Plant Evolution

Plant Evolution

An Introduction to the History of Life

KARL J. NIKLAS

The University of Chicago Press

Chicago and London

Karl J. Niklas is the Liberty Hyde Bailey Professor of Plant Biology and a Stephen H. Weiss Presidential Fellow in the Plant Biology Section of the School of Integrative Plant Science at Cornell University. He is the author of *Plant Biomechanics*, *Plant Allometry*, and *The Evolutionary Biology of Plants*, and coauthor of *Plant Physics*, all published by the University of Chicago Press.

The University of Chicago Press, Chicago 60637
The University of Chicago Press, Ltd., London
© 2016 by The University of Chicago
All rights reserved. Published 2016.
Printed in the United States of America

25 24 23 22 21 20 19 18 17 16   1 2 3 4 5

ISBN-13: 978-0-226-34200-9   (cloth)
ISBN-13: 978-0-226-34214-6   (paper)
ISBN-13: 978-0-226-34228-3   (e-book)
DOI: 10.7208/chicago/9780226342283.001.0001

Library of Congress Cataloging-in-Publication Data

Names: Niklas, Karl J., author.
Title: Plant evolution : an introduction to the history of life / Karl J. Niklas.
Description: Chicago ; London : The University of Chicago Press, 2016. | Includes
    bibliographical references and index.
Identifiers: LCCN 2016002294 | ISBN 9780226342009 (cloth : alk. paper) | ISBN
    9780226342146 (pbk. : alk. paper) | ISBN 9780226342283 (e-book)
Subjects: LCSH: Plants–Evolution. | Evolution (Biology)
Classification: LCC QK980 .N556 2016 | DDC 581.3/8–dc23 LC record available at
    http://lccn.loc.gov/2016002294

♾ This paper meets the requirements of ANSI/NISO Z39.48-1992 (Permanence of Paper).

CONTENTS

PREFACE

The thought we read is related to the thought which springs up in
ourselves, as the fossil-impress of some prehistoric plant to a plant as
it buds forth in springtime.
—ARTHUR SCHOPENHAUER, *Parerga und Paralipomena* (1851)

It is nearly two decades since the publication of *The Evolutionary Biology of Plants* and much has been learned in each of the scientific disciplines that contribute to our understanding of evolution and evolutionary theory. By necessity, therefore, *Plant Evolution: An Introduction to the History of Life* is a new book and not a second edition of my earlier work. To be sure, some of what is contained in the present volume draws heavily from my previous book. Although extensively rewritten, chapters 1 and 2 present the same general overview of the major evolutionary transformations that have shaped our planet's biosphere. On the other hand, the rest of this book is new and reflects recent advances. Chapters 3 and 4 review population genetics and developmental biology. Chapters 5 and 6 review the processes of speciation and macroevolution. Chapters 7 and 8 discuss the evolution of multicellularity, the plant body plans, and how evolution and biophysics are interrelated. Finally, chapter 9 provides a perspective on evolutionary ecology.

Like all of my other books, this one is personal and eclectic. Every book, no matter how technical, is a reflection of the author's mind, and *Plant Evolution* is no exception, since it is a distillation of what I consider to be significant concepts, principles, and facts. This book was written with a single purpose in mind—to show why the study of evolution is important and why the study of plants is essential to understanding evolution. To paraphrase Theodosius Dobzhansky (1900–1975), if nothing in biology makes any sense except in the light of evolution, then nothing in evolutionary biology makes sense if not illuminated by studying plant biology. This claim rests on the fact that plants are very different from animals or fungi, yet much of the standard evolutionary literature is zoocentric. It is ironic to a botanist such as myself that biology textbooks sometimes begrudgingly treat plants as decorations on the tree of life rather than as the forms of life that allowed that tree to survive and grow. This is a blue planet, but it is a green world.

Plant Evolution does not provide a comprehensive overview of evolution. It is a collection of nine essays that attempt to cover the major lines of evidence that prove that evolution is fact—that is, evidence from the fossil record, molecular biology, comparative embryology, comparative anatomy, comparative physiology, the experimental manipulation of species, and biogeography. No author can survive the pretense of claiming to understand everything about evolution, and no single book can delve into every aspect of evolution, even superficially. Each book can only bring a particular perspective to illuminate some corner of the history of life. My goal was to cover the fundamental processes of evolution from the perspective of a botanist who has enjoyed studying plants and who has tried to comprehend the breathtaking discoveries made about them over the past two decades. By way of inclinations and biases, you should know that I was trained first as a mathematician and second as a paleobotanist. As a student of numbers, I know that understanding evolution depends in part on understanding mathematics. I hold to the idea that if we cannot quantify something, we do not truly understand it. In addition, probability theory, the dynamics of changing gene frequencies, and many other aspects of evolutionary theory are mathematics in different guises.

But I am also a student of the dead. Knowing the fossil record is just as important as knowing about living organisms. If we neglect the past and study only the living, we would be totally ignorant of over ninety percent of all the species that ever lived and we would have no real insight into the causes and consequences of phenomena such as global climate change and mass extinctions. The study of the fossil record requires no apology—it is vital and important to every aspect of modern biology.

This book is intended to serve as the text for an upper-level undergraduate course, or as a text for an early postgraduate course. I freely admit that it is my intention to steer students toward studying plants professionally. For this reason, I have tried to keep the writing style of this book simple (albeit without sacrificing accuracy) and to use jargon sparingly. Jargon is unquestionably necessary at times. But technical words spoken glibly in the classroom or released on paper tend to sound like explanations rather than descriptions, and jargon in any form can easily mystify and alienate a student. Jargon has yet another negative aspect—it can build walls around disciplines and thereby hide a discipline's relevance to other fields of study. Unfortunately, biology has more than its fair share of terminology. Consider for example that the basal body is called a kinetosome by the protozoologist, who calls the cell biologist's undulipodium a flagellum, which is a very different structure from the flagellum studied by the bacteriologist. Likewise, what a phycologist calls siphonocladous, a bryologist calls symplastic, which is not to be confused with what the zoologist calls a syncytium. As these few but perhaps painful examples of terminology attest, the study of evolution, which employs the vocabulary of all of the scientific disciplines, involves learning the meaning of many technical terms. I have done my best to inflict as little pain as possible, and I have provided a glossary for some of the more troublesome technical terms (or those that I may have used idiosyncratically).

This book is intended as a textbook, or as a reading for a seminar; it was not written to cite the primary literature wherever a declarative sentence appears in the text. A few citations appear from time to time, but only when necessary to acknowledge the contributions of particular individuals. This tactic was predicated on the notion that libraries and Internet search engines are available to students. It was also based on the fact that the most recent literature is easily found on the Internet, whereas the older literature is more difficult to find electronically. A few key words, such as the name of a researcher, typed into the Internet's vast resource network can locate the required new literature easily. Whether the literature found in this manner provides reliable information is another matter. Each of us must exercise great judgment when reading anything, including this book. In lieu of references in the text, a suggested reading list is provided at the end of each chapter. An emphasis is placed on the older literature, which is often neglected or forgotten. Even though one of my goals was to integrate the new discoveries made about plants with the literature published one or more hundred years ago, re-reading this older literature gives credence to the proclamation *nihil sub sole novum.*

A few words are needed to explain the use of the word "we" throughout the text. As in my classroom when lecturing, I want you the reader and me the author to have a conversation as we take a journey through the corridors of time and space. I want us to engage in the joys of discovering new ideas or facts together. My copy editor will detest this practice, but I find the use of the word "we" singularly pleasing and appropriate.

Finally, I want to thank the wonderful people who have made this book possible. Everything that I am is a consequence of my parents' nurturing and genetics. Although they had little themselves, they valued an education, and they lived long enough to see me become a teacher, but not alas long enough to see this book, which is dedicated to their memory. My teachers and role models in school are also responsible for who I am. Lawrence J. Crockett showed me the wonders of plant life and how I could make mathematics literally come to life. Tom L. Phillips opened my eyes to the fossil record and served as the best model scientist and scholar. When speaking of my teachers, I must also thank all my students who, by asking seemingly simple questions, taught the teacher that he had much more to learn (and still does). I am also deeply grateful to Michael Christianson (University of California, Berkeley), Randy Wayne (Cornell University), and Bruce Tiffney (University of California, Santa Barbara), who offered many suggestions and comments while reading preliminary drafts of all the chapters. I also extend my deepest gratitude to my editor, Christie Henry, who provided unyielding encouragement throughout my pre- and post-publication anxieties; Mary Corrado, who provided brilliant re-edits of a very imperfect manuscript; Logan Smith, who attended to details too numerous to recount; and Julie Shawvan, who provided a beautifully detailed index. Finally, I acknowledge my 37-year-long collaboration with Edward Cobb, who took the majority of the photographs that adorn the text (the artwork with all its faults is mine) and who has given me more support than anyone truly deserves.

Introduction

Much has been written about evolution from the perspective of the
history and biology of animals, but significantly less has been written about the evolutionary biology of plants. Zoocentricism in the
biological literature is understandable to some extent because we are
after all animals and not plants and because our self-interest is not
entirely egotistical, since no biologist can deny the fact that animals
have played significant and important roles as the actors on the stage
of evolution come and go. The nearly romantic fascination with dinosaurs and what caused their extinction is understandable, even
though we should be equally fascinated with the monarchs of the
Carboniferous, the tree lycopods and calamites, and with what caused
their extinction (fig. 0.1). Yet, it must be understood that plants are
as fascinating as animals, and that they are just as important to the
study of biology in general and to understanding evolutionary theory
in particular. Consider, for example, that the fossil remains of the tree

1

Figure 0.1. A suggested reconstruction of the Carboniferous (359–300 Mya) flora dominated by tree-sized (arborescent) lycopods such as *Lepidodendron* (right foreground) and arborescent calamites such as *Calamites* (left rear). This type of vegetation grew in geographically expansive, swampy environments throughout Europe and North America. Its fossil remains constitute most of today's commercial grade coal. The extinction of the Euramerican lepidodendrids and calamites toward the end of the Westphalian stage of the Carboniferous (≈312–299 Mya) is attributed to climate changes and to tectonic activity that reduced the geographical expanse of the coal-swamp ecosystems. Courtesy of The Volk und Wissen Volkseigener Verlag, Berlin.

Table O.1. Formal and informal names of some of the living plant groups mentioned in the text

Prokaryota (polyphyletic)

    Eubacteria
    Archaea

Eukaryota (eukaryotes)

    algae (polyphyletic)

            Class Charophyceae (charophytes)[a]
            Class Chlorophyceae (green algae)[a]
            Class Phaeophyceae (brown algae)
            Class Rhodophyta (red algae)

    Embryophyta (monophyletic) [a]

        bryophytes (paraphyletic)

                Phylum Bryophyta (mosses)
                Phylum Marchantiophyta (liverworts)
                Phylum Anthocerotophyta (hornworts)

        vascular plants/tracheophytes

            seedless vascular plants

                Phylum Lycopodiophyta (lycopods)
                Phylum Monilophyta (ferns and horsetails)[b]

            seed plants

                gymnosperms (polyphyletic)

                Phylum Cycadophyta (cycads)
                Phylum Ginkgophyta (*Ginkgo*)
                Phylum Coniferophyta (conifers)
                Phylum Gnetophyta (gnetophytes)

              Flowering plants (monophyletic)

                Phylum Anthophyta (angiosperms)

---

[a] The green algae (Chlorophyceae and Charophyceae) and the Embryophyta are a monophyletic group of plants that are collectively called the Viridiplantae. The Charophyceae and the Embryophyta are collectively referred to as streptophytes.

[b] Although the monilophytes have been given formal taxonomic status and evolved from a last common ancestor (trimerophytes), the horsetails evolved independently from the ferns, and there is ample evidence that modern-day ferns had independent origins. Thus, the monilophytes should be considered a paraphyletic group of plants.

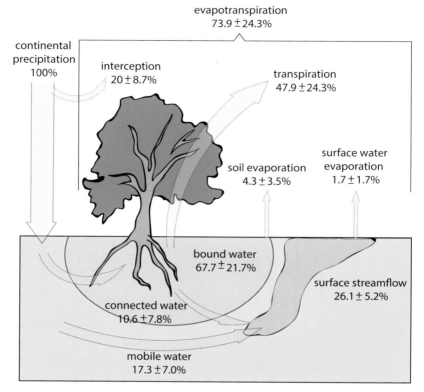

Figure 0.2. Schematic of global hydrological fluxes (expressed as percentages of continental precipitation, 100%) based on a model using isotopic data and estimates of terrestrial plant gross primary productivity. The model assumes that plants lose ≈300 water molecules per CO_2 fixed by photosynthesis, which predicts that plant transpiration equals 55,000 kn³/yr., and that gross primary productivity equals ≈120 Pg C/yr. Note that one petagram (Pg) equals 10^{15} grams, or one billion metric tons. Note further that transpiration accounts for ≈47.8% of the total continental precipitation, which is ≈65% of total evapotranspiration. These data emphasize the important roles land plants play in influencing Earth's hydrological cycles that in turn influence the movement of nutrients and soil contaminants. The schematic is based on data reported by Good, Noone, and Bowen (2015).

lycopods and tree horsetails produced much of the coal that fueled the early days of the industrial revolution (table 0.1). Consider also how important plants are to the Earth's ecosystem (fig. 0.2).

   The introduction to a book about evolution can serve many purposes as for example to disabuse the notion that evolution has directionality or purpose, which is a common misconception that can lead to heated debates where none should exist. Nevertheless, the misconception emerges for a number of reasons. Clearly, time has direction, and the fossil record preserves the long history of evolution in chronological order that reveals many clear-cut trends as for example a trend toward increasing body size in some, but not all, lineages. Likewise, our species has a predilection for pareidolia—the tendency to see patterns where none exist, as for example "the man in the moon." However, none of these phenomena justify the assumption that evolution has a prefigured pattern, or some sort of goal. Evolution must abide by many rules, but these are prefigured in the laws of physics and chemistry, and the overarching laws of chance.

Why Study Plants?

But first, why study plants? The next time you walk through a forest, park, or garden, consider how alike and yet unalike you are from the plants that surround you. You and they are made of cells, each of which contains organelles called mitochondria that consume oxygen to power cellular metabolism. Like plants, our cells also contain copies of the remarkable molecule called DNA (deoxyribonucleic acid) that contains most, albeit not all, of the information needed to keep you alive. Perhaps even more astounding is the fact that we and every plant around us are distantly related, albeit at a time when life first started to evolve billions of years ago. As surmised by Charles Darwin (1809–1882), all forms of life are related because, with the exception of the very first living things, organisms can evolve only from preexisting organisms. To be specific, Darwin vigorously proposed and defended five propositions in his magnum opus, *On the Origin of Species*:

(1) All life evolved from one or a very few simple, unicellular organisms.
(2) All subsequent species evolve from preexisting species.
(3) The greater the similarities between taxa, the more closely they

are related to one another and the shorter their evolutionary
divergence times.
(4) The process giving rise to species is gradual and of long duration.
(5) Higher taxa (genera, families, etc.) evolve by the same evolutionary mechanisms as those that give rise to new species.

As we will see throughout this book, propositions (4) and (5) are
problematic for certain species and some higher taxa. However, propositions (1)–(3) have received extensive experimental validation, both
in terms of molecular analyses and classical comparative anatomy and
morphology. There is no doubt that each of us is related to every other
living thing as a consequence of uncountable ancestor-descendant relationships comprising a genealogy that extends back to the dawn of life.

Yet, consider too that we are very unlike plants. Most of our cells
are held together primarily by glycoproteins called cadherins, whereas
most of the cells in land plants are held together with the help of
multifunctional pectic polysaccharides. Likewise, with only a few exceptions, plant cells have cell walls that provide mechanical support
by virtue of one of the strongest naturally occurring polymers on the
planet, cellulose. In addition, green plant cells contain organelles
called chloroplasts that, in the presence of sunlight, convert carbon
dioxide, water, and a few essential elements into new living cells. As-
tronomers like to tell us that we are made of stardust—because the
elements in our bodies were fabricated in the hearts of stars now long
vanished from the night's sky. If this is true, it must also be said that
we are made of starlight—because plants provide all animals, either
directly or indirectly, with food thanks to the evolution of a process
called photosynthesis.

Even if plants were not the foundation of almost every food chain
on our planet, they deserve our unwavering attention because they
have done far more than feed the world over the course of evolutionary history. Consider two facts. Most extant organisms require
oxygen to live. Yet, Earth's first atmosphere lacked oxygen. Indeed,
oxygen was probably toxic to many of the first forms of life on this

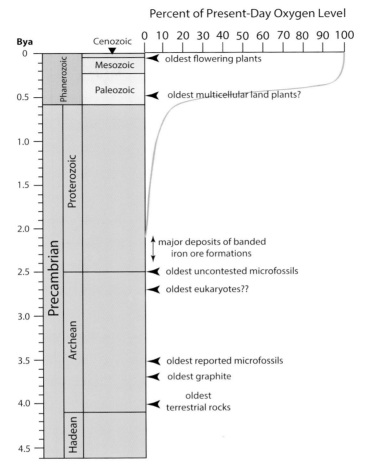

Figure 0.3. Estimates of the percent of present-day levels of atmospheric oxygen (100% denotes current oxygen level) plotted as a function of geological time (in billions of years before present). A few evolutionarily important events, such as the appearance of the first cells containing organelles (eukaryotic cells), are concurrently plotted. The horizontal line measures our uncertainty about the precise timing of each of these events.

planet. So, how did the majority of organisms come to require oxygen? The answer requires knowing that plant photosynthesis splits water molecules and releases oxygen. Once this metabolic process evolved, Earth's atmosphere changed from one composed of methane, ammonia, carbon monoxide, and other reducing gases into an oxidizing atmosphere like the one we breathe today (fig. 0.3). The evidence for

Table 0.2. Six examples of how plant evolution changed the physical and biological world

(1) Evolution of Photosynthesis	→	Transformed a reducing atmosphere into an oxidizing atmosphere; provided heterotrophs food.
(2) Evolution of Land Plants	→	Ameliorated the terrestrial landscape; paved the way for the colonization of the land by animals; shaped water and nutrient soil cycles.
(3) Evolution of Wood	→	Sequestered carbon dioxide; provided lightweight building material that amplified the three-dimensionality of terrestrial communities; shaped ecosystems by virtue of forest fires.
(4) Evolution of Flowering Plants and Endosperm[a]	→	Permitted the storage of seeds by early humans thereby fostering the transition from a hunting-gathering society to an agrarian society.
(5) Fossilization of Plants and Coal Formation	→	Fostered the Industrial Revolution.
(6) Diversification of Secondary Plant Metabolic Products	→	Continues to provide numerous pharmaceuticals.

[a] Endosperm is a specialized tissue produced in the developing seeds of flowering plants. It provides nutrients to the developing embryo within the seed, which typically dehydrates and undergoes a dormancy period. This developmental pattern allows seeds to be dried and stored for long periods.

this claim is extensive and will be presented in greater detail when we discuss the origin and early evolution of life (see chapter 1). For now, it is sufficient to recognize that the evolution of plants has literally changed the world (table 0.2), and that no one can claim to understand evolution unless they understand plant biology.

What Does Evolution Mean?

But what is evolution? What does the word really mean? To be sure, definitions of complex things are difficult to construct in ways that are acceptable to everyone. This generalization holds true for the concept of evolution, which helps to explain why different authors have defined

Figure 0.4. Sassafras leaves taken from the same branch illustrate phenotypic plasticity. The differences in shape result from developmental responses to the effects of gravity on developing leaves. Leaves developing on the upper sides of branches tend to be unlobed. Leaves developing on the sides of branches tend to be mitten shaped. Leaves developing on the lower sides of branches tend to have three lobes.

evolution in slightly different ways. Yet, most definitions adopt the phrase *descent with modification* or contain language that says much the same thing. Evolution is a record of the heritable changes in the characteristics (traits) of organisms over a few or many generations. Notice that this definition does not speak to *how* evolution occurs. Rather, it merely *describes* a process. Also notice the use of the word *heritable*. The changes that occur across successive generations must be the result of genomic modifications and not the result of developmentally reversible responses of individual organisms to their environmental conditions. The leaves developing on the same branch of a tree can differ in size, shape, or other traits in response to differences in light or the effects of gravity (fig. 0.4). The capacity for this developmental plasticity is an inherited feature, and it is nowhere better expressed than in sedentary organisms like the land plants who must continue to grow in one place where environmental conditions can change, often dramatically over the course of a few or many years. However, the particular differences among individual leaves growing in the sun or in the shade are not inherited traits that can be passed down to the next generation a tree produces. If they were, each tree would be capable of producing leaves of only one shape and size. Rather, leaves differing in shape but drawn from the same plant illustrate that a single genotype (the combined genome of an organism, which in the case of plants includes the genetic information stored in the nucleus, mitochondria, and chloroplasts) can produce different phenotypes (the physical manifestation of all of an organism's traits) in response to different environmental conditions. Consequently, the word evolution is not applied to changes in an individual organism, but rather to modifications in the traits of descendants with respect to ancestral traits.

As mentioned earlier, a regrettable misconception about evolution is that it has purpose—some grand design. This misconception rests in part on the notion that heritable changes cannot revert back to the ancestral condition. Yet, this is demonstrably wrong—evolution does not have a prefigured direction and reversals are not uncommon as for example the evolution of vestigial leaves in the relatives of plants that possessed large leaves (as for example, the leaves of the herbaceous horsetail *Equisetum* and the arborescent horsetail *Calamites*; fig. 0.5). Reduction is particularly evident in instances of the evolution of parasitism as for example the Indian Pipe (*Monotropa uniflora*), which has highly reduced, nonphotosynthetic leaves (fig. 0.6).

Nevertheless, most biologists agree on the existence of major evolutionary transitions that have collectively established what appear to be trends in the fossil record. For example, prebiotic replicating molecules preceded the appearance of membrane-bound protocells in which originally independent genes subsequently became aggregated into chromosomes (table 0.3). Yet, at finer levels of resolution, each of these transitions must be seen as the statistical summation of numerous smaller events, some of which involved gains, losses, or reversals of previous events. For example, depending on the group of organisms (or the time interval) examined, body size may increase or decrease in

Figure 0.5. Comparison of the vestigial leaves of the horsetail (*Equisetum*) shown on the left panel and the larger leaves of the organ genus for the leafy shoots of the tree-sized calamites (*Annularia*) shown on the right panel (scale is in millimeters). The leaves of the horsetail are highly reduced in size and fused together along their margins to form a crown-like whorl. Only their tips are individually recognizable, both on main and lateral branches (arrows). Unlike the leaves shown here, most mature horsetail leaves are not photosynthetically functional. The leaves of calamites were likewise arranged in a whorl, but they were unfused at their margins, larger, and photosynthetic.

the fossil record of a particular lineage just as the degree of ecological specialization may increase or decrease over the long course of the history of a lineage or clade. Consequently, claims for the existence of evolutionary trends depend on our particular taxonomic and temporal foci. Indeed, in a very real sense, what appear to be broad patterns in evolutionary history are fractal-like in the sense that their existence depends on our scale of measurement (much like the length of a coastline depends on the length of the yardstick used to measure it).

Figure O.6. Colorless flowers and vestigial leaves (left) and developing fruits (right) of the Indian Pipe (*Monotropa uniflora*), a parasitic angiosperm placed in the Blueberry family (Ericaceae).

Table O.3. Six examples of evolutionary transitions (in approximate chronological order of occurrence; top to bottom) that collectively appear to constitute an evolutionary trend of increasing complexity

(1) Replicating Molecules	→	Compartmentalized Replicating Molecules
(2) Independent Genes	→	Chromosomes
(3) Unicellular Prokaryotes	→	Multicellular Prokaryotes
(4) Multicellular Prokaryotes	→	Cellular Specialization
(5) Unicellular Eukaryotes	→	Multicellular Eukaryotes
(6) Aquatic Multicellularity	→	Terrestrial Multicellularity

Patterns and Trends

The coastline-yardstick analogy helps us to understand why the interpretation of some patterns in the fossil record has proven contentious. Consider the contrasting perspectives of Christian De Duve (1917–2013) and Stephen Jay Gould (1941–2002). De Duve observes a clear directionality in a trend going from functionally general and inefficient biochemical reactions to progressively more specific and efficient reactions during the molecular transition from an abiotic to a biotic world. According to this view, evolutionary patterns emerge from orderly molecular modifications and adaptive innovations that translate ultimately into complex molecules such as DNA. Gould also sees patterns in life's macroscopic history, but argues that most are largely the result of unpredictable contingent events ranging from developmental quirks early in the ontogeny of ancestral organisms carried forth into their descendants to global catastrophes such as the asteroid collision that resulted in the Cretaceous-Paleogene mass extinction (also called the K-T event; see fig. 9.19). However, these two worldviews arise because De Duve and Gould are viewing different coastlines and using very different yardsticks with which to measure it. De Duve's coastline is constructed by the unalterable laws of physics and chemistry. His yardstick is a molecule. Gould's coastline is constructed out of macroscopic morphological transformations preserved in the fossil record. His yardstick is the observable phenotype. De Duve sees patterns because of predictable molecular verities. Gould sees patterns resulting from seemingly random historical events that are refined subsequently by the operation of natural selection. Both worldviews are real, but the two are very different. One emerges from necessity. The other comes largely from chance.

Necessity and Chance

The tension between necessity and chance lies at the heart of many aspects of biology, but none more so than in evolutionary biology because of the roles played by selection and genomic variation. Physics certainly encompasses the determinism of classical Newtonian mechanics (which describes the behavior of billiard balls and planets) and the randomness of quantum mechanics (which describes the behavior of quarks and electrons). However, these two contrasting paradigms operate at such different physical scales that one paradigm rarely affects the other in ways perceptible to us. This does not hold true when we examine classical Darwinian evolutionary dynamics. Consider the

theory of natural selection as proposed independently by Charles Darwin and Alfred Russel Wallace (1823–1913), which makes five major
assertions:

(1) The number of individuals in a population should increase
    geometrically.
(2) However, the number of individuals tends to remain constant.
(3) Therefore, only a fraction of the offspring that are produced
    survive because the environment provides limited resources.
(4) Those offspring that survive and reproduce differ from those
    that die because the individuals in a population differ owing to
    heritable variation; and
(5) the struggle to survive and reproduce determines which variants
    will perpetuate the species.

According to this theory, the *necessity* of natural selection results
in the accumulation of favorable heritable traits in successive generations by means of the elimination of individuals bearing traits that
are less favorable to survival and reproductive success. The result is a
macroscopic evolutionary pattern that can appear to have direction
(and, to some people, even design and purpose).

However, heritable differences in traits are the result of chance
molecular changes in an organism's genome, changes that result from
spontaneous random mutations (table 0.4), or from genetic recombination during meiosis and sexual reproduction (fig. 0.7). Most mutations are lethal, or at best neutral, in their effects. Those that are
not lethal introduce heritable changes in the next generation without
the benefit of sexual reproduction. Genetic recombination results as
a consequence of chromosome pairing and crossing-over during meiosis, a process that will be described in detail in chapter 3. To be very
clear, organisms have evolved extremely sophisticated mechanisms
to proofread their DNA and repair or purge modifications from their
genomes. Likewise, mutations and crossing-over do not occur with
equal probability throughout an organism's genome. Some DNA se-

Table 0.4. Examples of mutations that alter DNA sequences and thus protein function

(1) Deletion	The removal of DNA nucleobases,[a] e.g., CTGGAG → CTGGA.
(2) Duplication	The formation of a DNA sequence that is copied more than one time.
(3) Insertion	The addition of DNA nucleobases, e.g., CTGGA → CTGGAT.
(4) Frameshift mutation	A deletion or insertion of one or more DNA nucleobases that shifts the type(s) of amino acid(s) encoded for a protein, for example The Fat Cat Ate Fat → heF atC atA teF at.
(5) Missense (substitution) mutation	A change in one DNA nucleobase triplet that results in the substitution of one amino acid for another amino acid in a protein sequence, e.g., GAG→GTG in the β-globin gene results in sickle cell anemia.
(6) Nonsense mutation	A change in one DNA nucleobase pair that truncates protein construction and results in a shortened protein.
(7) Repeat expansion	Short DNA sequences that are repeated one or more times in a row, e.g., CTGGAG → CTGGAG CTGGAG CTGGAG.

[a] The four DNA nucleobases are adenine (A), cytosine (C), thymine (T), and guanine (G).

quences are more prone to mutation and crossing-over, while others are not. Consequently, in this context, the word chance does not mean random.

Nevertheless, mutations and recombination involve elements of chance. Mutations are random in the sense that an organism cannot instigate or specify what part of its genome will mutate or what a mutation will produce. Likewise, with the exception of human medical intervention, an organism has no direct control over which sperm cell will fertilize a particular egg. Viewed in the most simplistic of ways, mutations and genetic recombination are genomic accidents that provide the heritable variation that opens the possibility that selection will subsequently influence which variants die and which prosper.

It is important to not lose sight of one of the great insights pro-

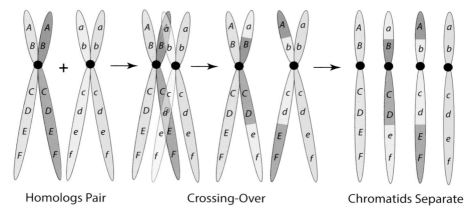

Homologs Pair   Crossing-Over   Chromatids Separate

Figure 0.7. Genetic recombination results in progeny with combinations of genetic information differing from those of either parent. It is the result of genetic materials shuffled between parents when sperm and egg fuse to form a zygote and from a phenomenon called crossing-over wherein homologous chromosomes (homologs), each consisting of two chromatids (here shown in different shades of blue and different shades of yellow) pair during meiosis (to produce sperm or egg) and transmit physical portions from one chromatid to the corresponding portions of the other chromatid (diagrammed from left to right). In the process sister chromatids will differ in allelic forms of genes (shown as a series of letters; dominant alleles in capitals and recessive alleles in lower cases). The chromatids of each chromosome are separated during meiosis to produce four chromosomes that in this case differ in each of their genetic makeup. The exchange of genetic information need not involve chromosome breakage; it can result from the transfer of copies of portions of chromosomes (not shown).

vided by the Darwin-Wallace theory of natural selection—an insight that significantly colors our perception of what we mean when we speak of adaptation. Correlated variables have meaning only in relation to one another such that one variable cannot be conceived of as cause or effect. This is a subtle but profoundly important insight. Organisms evolve, and by doing so they change their environments. Reciprocally, when environments change, organisms must evolve if they are to survive and reproduce successfully. This reciprocity drives a process that gives the appearance of progress because competition among individuals necessitates adaptive invention and novelty, or extinction. The theory of natural selection tells us that each species must either evolve to survive the gauntlet of changing environmental conditions, or suffer and ultimately perish. The fossil record also tells us that the end game of evolution is death. Well over ninety percent of all previous forms of life are extinct. This gruesome statistic shows that adaptions are never perfect. They are only temporarily effective.

Mendel, Planck, and Particulate Heredity

The theory of natural selection goes a long way to explain *why* organisms evolve, but it is silent about *how* they evolve. Charles Darwin mustered a remarkable amount of evidence for the physical manifestations of evolution, but he was unaware of hereditary mechanisms, including mutation and recombination. Darwin was remarkably clear about this. In his chapter on the "Laws of variation" (Darwin 1859, p. 170), he writes, "Whatever the cause may be of each slight difference in the offspring from their parents—and a cause for each must exist—it is the steady accumulation, through Natural Selection, of such differences, when beneficial to the individual, that give rise to all the more important modifications of structure, by which the innumerable beings on the face of this earth are enabled to struggle with each other, and the best adapted to survive." At the beginning of the same chapter (1859, p. 131), Darwin states that variation is "due to chance," but he goes on to say, "This, of course, is a wholly incorrect expression, but it serves to acknowledge plainly our ignorance of the cause of each particular variation." In this context, it is fair to say that the word chance has often been used to explain what we do not know or cannot explain.

This huge gap in knowing what chance means began to disappear with the rediscovery in 1900 of the seminal work of Gregor Mendel (1822–1884) on particulate inheritance, which was the same year that Max Planck (1858–1947) introduced his concept of quantum discontinuity. Curiously, the theories of Mendel and Planck had one important feature in common—both hypothesize discretized entities, traits in the context of Mendel's heredity theory and quanta in the case of Planck's black-body theory. In order to understand the depth of this coincidence, consider that Mendel selected peas (*Pisum sativum*) with which to explore heredity for two reasons. First, peas have non-opening, self-pollinating (cleistogamous) flowers, which allows plant breeders to know the source of the pollen used to produce the next generation of seeds, and, second, some of the more easily measured traits exhibited by peas have only two phenotypic states as for example seed color (yellow versus green) and seed shape (smooth versus wrinkled). The pollination syndrome and the "either or" genetics of peas allowed Mendel to discover the laws of inheritance using seven traits: plant height, pod shape and color, seed shape and color, and flower position and color. Over the course of his studies, Mendel discovered that some phenotypes were dominant, whereas others were recessive. For example, when a yellow pea plant is pollinated with pollen from a plant with green peas, all of the peas in the next generation are yellow (thus yellow is dominant, whereas green is recessive). However, in the following generation of plants that were allowed to self-pollinate, green peas reappeared at a ratio of 1:3. A graphical technique, formulated by Reginald Punnett (1875–1967) and named in his honor as Punnett squares, diagrams these relationships efficiently (fig. 0.8).

In contemporary terminology, the molecular domains of DNA that code for a trait are called genes, whereas alternative DNA sequences in the same DNA segment are called alleles (that is, alleles are alternative forms of the same gene). In the foregoing example of Mendelian genetics, the gene for pea color has two allelic forms (yellow and green). Diploid organisms such as peas inherit one allele for each gene from each parent. An individual that has two copies of the same allelic form of a gene (as for example *YY* in fig. 0.8) is said to be homozygous for that gene, whereas an individual that has two different allelic forms of a gene (*Yg* in fig. 0.8) is said to be heterozygous for that gene.

The "Modern Synthesis" That Was Neither Modern nor Synthetic

Unfortunately, Mendel's brilliant insights were not understood by those who initially read his work. Perhaps worse, Mendel's work was

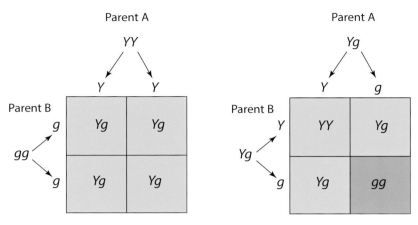

Figure 0.8. Punnett squares illustrating what happens when a yellow pea is crossed with a green pea (left) and when the progeny of this cross are allowed to self-pollinate (right). When a yellow pea (*YY*) is crossed with a green pea (*gg*), all of the progeny are yellow peas (*Y* is dominant), despite the fact that the allele for green is present in each of the four genotypes (*g* is recessive). When the progeny produced by the first cross are allowed to self-pollinate, three genotypes are produced, one of which is homozygous for green (*gg*) and two of which produce the yellow phenotype (one that is homozygous, *YY*, and another that is heterozygous, *Yg*). Statistically, the result is one green phenotype for every three yellow phenotypes (1:3). Note: It is conventional to denote genes in italics and to use upper- and lower-case letters for dominant and recessive genes, respectively.

wholly unknown to Darwin. Had the latter learned of the laws of Mendelian inheritance, genetics might have prospered earlier than it did and Darwin would never have invented pangenesis as a mechanism for inheritance. Fortunately, Mendel's work was independently duplicated and rediscovered by Hugo de Vries (1848–1935) and Carl Correns (1864–1933), both of whom published their work within a two-month period in the spring of 1900. The curious initial result was that biologists quickly accepted Mendel's ideas, but supposed them to be largely incompatible with Darwinian evolution for the simple reason that Darwin's theory emphasized the effects of selection on traits manifesting continuous rather than "either or" variation. In contrast, Mendelian genetics was particulate (either yellow or green, either wrinkled or smooth, etc.) with no intermediates. Notice that the example of Mendelian genetics illustrated in fig. 0.8 can never achieve more than three genotypes (*YY*, *Yg*, and *gg*) and never more than two

seed color phenotypes (yellow and green). Barring some sort of mutation, there are no possible intermediates upon which selection can act
because the genes underlying seed color are qualitative in nature. The
impasse between Darwin's theory and Mendel's theory was resolved
when the existence and behavior of quantitative genes were fully
recognized (box 0.1). Quantitative genes typically act in concert and
result in phenotypic traits that vary by degrees. Quantitative traits,
such as body mass or height, are those that can vary continuously and
that depend on the cumulative actions of more than one gene and
their interaction with the environment.

The comfortable merger of Darwinian evolution with Mendelian

Box 0.1. Quantitative Traits and the Length of Tobacco Corollas

Mendelian genetics was described in the text as "particulate" because the traits originally studied by Gregor Mendel were discontinuous discrete traits, as for example green
or yellow peas. However, many traits are continuous traits, such as body length or plant
height. These attributes are called quantitative traits, many of which are the result of
the cumulative interactions among two or more genes and interactions among these
genes and the environment. A quantitative trait locus (QTL) is a polygenic portion of
DNA that correlates with and participates with the regulation of the phenotypic variation
in a quantitative trait. Early in the twentieth century, after the rediscovery of Mendel's
work, it was not immediately obvious to geneticists how Mendelian (particulate) genetics
could be reconciled with quantitative traits. The American geneticist William E. Castle
(1867-1962) is generally credited with making the first attempt to reconcile Mendelian
genetics with Darwin's theory of speciation. Castle argued that the appearance of novel
traits complying with Mendelian genetics resulted in new species—that is, the evolution
of new discontinuous traits is the basis for phenotypic discontinuity and thus speciation.
This speculation did not address the mechanisms responsible for QTL. However, it did
help to shift attention to the genetics of QTLs.

One of the early pioneers studying quantitative traits was the American plant geneticist Edward M. East (1879-1938), who studied tobacco and corn. One of his seminal
papers dealt with the inheritance of the style and corolla length of tobacco (*Nicotiana*).
He made crosses between *N. alata grandifolia* and *N. forgetiana*, which differ phenotypically only in the size and color of their flowers, and measured the lengths of styles and
corollas of the parental plants, their progeny (F_1), and the second generation of plants
(F_2). The mean corolla lengths of these two species were found to differ by more than 53
mm, whereas the frequency distribution of the corolla length of the F_2 generation was
continuous, albeit positively skewed (fig. B.0.1). From these measurements, East devel-

genetics along with the contributions of biometricians, such as Ronald Fisher (1890–1962) and Sewall Wright (1889–1988), lead to what is popularly called the Modern Synthesis. We will examine some of the historical details of this epoch in chapter 3. For now, it is sufficient to enumerate a few of the major concepts that emerged when evolutionary theory was invigorated by the insights of population genetics (table 0.5), and to juxtapose some of these concepts with those of Darwin.

For example, Darwin as well as most of the major contributors to the Modern Synthesis conceived of speciation as a comparatively slow process. However, this is not necessarily always true. Although

oped a genetical model and concluded, "the difference in corolla length shown by these two species [was] represented by the segregation and recombination of four cumulative but independent pairs of unit factors [genes], dominance being absent" and that "the Mendelian notation . . . to describe complex qualitative inheritance . . . is similarly useful in describing the inheritance of quantitative characters." This seminal conclusion set the stage for a true synthesis of genetics and evolutionary theory.

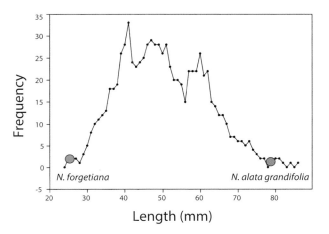

Figure B.0.1. The frequency distribution of the corolla length of the second generation (F_2) crosses between *Nicotiana forgetiana* and *N. alata grandifolia* illustrates what is meant by a quantitative (continuous) trait. The mean corolla length for each of the two parental species is shown as colored circle. Data taken from East (1913).

Table 0.5. Eight major concepts and conclusions characterizing the Modern
Synthesis

(1) Evolution is the change of allele frequencies in the gene pool of a population
over many generations.
(2) The gene pools of different species are isolated from one another, whereas the
gene pool of a species is held together by gene flow.
(3) Each individual of a sexually reproductive species has only a portion of the
alleles in the gene pool of its species.
(4) The alleles and allelic combinations of the individual are contributed by two
parents (and arise from independent assortment) that may be modified by chromosomal or genic mutations. Mutations are the ultimate source of new alleles
and genes.
(5) Individuals favored by natural selection will contribute larger portions of their
genes or gene combinations to the gene pool of the next generation.
(6) Changes in allelic frequencies in populations come about primarily by means of
natural selection, even though random mutations occur frequently.
(7) Barriers that restrict or eliminate gene flow between the subpopulations of a
species are essential for genetic and phenotypic divergence of the subpopulations of a species.
(8) Speciation is complete when gene flow does not occur between a divergent
population and the population of its parent species.

monogenic (single-gene) mutations resulting in speciation are rare, it
is increasingly clear that phenotypes can be altered dramatically as a
result of just one or a few mutations. Indeed, there is good evidence
that phenotypes can diverge rapidly by virtue of single allele differences. For example, a mutation in the *AFILA* allele in pea results in
a leaf composed entirely of tendrils (fig. 0.9). Likewise, the effects of
mutation on flower structure can affect pollination syndromes and
thereby limit or eliminate gene flow among neighboring populations
of plants. For example, flowers lacking petals (apetalous flowers) are
typically wind pollinated or self-pollinated, while flowers with large
numerous petals (polypetalous flowers) are generally pollinated by
animals. Single-gene mutations resulting in apetalous, fertile flowers
are known for mountain laurel (*Kalmia latifolia*), evening primrose
(*Oenothera parodiana*), tobacco (*Nicotiana tabacum*), and a variety of
annual chrysanthemum species. Conversely, monogenic mutations

Figure O.9. Representative leaves of eight genotypes of peas differing in their *af*, *st*, and *tl* allelic composition (see inserts for genotypic compositions). A mutation in a single gene can result in dramatic differences—for example, the wild type of pea is *AFAF STST TLTL* (shown at the upper left), whereas the *afaf STST TLTL* genotype leaf is all tendrils (shown below the wild type). Each of the three recessive allelic mutations is a naturally occurring mutation on three separate chromosomes that alter leaf architecture. In the examples shown here, each combination of alleles has been introduced into otherwise isogenic lines (i.e., all other genes in the diploid plants are homozygous). The use of isogenic lines reveals how each *af*, *st*, and *tl* allelic composition affects leaf shape.

resulting in flowers with supernumerary petals occur in mountain laurel, geranium (*Pelargonium hortorum*), soybean (*Glycine max*), gloxinia (*Sinningia speciosa*), garden nasturtium (*Tropaeolum majus*), and petunia (*Petunia* hybrids). Consider also monogenic mutations that affect floral organ identity. The mutations of the *AP3* or *PI* genes of the mouse-ear cress (*Arabidopsis thaliana*) or the *DEF* gene in snapdragon (*Antirrhinum majus*) cause petals to be replaced by sepals, and stamens to be replaced by carpels. None of these phenotypic alterations is known to have resulted in a new species. However, the structure and appearance of flowers are extremely important to attracting specific animal pollinators, and changes of the types just described can reduce or even eliminate gene flow between populations of wild-type and mutated plants that can in turn be the prelude to speciation.

Inspection of table 0.5 also reveals a serious omission in the Modern Synthesis—a failure to incorporate the insights of developmental biology when conceptualizing evolutionary mechanisms. Indeed, the Modern Synthesis was not a *synthesis* in the true meaning of the word. It did little to bring the different fields of biology together except to say "nothing in biology makes sense other than in light of evolution." Rather, it offered a reductionist approach to understanding evolution, one that abridged the mechanics of evolution to the level of population genetics. This claim may seem unwarranted. However, no less an important architect of the Modern Synthesis than Theodosius G. Dobzhansky (1900–1975) declared, "Evolution is a change in the genetic composition of populations. The study of the mechanisms of evolution falls within the province of population genetics." This perspective was grounded on a number of assumptions, some of which are extremely problematic. Four of these assumptions are particularly notable:

(1) Evolution proceeds gradually in small steps ("gradualism prevails").

(2) The mechanisms responsible for the appearance of new species are the same as those that give rise to higher taxa ("microevolution explains macroevolution").

(3) It is possible to directly map an organism's phenotype directly onto its genotype ("the genotype explains everything").

(4) Taxonomically widely separated organisms lack genetic similarities ("there are no widely shared 'old' genes").

As noted, assumptions (1) and (2) directly mirror those of Darwin—speciation is slow and thus of long duration, and the appearance of higher taxa involves the same mechanisms as those responsible for speciation. Assumptions (3) and (4) emerge directly from a single-minded focus on population genetics. Importantly, all four assumptions are incomplete at best. The monogenic mutations mentioned earlier have profound biological effects on morphology in just one generation, and we know of examples in which new plant species make their appearance over the course of a few generations, or, in the case of hybrids, one generation (see chapter 5). Likewise, epigenetic phenomena, microRNA gene silencing, and many other phenomena refute the notion that the phenotype emerges purely and simply from the genotype. It is also apparent that the co-option of "old genes" to do new things is ubiquitous in evolutionary dynamics. The mind-set of the Modern Synthesis emerged from a philosophy that failed to recognize that the developmental *arrival* of a novel phenotype is as important as the *survival* of the phenotype. This serious mistake had a number of consequences that will be explored in chapter 3.

What Is a Theory?

Before we proceed to examine evolution in the following nine chapters, it is useful to understand what is meant by "a scientific theory" such as the theory of evolution. The word "theory" has many colloquial meanings as for example "a hunch" or "an idea." However, in the sciences, the word has a much more focused and precise meaning, as for example *a predictive set of interrelated hypotheses that integrates facts to provide a broad explanation for one or more naturally occurring phenomena*. The theory of evolution is predictive (it expects adaptation, speciation, extinction, etc.) on the basis of a comparatively small set of hypotheses (heritable variation, natural selection, etc.) that integrates facts (empirical observations of living organisms and the fossil record) to provide a broadly applicable explanation of naturally occurring phenomena (how living things change and adapt to the world around them). Importantly, a scientific theory employs the scientific method of hypothesis building and testing, and as such it can and must be modified as new facts are brought to light. Indeed, we shall see that Darwin's theory of evolution was not complete. In fact, he got some things wrong. This is a characteristic of science because we are always learning new things and because our theories are constantly changing as new facts are learned. This is not a sign of weakness or failure. It is a sign of intellectual vigor and honesty. It is also a sign that our universe is extraordinarily complex.

Although the Darwinian theory of evolution has been modified and amplified over many decades of research, its predictive powers nevertheless remain impressive. Consider the case of Darwin's orchid, *Angraecum sesquipedale*. Early in the year 1862, the English horticulturalist James Bateman (1811–1897) sent Darwin a shipment of orchids collected in Madagascar containing an orchid bearing a beautiful star-shaped, white flower with an exceptionally long spur measuring as long as 30 cm (fig. 0.10). Inspection revealed a nectary within the tip of the spur, which prompted Darwin to hypothesize that the orchid must be pollinated by a moth with an exceptionally long proboscis (Darwin predicted a moth rather than a butterfly because the flower of *A. sesquipedale* is white rather than pigmented; see fig. B.6.1 in box 6.1 in chapter 6). On the basis of this hypothesis, which rested on the hypotheses called selection and adaptation, Darwin predicted the existence of an unknown insect (most probably a moth). In 1907, more than 20 years after his death, Darwin's hypothesis was vindicated by the discovery of a large Madagascar moth bearing a proboscis that measured on average 20 cm in length. The moth was named *Xanthopan morganii praedicta* in honor of Darwin's prediction.

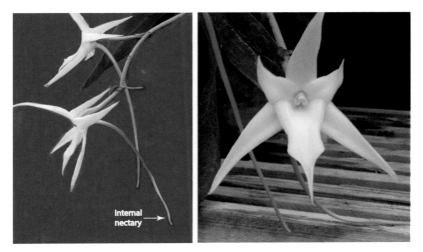

Figure 0.10. Representative flowers of "Darwin's Orchid," *Angraecum sesquipedale*. On the basis of his theory of evolution and his familiarity with pollination syndromes, Darwin predicted that the white flowers of this species would be pollinated by a nocturnal moth. This prediction was vindicated more than 110 years after his death when a large Madagascar moth with a 20 cm long tongue was observed under field condition pollinating this orchid. This example typifies what a scientific *theory* means.

Nevertheless, at that time there was no direct evidence that the moth fed on the nectar of *A. sesquipedale*, or that the moth was the orchid's pollinator. The scientific method required proof. It was not until 1992, more than 110 years after Darwin's death, that *X. morganii praedicta* was *directly* observed to feed on the orchid's nectar and to transport pollen from one flower to another.

The mutualistic nature of Darwin's orchid and *X. morganii praedicta* has become a classic example of plant-insect coevolution. Perhaps more important, it epitomizes what is meant by the ability of a *scientific* theory to explain the world around us in a rational, coherent, and empirically testable way. If some one asks you, "Do you believe in evolution?" answer, "Do you believe in the sun?" We can see and measure the sun. We can see and measure evolution. The phrase *I believe* is irrelevant to the scientific method or the scientific community. The sun is a fact. Evolution is a fact as well as an idea.

Knowledge requires us to possess both Facts and Ideas;—that every step in our knowledge consists in applying the ideas and the conceptions furnished by our minds to the facts which observation and experiment offer us. When our conceptions are clear and distinct, when our facts are certain and sufficiently numerous, and when the conceptions, being suited to the nature of the facts, are applied to them so as to produce an exact and universal accordance, we attain knowledge of a precise and comprehensive kind, which we may term *Science*.

—**WILLIAM WHEWELL**, *The Philosophy of the Inductive Sciences*, Part II, Book XI (1847)

Suggested Readings

Conway Morris, S., ed. 2008. *The deep structure of biology: Is convergence sufficiently ubiquitous to give a directional signal?* West Conshohocken, PA: Templeton Foundation Press.

Darwin, C. 1859. *On the origin of species by means of natural selection.* London: John Murray.

East, E. M. 1913. Inheritance of flower size in crosses between species of *Nicotiana*. *Bot. Gazette* (now the *Int. J. Plant Sci.*) 60: 177–88.

Futuyma, D. J. 1979. *Evolutionary biology.* Sunderland, MA: Sinauer.

Good, S. P., D. Noone, and G. Bowen. 2015. Hydrologic connectivity constrains partitioning of global terrestrial water fluxes. *Science* 349: 175–77.

Gould, S. J. 2003. *The structure of evolutionary theory.* Cambridge, MA: Harvard University Press.

Hull, D. L. 1973. *Darwin and his critics.* Chicago: University of Chicago Press.

Niklas, K. J. 1997. *The evolutionary biology of plants.* Chicago: University of Chicago Press.

Mayr, E. 1982. *The growth of biological thought: Diversity, evolution, and inheritance.* Cambridge, MA: Belknap Press.

Ruse, M. 1996. *Monad to man: The concept of progress in evolutionary biology.* Cambridge, MA: Harvard University Press.

Ruse, M. 2003. *Darwin and design: Does evolution have a purpose?* Cambridge, MA: Harvard University Press.

Origins and Early Events

On looking a little closer, we find that inorganic matter presents a
constant conflict between chemical forces, which eventually works to
dissolution; and on the other hand, that organic life is impossible without continual change of matter, and cannot exist if it does not receive
perpetual help from without.
—ARTHUR SCHOPENHAUER, *Parerga und Paralipomena* (1851)

This chapter begins our exploration of plant evolution with an outline
of plant history and a focus on broad physiological and morphological
innovations. This outline is difficult to sketch because the history of
plant life is extremely complex and begins with the evolution of the
first living things, estimated to be as old as 3.6 billion years (fig. 1.1).
Nevertheless, eight events in the history of plant life loom as mon-
umental in their consequences on all forms of life. These are (1) the
evolution of the first cellular forms of life, (2) the origin and development of photosynthetic cells, (3) the evolution of multicellularity in
prokaryotic microorganisms, (4) the first appearance of eukaryotic
cells (organisms with organelles and the capacity of sexual reproduction), (5) the ecological assault of plants on land and their growth in
air, (6) the evolution of conducting tissues, (7) the evolution of the

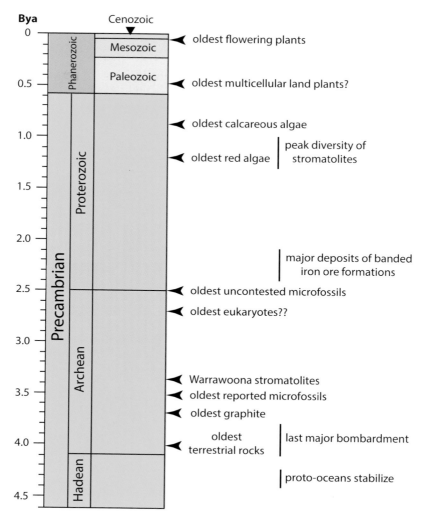

Figure 1.1. A schematic of the geological column showing its chronology in billions of years (to the left) and highlighting some major evolutionary events (to the right). The vast majority of the geological column is occupied by the Precambrian, and many of the most important historical events are recorded in this span of time. For example, the origins of living things are indicated by the oldest known deposits of graphite and stromatolites. The major deposits of banded (oxidized) iron ores indicate the presence of free oxygen (and thus the evolution of photosynthesis). The appearance of the red algae in the fossil record is evidence for the evolution of multicellular eukaryotic organisms, and the fossil remains of land plants indicate that these organisms evolved features that permitted growth in the air. It is important to bear in mind that the oldest evidence for any of these evolutionary events is likely not the earliest appearance of these features since future discoveries may reveal still older occurrences.

seed habit, and (8) the rise of the flowering plants. The first four of these events are treated in this chapter. The last four are treated in the chapter 2. The division is somewhat arbitrary, since history is a continuous process. However, the eight events selected for review are arranged in chronological order, and partitioning them into two equal-sized groupings is parsimonious.

Compared with the origins of life, subsequent historical events may seem pale. But each of the eight major events is of comparable importance to life's story because without them the world as we know it would not exist. Perhaps the most obvious of these events is the evolution of metabolism (such as photosynthesis, the chemically intricate process that converts light energy into chemical energy). Living things are characterized by having the ability to process energy and matter in ways that permit them to survive, grow, and reproduce. Some of life's basic and generalizable metabolic systems are the pentose phosphate pathway, glycolysis, and photosynthesis. In one way or another, nearly all forms of life depend on these processes. For example, the evolution of photosynthetic organisms, particularly those that release oxygen as a by-product, dramatically changed the physical and biological environment of the ancient Earth (box 1.1). Equally important to the history of all life was the evolution of cells containing the microscopic membrane-bound objects called organelles, whose internal structure and chemical composition are highly specialized to participate in life's various biochemical and genetic processes. The absence or presence of organelles in cells divides all living things into two great organic camps: the prokaryotes, organisms whose cells lack organelles, and the eukaryotes, whose cells contain them. The prokaryotes comprise two large and distinctively different groups of prokaryotes—the Archaea and the Eubacteria (fig. 1.2). The eukaryotes encompass all other forms of life—plants, animals, and fungi. The prokaryotes are the more ancient and undoubtedly the more physiologically resilient and ecologically successful forms of life. Current evolutionary theory states that the first eukaryotes evolved from ancient loose symbiotic confederations of prokaryotic-like forms of life. Thus, the "great organis-

Box 1.1. Banded Iron Formation

Banded iron ore formations consist of comparatively thin layers of iron oxides (either hematite or magnetite) that are typically red in color alternating with iron-poor shales and cherts containing very think layers of iron oxides (fig. B.1.1). Formations that are particularly rich in iron were deposited between 2.4 and 1.9 Bya during a geological interval referred to as the great oxygenation event, so called because it was in this time interval that significant quantities of oxygen were released by photosynthetic cyanobacteria-like organisms. Iron oxides formed as a result of oxygen combining with iron dissolved in Earth's ancient oceans. Cycles of iron oxide precipitation resulted in thin layers of the ocean floor that when anoxic would have resulted in the formation of shales and cherts. Each band (referred to as a varve) reflects periods of oxygen availability followed by a return to anoxic conditions, possibly reflecting cycles of cyanobacterial "blooms." Additional geochemical data indicate that Earth's oceans were acidic with large amounts of dissolved nickel and iron, which served as sinks for dissolved oxygen. With continued oxygenation, however, these minerals became oxidized and precipitated out of the waters. Over time, the oxygen released by photosynthetic organisms reached Earth's surface waters and eventually changed the atmosphere from a reducing atmosphere to an oxidizing atmosphere.

Figure B.1.1. The author with an example of banded iron ore, Dresden, Germany.

mic wedge" between prokaryotes and eukaryotes was not as sharply defined in the distant past as it is now. Indeed, many prokaryotes manifest a process called pseudo-sexuality in which genetic materials are exchanged between neighboring cells in a seemingly disorganized manner. As in the case of eukaryotes, the introduction of novel genetic information into prokaryotic populations relies on chance mutations.

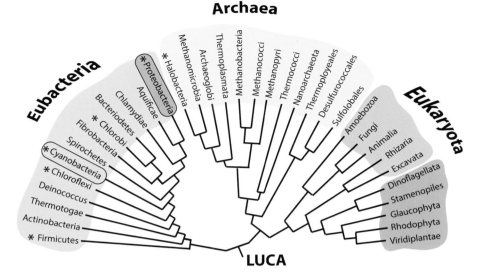

Figure 1.2. Schematic showing the broad phylogenetic relationships among the prokaryotes (Eubacteria and Archaea) and the eukaryotes (Eukaryota consisting of algae, plants, animals, and fungi) that descended from a last universal common ancestor (LUCA). The cyanobacteria and the proteobacteria are highlighted because, according to the endosymbiotic theory, cyanobacteria-like and proteobacteria-like organisms respectively evolved into modern day chloroplasts and mitochondria. Clades within the Eubacteria and Archaea that have photosynthetic representatives are denoted by an asterisk.

Organisms capable of sexual reproduction likely appeared collaterally with or shortly after the first eukaryotes evolved, although some believe a significant time gap exists between these two events. What can be said with certainty is that sexual reproduction (as defined with eukaryotes in mind) required the evolution of meiosis, the "two stage" type of cell division producing cells with half the chromosome number of the cells from which they arise. Meiosis results in genetic recombination, whose evolutionary consequences are treated in chapter 3. The eukaryotes can mix and blend the traits of different parents through sexual reproduction and meiosis. Although it is somewhat surprising, biologists continue to debate the importance of sexual reproduction to evolution, and there is no general agreement about the benefits conferred on organisms by genetic recombination (we will explore this topic again in chapters 3 and 7). The bacteria continue to be the most successful forms of life without benefit of sexual reproduction, but genetic recombination continues to be an effective way of adapting organisms to new habitats and adopting new and different forms of biological organization. Sexual reproduction and the sequestration of reproductive cells from somatic cells may have evolved to deal with the decreasing mutation rates associated with increasing organismic size. Larger organisms reproduce at a slower rate than smaller organisms, but they tend to accumulate genetic mistakes the more times cells divide. Arguably, the isolation of reproductive cells (the germ line) from somatic cells, meiosis, and sexual reproduction may have been required to engender sufficient genetic variation as the number of mutations increased in proportion to the increasing body size of eukaryotes. The isolation of a germ line with a concomitant reduction in cell divisions would also have reduced the accumulation of potentially deleterious genetic anomalies in reproductive cells. (This topic is taken up in chapter 6, wherein we will also discuss the possibility that meiosis may have also evolved before sexual reproduction as a method of correcting for the spontaneous doubling of chromosome numbers, called autopolyploidy.) What can be said without doubt is that the appearance of unicellular eukaryotes presaged multicellular organisms and the evolution of the most visibly conspicuous forms of aquatic and terrestrial life.

Life's Beginnings

Precisely how life began is still a mystery, particularly since the prevailing conditions of the early Earth were inimical to life as we know it today. The embryonic Earth was a protoplanet devoid of atmosphere, cracked by volcanic upheaval and lightning, flooded by lethal ultraviolet light, and rained on by the rocky debris of a juvenile solar system still actively in the violent process of birth—a world whose original surface was molten rock and whose first atmosphere, largely composed of hot hydrogen gas, was vented into space almost as quickly as it emerged from the Earth's radioactive core. Although hard to conceive of, these were the conditions when Earth formed roughly 4.6 billion years ago (see fig. 1.1). Earth's first age is called the Hadean, reflecting the netherworld status of our juvenile planet. Nonetheless, with the passage of a comparatively short time, Earth's version of Hades cooled sufficiently to retain the first proto-oceans formed from torrential rains when water condensed in the atmosphere. Earth continued to be bombarded by fragments of a still forming solar system, and, although possibly diminished, sources of energy like lightning, volcanoes, and ultraviolet light continued to inspire molecular upheaval, breaking and forming molecular bonds in the atmosphere and the shallow proto-oceans.

On the basis of the gases vented from currently active volcanoes, Earth's early atmosphere likely consisted of a mixture of water vapor, carbon dioxide, carbon monoxide, nitrogen, methane, and ammonia, but its precise composition is still debated. Most textbooks categorically state that Earth's first atmosphere was a strongly reducing one, in large part because the first successful laboratory experiments to abiotically synthesize organic molecules used a strongly reducing atmosphere (see below) and because the energies required to abiotically synthesize simple organic molecules like amino acids from raw ingredients such as carbon dioxide and ammonia are much lower in a strongly reducing atmosphere (such as one composed of methane, ammonia, water vapor, and hydrogen) than in a strongly oxidizing one (composed of carbon dioxide, water vapor, nitrogen, and oxygen). Nevertheless, it is worth noting that geochemical and astronomical evidence opens the possibility that the ancient atmosphere was not as reducing as formerly believed but may have been very much like that of today without, naturally, the effects of life (an atmosphere dominated by N_2 and CO_2, with a small ration of noble gases along with some H_2 and CO).

For convenience, we will discuss the continuous evolution of early life in four stages:

(1) The accumulation of abiotically synthesized small organic molecules (monomers) such as amino acids and nucleotides.

(2) The abiotic joining of these monomeric chemicals into polymers such as proteins, lipids, and nucleic acids.

(3) The origin of the heredity molecules ribonucleic acid (RNA) and deoxyribonucleic acid (DNA).

(4) The aggregation of abiotically synthesized molecules into cell-like membrane-bound droplets (called protobionts) that had an internal chemical environment differing from the external chemical environment.

Notice that evolution technically does not enter into the first two stages because evolution requires heritability that in turn requires heredity molecules. Although some may talk about "the evolution of the solar system" or "the evolution of the Earth's atmosphere," the use of the word "evolution" in these contexts must be perceived as having a different meaning from that of a biologist's.

Of these four stages, the first is the most easily simulated in the laboratory. Organic monomers can be abiotically synthesized with comparative ease using a variety of energy sources and a range of atmospheric compositions believed to mimic early Earth conditions. This simulation was first widely publicized with the landmark experiments in 1952 of Stanley L. Miller (1930–2007) and Harold Urey (1893–1981), who created a strongly reducing "mock" atmosphere of water vapor, hydrogen gas, methane, and ammonia (probably too strongly reducing, according to recent thinking) suspended over a mock ocean of heated water (fig. 1.3). After continuously subjecting the atmosphere to an electrical discharge using tungsten electrodes (to simulate lightning), Miller and Urey chemically analyzed the aquatic residue in their experimental system and found a variety of organic compounds (fig. 1.4). The most abundant of these was glycine, a common amino acid. Additionally, many of the other standard twenty amino acids were identified, although in lesser amounts than glycine (box 1.2). Subsequent experiments using different recipes for Earth's

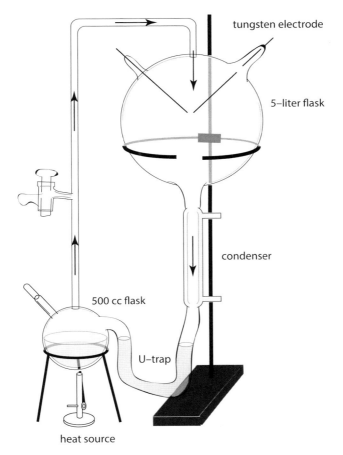

tungsten electrode

5-liter flask

condenser

500 cc flask

U-trap

heat source

Figure 1.3. Schematic of the reaction vessel and ancillary equipment used in the Urey-Miller experiments. Distilled water was placed in a 500 cc flask that was heated to release water vapor that was cycled into a 5 liter flask containing a mixture of gases mimicking Earth's ancient reducing atmosphere. The gas mixture (water, methane, ammonia, and hydrogen) was exposed to continuous electrical discharge by means of two tungsten electrodes to mimic lightening. The chemical reactants occurring in the atmosphere precipitated as "rain" and were collected in a U-trap. The arrows placed in various parts of the schematic show the direction of the cycling reactants and products.

ancient atmosphere have abiotically synthesized several sugars and lipids as well as the compounds generated in the original Urey-Miller experiment. Equally important was the discovery that the purine and pyrimidine bases found in RNA and DNA can also be synthesized under comparatively mild chemical conditions and by simulating the

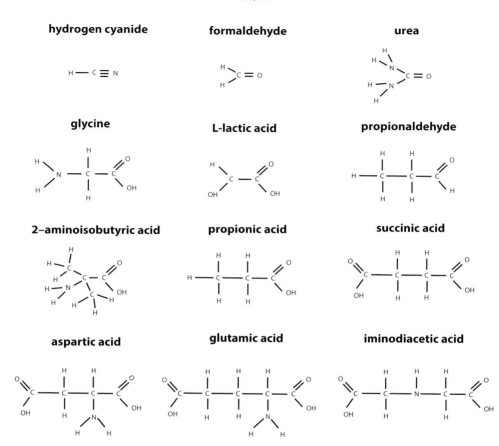

Figure 1.4. Diagrams of a few of the chemicals produced abiotically in the Urey-Miller experiments. In addition to toxic compounds such as hydrogen cyanide and formaldehyde, primary metabolites such as glycine, L-lactic acid, and propionaldehyde are produced. The most abundant product produced in the original experiments is glycine (the smallest of the 20 amino acids normally found in biotically produced proteins).

extraordinarily violent effects of asteroid collisions with earth. Also, when phosphate was added to a mock ocean, adenosine triphosphate (ATP) could form spontaneously. ATP is the major energy-transferring molecule in living cells. The demonstration of the abiotic synthesis of ATP under conditions mimicking those of early Earth is important to any theory of the origin of life.

More recent experiments by Markus A. Keller, Alexandra V. Turchyn, and Markus Ralser have shown that the reaction pathways cen-

Box 1.2. Why H, O, N, and C?

Living things on Earth are composed almost entirely of 16 to 21 elements, 16 of which occur in almost all organisms and 5 of which are found in particular groups. The elements found in most living things are H, O, N, C, P, S, Na$^+$, K$^+$, Mg^{2+}, Ca^{2+}, CL$^-$, Mn, Fe, Co, Cu, and Zn. Those found at low concentrations in special groups of organisms are B, Al, V, Mo, and I. With the exception of molybdenum and iodine, most of the common *bio*elements occur within the lightest 30 of the 92 natural elements. These tend to be the most abundant in the universe. However, an argument for their occurrence in living things based on their relative abundance is not consistent with the occurrence of elements such as carbon, nitrogen, and sulfur. The special distinction of these elements is that many of them are small and achieve stable electronic configurations by gaining and sharing a few electrons with other elements. For example, hydrogen, oxygen, nitrogen, and carbon gain 1, 2, 3, and 4 electrons, respectively. In addition, the smallest elements also make the strongest bonds. Note further that carbon, nitrogen, and oxygen are the only elements that regularly form double and triple bonds, which makes them chemically versatile. By the same token, calcium is a ubiquitous second messenger in calmodulin-mediated signal-perception across diverse taxa. Certainly, it is abundant in the environment accessible to living things. However, it is a cytotoxin at the millimolar quantities found outside the cell. Because phosphate-based energy metabolism would be severely inhibited if intracellular Ca^{2+} concentrations approached the concentrations found outside the cell, cells have evolved efficient methods for removing Ca^{2+} from the cytosol (lowering its concentration to approximately 0.1 µM). Whatever the circumstances, because the concentration of free Ca^{2+} in cells is approximately 10,000 times lower than that in the environment, cellular metabolism is intrinsically very sensitive to any change in cytosolic Ca^{2+} concentration, which makes it an extremely useful signaling element.

It is suggested from time to time that silicon could replace carbon in some forms of life on another planet because, like carbon, it can combine with itself to form long chains and because it is roughly 135 times more abundant than carbon in the Earth's crust. However, a silicon-based life form has two drawbacks: (1) silicon chains can be attacked by molecules with lone pairs of electrons, and thus cannot exist in the presence of oxygen, ammonia, or water, and (2) silicon cannot make double bonds to oxygen (as can carbon) and thus, in the presence of oxygen, silicon forms huge silicon dioxide polymers (such as quartz) that can participate in additional chemical reactions only by breaking covalent bonds. Therefore, judging from Earth's earliest surface chemistry, a silicon-based life form on another planet is a highly unlikely possibility.

tral to cellular metabolism, the pentose phosphate pathway and glycolysis, can be reconstructed nonenzymatically by Earth's Archean ocean chemistry. The pentose phosphate pathway (also called the phosphogluconate pathway) is a reaction sequence parallel to glycolysis. It generates nicotinamide adenine dinucleotide phosphate (NADP$^+$),

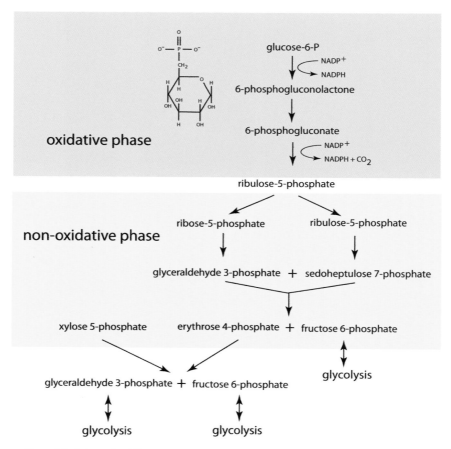

Figure 1.5. Schematic of the pentose phosphate pathway wherein NADPH is synthesized during an oxidative phase of enzymatic reactions and pentoses are synthesized during a non-oxidative phase of reactions. Unlike glycolysis (see fig. 1.6), this pathway is primarily anabolic (it constructs rather than breaks down chemicals). In plants, the pentose phosphate pathway takes place in plastids. In most other organisms, these reactions occur in the cytosol.

which serves as a cofactor in anabolic reactions such as lipid and nucleic acid synthesis, and pentose (5-carbon) sugars (fig. 1.5). Like the pentose phosphate pathway, glycolysis is extremely ancient and occurs in nearly all kinds of organisms, both aerobic and anaerobic. However, unlike the pentose phosphate pathways, which constructs molecular entities, glycolysis sequentially converts glucose into pyruvate and releases energy to form ATP and NADH$^+$ (fig. 1.6). In tandem, these two pathways may be thought of as the yin and yang of life's energy flow. Keller and coworkers were able to simulate 29 abiotic reactions including the formation and/or interconversion of glucose, pyruvate, the nucleic acid precursor ribose-5-phosphate, and the amino acid precursor erythrose-4-phosphate, antedating reaction sequences similar to that used by the metabolic pathways (see fig. 1.5). Moreover, the Archean ocean mimic containing iron increased the stability of the phosphorylated intermediates and accelerated the rate of intermediate reactions and pyruvate production. The catalytic capacity of the reconstructed ocean was attributable to its metal content and was particularly sensitive to ferrous iron Fe(II), which is understood to have had high concentrations in the Archean oceans.

Regarding the second stage in the evolution of life, Sidney W. Fox (1912–1998) in 1958 was able to abiotically synthesize polypeptides, called proteinoids, by heating amino acid mixtures on hot sand, clay, or rocks and subsequently flooding them with water. Polypeptides and other organic polymers may have been synthesized during the early history of the Earth when mixtures of monomers carried by ocean waves were repeatedly dehydrated and rehydrated on the surfaces of a cooling volcanic shoreline. The abiotic synthesis of complex polymers may also have occurred on the surface of cool clay, which has the capacity to concentrate amino acids and other monomers from dilute aqueous solutions. Metal atoms, such as iron and zinc, found in a variety of clays, are known to expedite dehydration reactions that precede the polymerization of monomers into larger, more complex chemicals. Other experiments show that clays can also mediate the abiotic synthesis of RNA from its constituent monomers.

Clearly, from a genetic perspective the most complex and important biopolymers are the two nucleic acids, RNA and DNA. The evolution of the genetic code was obviously a pivotal event, since evolution is technically impossible unless traits are inherited. Because all living organisms carry their genetic codes in the form of RNA and DNA, there is every reason to believe that these heredity molecules are ex-

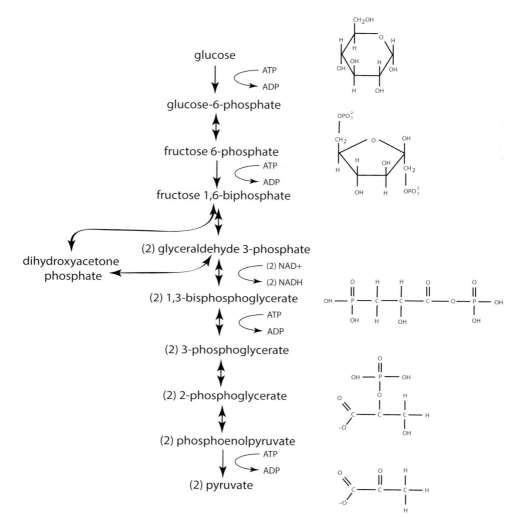

Figure 1.6. Schematic of glycolysis wherein glucose is enzymatically broken down to yield two molecules of pyruvate, indicated by (2). Unlike the pentose phosphate pathway (see fig. 1.5), this pathway is primarily catabolic (it breaks down chemicals). Glycolysis occurs in anaerobic organisms as well as aerobic organisms. Like the pentose phosphate pathway, glycolysis is extremely ancient.

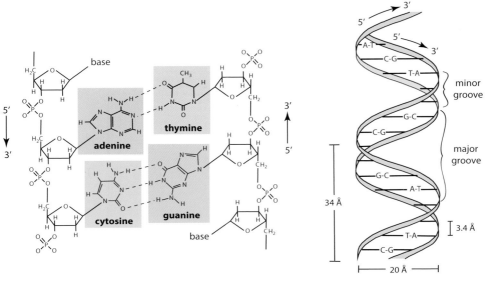

Figure 1.7. Schematic of the molecular structure of DNA (left) and the dimensionality (in angstroms) of the double-stranded DNA molecule (right). The two categories of the four DNA nucleobases (purines and pyrimidines) are shown in blue and green, respectively. The fifth nucleobase, which substitutes for thymine in single-stranded RNA molecules, is uracil (U), which is similar to thymine but lacks the 5' methyl group ($-CH_3$).

tremely ancient and can be traced back to the last common ancestor to all living things. Among current organisms, DNA is the master heredity molecule of the two nucleic acids. It transcribes genetic information into RNA, which subsequently translates this information during the biosynthesis of specific enzymes and other proteins that regulate and participate in the chemical reactions within cells. Despite the deferential role RNA plays in current cells, DNA and RNA essentially use the same genetic code—sequences of nucleotides arranged in strands. Each nucleotide in RNA and DNA consists of a sugar, a phosphate group, and one of five nitrogen-containing bases. In DNA the sugar is deoxyribose and the four bases are adenine (A), guanine (G), cytosine (C), and thymine (T) (fig. 1.7). In RNA the sugar is ribose and the base uracil (U) substitutes for thymine. The four bases pair very specifically in the double-stranded molecular architecture of DNA. Specifically, adenine pairs with thymine (A-T), while guanine pairs with cytosine (G-C). Adenine on the DNA molecule is transcribed as uracil on the RNA molecule (A-U). The bases in DNA constitute the genetic alphabet, triplets of bases forming the genetic words that encode the information for protein structure. For example, the triplet CTT in DNA is transcribed as CUU in RNA, where it designates the information to add the amino acid leucine to the growing strand of a polypeptide. The universality of these DNA and RNA features indicates that the last common ancestor of all living things stored its genetic information in some kind of nucleic acid polymer that could specify all of life's intricate biochemical machinery by encoding the information for protein structure.

Given that there are two nucleic acids, DNA and RNA, an intriguing question is whether DNA was the first heredity molecule or whether RNA served this function only to become displaced by DNA later in the history of life. Many scientists favor the latter hypothesis and speak of an ancient RNA world (Gesteland and Atkins 1993; Orgel 1994). The concept of the RNA world has gained considerable credibility. For example, although Urey-Miller type experiments show that the purine nucleic acid bases (adenine and guanine) form readily under laboratory-simulated early Earth conditions, an efficient, plausible chemical pathway for the abiotic synthesis of the pyrimidine bases remained elusive until the experiments by Michael P. Robertson and Stanley L. Miller in the latter part of the past century showed that as its concentration increases in water, urea can react with cyanoacetaldehyde to form cytosine, which in turn forms uracil. These authors speculate that urea and cyanoacetaldehyde, which are easily synthesized under Earth's early conditions, may have reacted in pools of water slowly evaporating under a hot sun to spawn the chemical reactions requisite to abiotically synthesize cytosine and uracil or perhaps even reacted in between ice crystals in periods of glaciation. The Urey-Miller-Robertson experiments show that Earth's ancient conditions were likely very conducive to the abiotic synthesis of all the bases required to make RNA and ultimately DNA.

The RNA-world hypothesis receives further support with the discovery of ribozymes, enzymes composed of RNA. Until the discovery of ribozymes, the term enzyme was typically reserved for proteins that serve as catalysts. Now the concept must be expanded to include ribozymes, which join together short strands of RNA and can synthesize RNA strands up to forty nucleotides long when monomers of RNA are mixed with zinc in a test tube. We now know that the RNA in ribosomes acts as an enzyme rather than as an analog to a structural protein as formerly believed. Ribosomes consist of protein and ribosomal RNA (denoted rRNA) that move along strands of messenger RNA (mRNA) and link amino acids (specified by the mRNA trinucleotides) to assemble complex proteins. It is clear that rRNA can by itself catalyze some reactions, giving greater credibility to the hypothesis that RNA was one of the first autocatalytic molecules to appear in Earth's history. To be sure, short-chain polymers of RNA, called oligonucleotides, can be abiotically created in the test tube with comparative ease, and the in vitro evolution of the first RNA known to catalyze a bond-forming reaction on a substrate not containing nucleotides has been reported. If, in the ancient oceans of the Earth, some RNA molecules acquired the ability to link other oligonucleotides together, as ribozymes do today, and if some of these proto-ribozymes replicated autocatalytically, then they could have served as an extremely ancient genetic apparatus. The discovery of the catalytic properties of RNA demonstrates that a single molecular species can function simultaneously as a genome and an enzyme (doing much to resolve the chicken and egg paradox of which came first, functional proteins or information-carrying nucleic acids). In turn, this discovery indicates that the first genes may not have been on DNA molecules but on short strands of RNA capable of self-replication.

How the RNA world evolved into the DNA world with cells like those of today is not known, but researchers have provided a very plausible scenario based on chemical and mathematical laws. This scenario assumes that early Earth's oceans contained a jumble of different kinds of autocatalytic RNA molecules with different reaction rates. Over time, one molecular species of RNA eventually came to govern the reaction rates of its autocatalytic compatriots. In doing so, this molecular species assumed the role of the master autocatalytic molecule.

Nevertheless, all of these ancient RNA molecules were essentially naked genes—they replicated for no purpose but their own replication. Experiments and theory also show that very short strands of RNA cannot self-replicate with high fidelity without protein assistants (enzymes in the traditional sense). Consequently, for life to evolve, self-replicating molecules must have evolved specifying proteins assisting in the very process of self-replication. The solution to this seeming paradox is called the hypercycle. As first conceived by Manfred Eigen and Peter K. Schuster, a hypercycle consists of a network of mutually reinforcing chemical processes—crudely described as chemical A supports the synthesis of chemical B, chemical B supports the synthesis of chemical C, and chemical C supports the synthesis of chemical A—until the point is reached where the whole process creates a molecular environment that expands its synthetic capacity to produce more complex molecules (fig. 1.8). In order for it to expand, however, some kind of information-storage molecule must emerge from a hypercycle. In this respect DNA is thermodynamically far more stable than RNA, and so DNA is a superior molecular repository for the genetic code since it is less susceptible to random mutations. Theoretical treatments of hypercycles show that, once evolved from RNA, DNA will take over as the primary heredity molecule by virtue of its storage capacity and stability. The chemical differences between RNA and DNA are comparatively small and can be achieved by relatively small chemical mutations. Once DNA emerged as the heredity molecule, the length of its nucleotide strands could be increased, spelling out the alphabet of base pairs containing all the information required for the machinery of life.

The fourth stage in the early evolution of life was the aggregation of this machinery into a cell-like structure, call a protobiont, whose interior differed chemically from its external surroundings. Protobionts have an essential property of life called boundary—the segregation

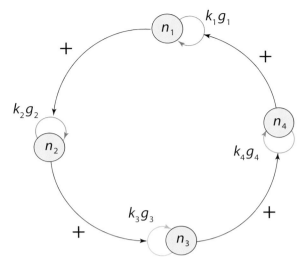

Figure 1.8. Schematic of a hypercycle consisting of four autocatalytic subcycles (n_1, n_2, n_3, and n_4). Each subcycle is a series of self-sustaining chemical reactions with reaction and degradation rates (k and g, respectively). In turn, each subcycle affects the rates of the next (n+1) subcycle, which links the reaction and degradation rates throughout the four-membered hypercycle. Other subcycles can be added to construct and integrate larger and larger hypercycles. This paradigm shows how metabolic pathways such as the pentose phosphate and glycolysis pathways can be integrated.

of chemical reactions from the external environment by some sort of organic container. Protobionts form spontaneously in the laboratory under early Earth conditions. They also can become coated with phospholipids, which are natural membrane builders. Likewise, cooling solutions of proteinoids, like those synthesized by Sidney W. Fox, can self-assemble into small spheres that, when coated with a lipid membrane, swell or shrink in solutions of different salt concentrations, store energy in the form of a membrane potential, and catalyze chemical reactions when enzymes are added to the solution (for example, maltose can be synthesized from glucosephosphate in the presence of amylase and phosphorylase). Protobionts and microspheres grow in size by absorbing proteinoids and lipids from their local environment, subsequently splitting into companion cell-like structures after reaching a critical, apparently unstable size.

A precise definition of life has eluded biologists. But it is clear that the protobionts and microspheres thus far created in the laboratory are not living things. They lack the sophisticated ensemble of biosynthetic pathways coordinated by enzymes synthesized from the genetic instructions encoded on DNA molecules by means of intermediary RNA molecules. Nevertheless, at some point in history the intricate machinery of life was assembled within a cell-like entity. This assembly need not have occurred all at once. The metabolism of the first living entities on Earth could have relied to some degree on externally supplied, abiotically synthesized monomers and simple polymers. Over time, however, the external supply of these abiotically synthesized chemicals would have gradually diminished, perhaps as they were used up by expanding populations of the first forms of life or as Earth's environment changed and the rates of abiotic synthesis in the proto-oceans declined. Under any circumstances, mutation and natural selection would have created and favored biosynthetic variants of cells that were metabolically more self-sufficient. In this way, over many hundreds of thousands of years, living matter progressively distanced itself from its ancestral physiological dependency on abiotic synthesis. This distancing of life required an amplification of the genetic library to encode for chemical pathways originally occurring outside the cell. In a very crude sense, the expansion of the genome allowed life to turn the chemical machinery of the abiotic world outside in. By doing so, early life acquired four prominent features shared by all contemporary living cells: the capacity to grow, self-replicate, inherit traits from ancestors, and mutate. With these four properties life emerged, crossing the thin, albeit profoundly important, line separating a purely chemical system from a strictly living and evolving one.

One plausible consequence of the passage from the inanimate to the animate world may have been viroids, virusoids, and viruses. Modern viroids and virusoids are the smallest known autonomously replicating molecules, and some of them have exactly the same enzymatic capabilities envisioned for the first self-replicating RNA molecules. Indeed, some workers suggest that viroids are living fossils that trace their ancestry back to the ancient RNA world. The viruses are extremely diverse in size, structure, and genomic organization. Nevertheless, every known virus is an obligate intracellular parasite whose genome consists of DNA or RNA, or both and encodes for an encapsulating protein coat. Three theories have been proposed for the origins of these life forms: (1) the regressive theory, which holds that viruses are degenerate forms of intracellular parasites; (2) the host-cell RNA theory, which argues that viruses arose from normal cellular components that gained the ability to autonomously replicate and thus evolve independently; and (3) the theory that viruses originated and evolved along with the most ancient molecules that had the ability to self-replicate. Whether one or more of these three contending theories is correct may never be known with utter confidence. Nevertheless, a reasonable case has been made that some RNA viruses are the evolutionary descendants of extremely ancient life forms that coevolved with the first self-replicating RNA molecules. This hypothesis argues that the antecedents of the RNA viruses were originally integrated parts of the normal metabolic machinery of the first prokaryotic cells but became obligate intracellular parasites later in their evolutionary course.

That life arrived at its present level of biochemical complexity through initial sequences of abiotic chemical reactions is remarkable. Even more remarkable is life's precociousness. Although geologists do not yet know the exact timing, the first atmosphere and oceans formed during the closure of the Hadean (see fig. 1.1). Paleontologists also tell us that the most ancient fossil cells are found in Archean rocks (fig. 1.9). However, it is worth remembering that most ancient means minimum age. Exploration of the fossil record conceivably could push the first occurrence of cells still further back in time. At most, therefore, only 300 million years separate nature's trial and error experiments with abiotic chemical synthesis and the first rudimentary forms of cellular life. True, by human standards this is a long time. But when dealing with evolution, human standards of time must be abandoned. We measure our lifetimes in decades. We measure recorded human

Figure 1.9. Examples of Precambrian fossils. a. Stromatolites in the Warrawoona succession, Australia, dated as 3.4 billion years old (scale is in centimeters). Stromatolites are the fossil remains of the biological activity of marine biofilms that when living precipitated minerals dissolved in water to form these large accretionary structures. b. Microorganisms from the Gunflint Formation, Canada, dated as 1.88 billion years old. The thick walled cell in the upper right is 10 microns in diameter. Courtesy of Dr. Andrew H. Knoll (Harvard University).

history in thousands of years. But the paleontologist typically measures time in hundreds of millions of years. Given the intrinsic complexity of biological metabolism and self-replication, the evolution of the inaugural forms of life occurred in a "brief instant."

Photosynthesis

Life gained tremendous physiological independence when it evolved the ability to convert light energy into chemical energy. This ability, called photosynthesis, is found in current representatives of both the eukaryotes and the prokaryotes. For the purposes of this book, the photosynthetic eukaryotes are considered plants and encompass the algae and the embryophytes. With few exceptions (in the Heliobacteria), most photosynthetic prokaryotes belong to the Eubacteria, a kingdom containing thirteen clades, five of which have one or more photosynthetic representatives as for example the purple bacteria, which include the sulfur and nonsulfur bacteria (in the proteobacteria), the green sulfur bacteria (in the Chlorobi), the filamentous green nonsulfur bacteria (in the Chloroflexi), the gram-positive bacteria (in the Firmicutes), and the cyanobacteria (see fig. 1.2). Of these, only the cyanobacteria release oxygen as a by-product of photosynthesis. In contrast to the oxygenic cyanobacteria, all other photosynthetic bacteria are non-oxygen-evolving (anoxygenic) photosynthetic organisms.

Although the first living entities may have been chemoautotrophs that evolved in hot, chemically active microenvironments (Bengtson 1994), the available evidence indicates that the photosynthetic prokaryotes evolved early in Earth's history. Fossils discovered in Archean rocks are morphologically very similar to present-day photosynthetic Eubacteria, and an isotopic signature characteristic of autotrophic carbon fixation as a consequence of photosynthesis is reported for rocks 3.8 billion years old (Schidlowski 1988). Whether these early Archean organisms were oxygenic photosynthetic prokaryotes, like the cyanobacteria, or anoxygenic photosynthetic prokaryotes, like most of the Eubacteria, is uncertain because of the morphological similarities among otherwise physiologically very diverse bacteria. Because the atmosphere of the early Earth was probably more reducing than the present atmosphere, and because the vast majority of present-day photosynthetic bacteria neither produce nor consume molecular oxygen—in fact, most are strict anaerobes that are poisoned by free oxygen—it is generally believed that anoxygenic photosynthetic prokaryotes evolved first. Furthermore, these organisms very likely were obligate anaerobes or extremely inefficient aerobic respirers. What can be said with assurance is that by Archean times, prokaryotes had physiologically diversified, some achieving a sophisticated level of metabolic organization capable of using the energy of light to manufacture organic compounds.

Attempts to reconstruct the evolution of photosynthesis necessarily focus on comparisons among the Eubacteria, because these organisms possess a number of features believed to be ancient rather than derived and because the fossil record does not preserve the chemical history of photosynthesis in sufficient detail. Among the Eubacteria, three features are almost universally shared among photosynthetic representatives: an antenna/reaction center organization of the photosynthetic unit; the use of chlorophyll-based pigments as the principal light-gathering molecule; and the presence of a heterodimeric protein core within the reaction center. These shared, presumably ancient features may inspire the incorrect belief that all photosynthetic prokaryotes evolved directly from a single common ancestor. However, the evolutionary origin and development of photosynthesis appears to have involved the lateral transfer of genes between different lineages of bacteria; that is, different parts of the photosynthetic apparatus at present found in the cyanobacteria and plants trace their ancestry to at least two prokaryotic lineages, each of which has an analogue in very different Eubacteria. This in turn lends compelling support to the hypothesis that the chloroplasts of eukaryotes evolved when a cyanobacteria-like organism gained entry and took up permanent residence within a heterotrophic prokaryotic host cell. This endosymbiotic hypothesis is treated in greater detail later. For now the important point is that the origin and fine-tuning of photosynthesis may not have proceeded along a linear evolutionary pathway.

The nonlinearity underlying the evolutionary history of photosynthesis is best appreciated by returning to the three universalities mentioned earlier. Photosynthesis involves the participation of pigment molecules that absorb light energy and give up electrons as a consequence of their elevated energy states. Photosynthesis also involves protein molecules that accept and transfer the electrons stripped from pigment molecules excited by light energy. These proteins and pigments are organized into a photosynthetic unit consisting of a light-gathering antenna system of pigment molecules attached to the electron transfer proteins arranged in what is called a reaction center. The antenna/reaction center organization of the photosynthetic unit is an almost universal feature of all photosynthetic organisms, both Eubacteria and Eukarya. The sole exceptions to this generality are the Halobacteria (in the Archaea; see fig. 1.2). These photosynthetic prokaryotes live in extremely saline habitats, some species growing in water ten times as salty as seawater. The Halobacteria are unique among the prokaryotic photoautotrophs because they use a retinal-containing protein, bacteriorhodopsin, to trap the energy of light. Bacteriorhodopsin is directly bound to membranes and is not organized in an antennae/reaction center configuration as in other photosynthetic organisms. Because there is no direct comparison between the photosynthetic apparatus of the Halobacteria and the other photosynthetic apparati of other photosynthetic prokaryotes, it is safe to say that photosynthesis, in the literal sense of the word, has independently evolved at least twice.

The second virtually universal feature of photosynthetic organisms is that they contain one or more types of the class of pigments known as the chlorophylls (fig. 1.10). Additional pigments, the carotenoids and bilin pigments, are also used to trap and channel light energy to the reaction center at the heart of the photosynthetic unit. These accessory light-harvesting pigments account for differences in the color of photosynthetic cells. But the role of the accessory photosynthetic

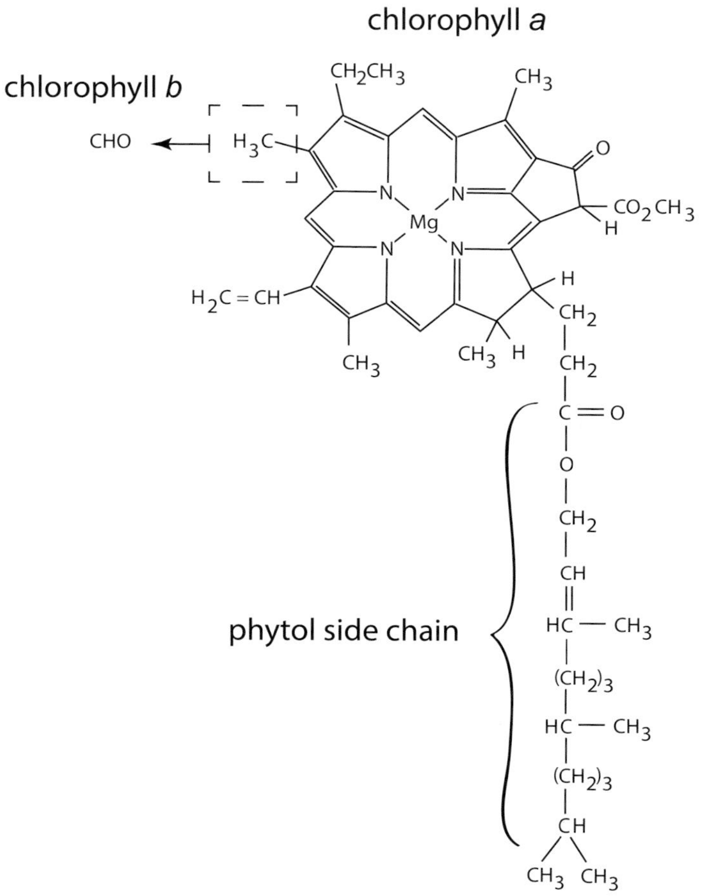

chlorophyll *a*

Figure 1.10. Schematic of the molecular structure of chlorophyll *a* and chlorophyll *b*. The two pigments differ in the functional groups attached to the location indicated by the dashed box. The functional group in chlorophyll *a* is a methyl group ($-CH_3$). The functional group in chlorophyll *b* is an aldehyde group ($-CHO$). Other forms of chlorophyll exist and provide one criterion for distinguishing the different algal lineages. All of the variants of chlorophyll molecular structure may be thought of as "evolutionary tinkering" with an ancient molecule that, once evolved, could not be evolutionarily abandoned in toto.

pigments does not color the fact that the principal underlying pigment in all photosynthetic organisms (other than the Halobacteria) is chlorophyll based—the chlorophylls and the bacteriochlorophylls. The type of chlorophyll-based pigment found in photosynthetic prokaryotes correlates with whether the organism is an anoxygenic or oxygen-evolving photoautotroph. All oxygen-evolving prokaryotes contain chlorophyll *a*; all anoxygenic prokaryotes lack chlorophyll *a* and instead use bacteriochlorophyll to trap the energy of light.

The presence of bacteriochlorophylls in four anoxygenic pro-karyotic clades, which differ in many other important respects, is evidence that these pigments are more ancient than chlorophyll. Cu-riously, however, chlorophyll *a* is an intermediary rather than a final step in the biosynthesis of the bacteriochlorophylls and chlorophyll *b* (fig. 1.11). Thus, the biosynthetic pathways of the chlorophyll-based pigments present something of a phylogenetic dilemma. If biosynthe-sis recapitulates phylogeny—that is, if intermediates in a biosynthetic pathway are the final products of biosynthetic pathways of earlier evo-lutionary forms—it follows that prokaryotes employing chlorophyll *a* evolutionarily preceded those using the bacteriochlorophylls as the principal photosynthetic pigment. Alternatively, the metabolic path-way for the chlorophyll-based pigments may have evolved backward. That is, if organisms typically evolve ways to synthesize essential chemical compounds from progressively simpler compounds as the available stocks of essential molecules are depleted in the environ-ment, then photoautotrophs with bacteriochlorophylls are truly more ancient than those using the chlorophylls as their photosynthetic pig-ments. Yet another possibility is that the earliest reductase enzyme for reducing the tetrapyrrole ring was nonselective.

Upon reflection, both the biosynthetic bootstrapping and the de-constructionist hypotheses seem wanting. For example, it is difficult to imagine that a molecule like chlorophyll (or its precursor, proto-chlorophyllide) could have been abiotically synthesized. On the other hand, the simplest precursors of the chlorophyll-based pigments are amino acids and other colorless compounds that could not have served as pigments (for example, 5-aminolevulinic acid). One plausible solu-tion is that the biosynthetic pathway for the chlorophyll-based pig-ments evolved from the middle outward (Blankenship 1992). The heliobacteria (in the Firmicutes), which contain the pigment bacte-riochlorophyll *g*, adds credibility to this middle outward biosynthetic

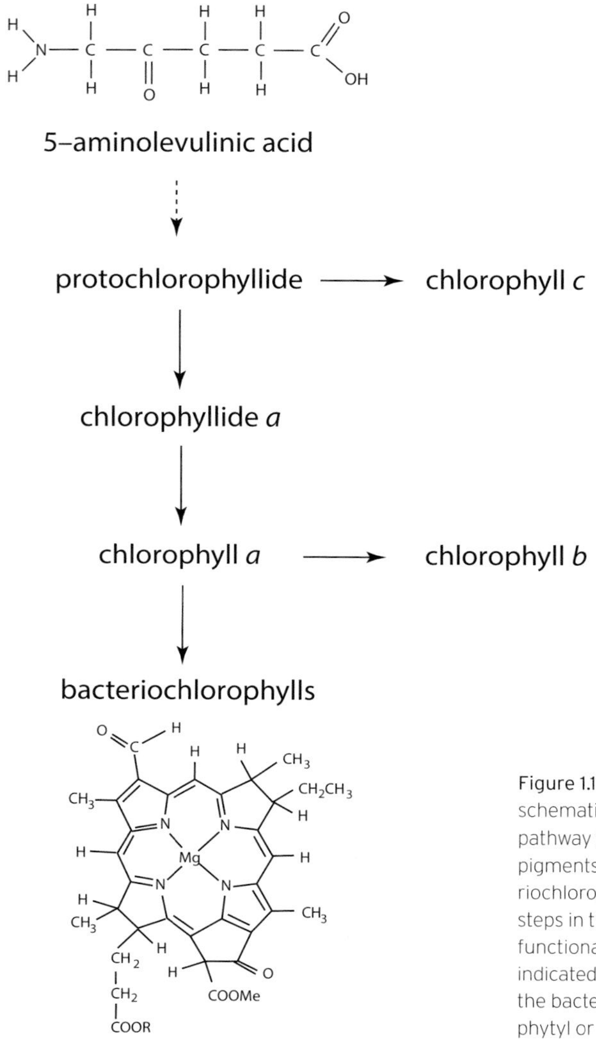

5–aminolevulinic acid

protochlorophyllide ⟶ chlorophyll *c*

chlorophyllide *a*

chlorophyll *a* ⟶ chlorophyll *b*

bacteriochlorophylls

Figure 1.11. A highly simplified schematic of the biosynthetic pathway to chlorophyll-based pigments. Note that the bacteriochlorophylls occupy terminal steps in the pathway. The functional group at the location indicated by R in the formula for the bacteriochlorophylls is either phytyl or geranylgeranyl.

hypothesis because bacteriochlorophyll *g* has the chemical properties of chlorophyll *a* and bacteriochlorophyll *b*. Therefore, it could have served as a possible intermediate pigment derived from an ancient biosynthetic pathway that was subsequently elaborated upon and fine-tuned to give rise to the chlorophylls on the one hand and the bacteriochlorophylls on the other. The heliobacteria are anoxygenic photosynthetic organisms and therefore fall in line with the generally held hypothesis that the first prokaryotes were obligate anaerobic bacteria.

An intriguing hypothesis has been advanced suggesting that bacteriochlorophyll evolved in ancient bacteria living near hydrothermal systems as an adaptation to finding and staying near their source of food (Nisbet, Cann, and Van Dover 1995). Molecular evidence supports the belief that the earliest Eubacteria were thermophilic organisms, well suited to living near hot, turbulent hydrothermal vents, which were widespread on Earth during the early Archean. Similar vents existing today provide a rich assortment of nutrients to chemotrophic bacteria, which risk starvation whenever they lose touch with their nutrient supply. Thus, organisms like the purple bacterium *Rhodospirillum centenum* are thermotactic and move toward the infrared light emitted by vents. Euan Nisbet and colleagues point out that the absorption spectra of bacteriochlorophylls match the thermal emission spectra of a hot body submerged in water. They speculate that the chemical precursors to the bacteriochlorophylls may have evolved in thermophilic, chemotrophic bacteria as an adaptation to locating and staying near hydrothermal vents. Ancient thermotactic bacteria could have evolved into phototactic and photosynthetic organisms using the infrared part of the spectrum of sunlight in shallow bodies of water and H_2 and H_2S provided by hydrothermal vents. For an organism that is already surviving as a chemotroph, a comparatively small evolutionary change to allow infrared phototaxis is an adaptive advantage. In turn, infrared phototaxis could preadapt an organism for optional infrared photosynthesis based on bacteriochlorophyll and, still later, the development of chlorophyll to use visible light.

Regardless of the evolutionary routes that gave rise to them, all the chlorophyll-based pigments absorb visible light efficiently because of their many conjugated bonds. Furthermore, these pigments can delocalize the energy absorbed from photons (the energy can be spread throughout their electronic structures). When seen in this light, the bacteriochlorophylls and chlorophylls appear to confer an advantage over other candidate pigments for photosynthesis. It is also worth noting that the absorption spectrum of chlorophylls falls between the ultraviolet and the infrared regions of the electromagnetic spectrum, and thus between forms of energy that can destroy chemical bonds or denature proteins. Yet, the chlorophyll-based pigments are far from being ideal light-absorbing compounds. Some of the energy they absorb is lost as either heat or fluorescence. Likewise, RUBISCO is not an ideal protein with which to fix CO_2, since it has a higher affinity for O_2 than CO_2 (fig. 1.12)—a reflection of its early evolution during a time when Earth's atmosphere was more reducing and CO_2-rich. Nevertheless, the chlorophyll-based pigments and RUBISCO are more than adequate as witnessed by the fact that they underpin Earth's food chain.

In terms of the evolution of the reaction center, Robert Blankenship points out that the reaction centers of all existing photosynthetic organisms fall into two general categories—those that contain pheophytin and a pair of quinones as the early electron acceptors, and those that use iron-sulfur clusters as early electron acceptors. For example, the green gliding bacteria and the purple nonsulfur bacteria have the pheophytin-quinone type of reaction center, and the green sulfur bacteria and the heliobacteria have the iron-sulfur type of reaction center. Despite their differences, the two types of reaction centers share some important features. Notable among these is that their protein cores are composed of two related yet distinct proteins. That is, the cores of both types of reaction centers contain a heterodimeric protein complex. It is interesting that the two types are found in very different kinds of Eubacteria, indicating perhaps that the ancestor of modern-day photosynthetic prokaryotes may have had a much simpler protein core, perhaps a monomeric protein core. Indeed, it has been suggested that gene duplication in this hypothetical ancestor may have led to a homodimeric protein core whose two protein components subsequently diverged to become heterodimeric (Blankenship 1992).

The notion that the otherwise diverse phyla comprising the Eubacteria are all derived from a common photosynthetic ancestor is

(a)

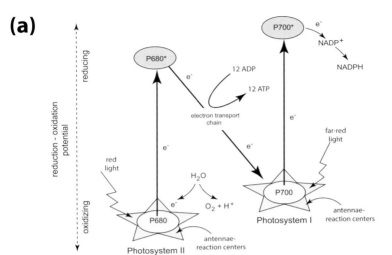

(b)

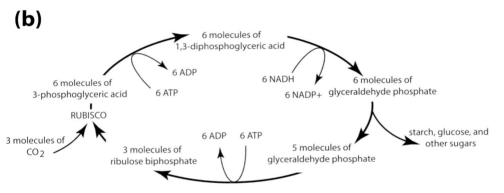

Figure 1.12. Highly simplified schematics of the light reactions (a) and the Calvin cycle (b). a. Schematic of the Z scheme of photosynthesis, which consists of two photosystems (PSI and PSII) connected by an electron transport chain. Red light that is absorbed by PSII produces a strong oxidant, which oxidizes water, and a weak reductant, which is an excited chlorophyll pigment (P680*); far-red light that is absorbed by PSI results in a weak oxidant and strong reductant, which reduces NADP+ to form NADPH. b. Schematic of the Calvin cycle in which the reducing power stored in ATP and NADPH is used to synthesize simple carbohydrates (sugars). CO_2 is added enzymatically by ribulose-1,5-bisphosphate carboxylase/oxygenase (RUBISCO) to a five-carbon sugar (ribulose biphosphate) to produce a six-carbon intermediate that splits into two molecules of 3-phosphoglyceric acid. Subsequent reactions utilize adenosine triphosphate (ATP) and nicotinamide adenine dinucleotide phosphate (NADPH) to form glucose, other sugars, and polymerized carbohydrates such as starch and cellulose.

also supported by another deep similarity between the pheophytin-quinone and the iron-sulfur types of reaction centers. In both types, a quinone serves as an early electron acceptor and is preceded by a chlorophyll-type pigment molecule in the electron transfer chain. Therefore, in addition to the heterodimeric protein core, the early steps in electron transfer within the two types of reaction centers are similar among all known photosynthetic prokaryotes. This similarity indicates that all photosynthetic prokaryotes likely shared a last common ancestor with a reaction center having a chlorophyll-type of pigment molecule preceding a quinone electron acceptor.

The pheophytin-quinone and iron-sulfur types of reaction centers are associated with two very different photosynthetic systems that coexist today only in the cyanobacteria and the photosynthetic eukaryotes. The simplest evolutionary scenario for the occurrence of the two types of reaction centers in the cyanobacteria, which are the simplest oxygen-releasing organisms, is that two kinds of bacteria, one possessing the pheophytin-quinone type of reaction center and another kind of bacteria possessing the iron-sulfur type, fused genetically to obtain a photosynthetically chimeric prokaryote. Initially the two photosystems may have operated independently, only later becoming functionally linked as they are in the cyanobacteria.

This hypothesis helps explain how the light reactions found in the cyanobacteria may have evolved (and via endosymbiosis, how the light reactions ended up in plants). The light reactions involve the participation of two photosystems, called photosystem I and photosystem II, each associated with its own chain of electron acceptors and donors (fig. 1.12). Photosystem I has an iron-sulfur type of reaction center, while photosystem II has a pheophytin-quinone type. With the exception of the cyanobacteria, prokaryotes possess photosynthetic apparati that are analogous to either photosystem I or photosystem II but not both. The most parsimonious hypothesis is that the cyanobacteria evolved from some kind of genetically chimeric prokaryote resulting from the fusion of two kinds of bacteria. Evidence indicates that photosystem I may optimize the efficiency of photosystem II under aerobic conditions. If so, then the fusion of two kinds of bacteria, each containing only one of the two photosystems, may have been particularly adaptive in a photosynthetic world with an atmosphere initially lacking much oxygen.

It is not known whether the chimeric ancestor to the cyanobacteria released oxygen as do present-day cyanobacteria or subsequently evolved the ability to use water as the electron donor molecule to reduce carbon dioxide and so release oxygen as a photosynthetic by-product. Among oxygenic photoautotrophs, water molecules are split to yield electrons, hydrogen ions (protons), and molecular oxygen. In chemical notation, $2H_2O \rightarrow 4e^- + 4H^+ + O_2$. The final electron-acceptor is carbon dioxide, which is fixed in the Calvin cycle to synthesize simple sugars, although the pyridine nucleotide $NADP^+$ serves as an intermediate electron acceptor from the light reactions (see fig. 1.12). The oxygen-releasing photosynthesis of the cyanobacteria, algae, and land plants is often summarized by the simple equation $nH_2O + nCO_2$ in the presence of light and chlorophyll yields $(CH_2O)n + nO_2$, where n is typically designated as 6 to correspond with the biosynthesis of simple sugars, like hexose ($C_6H_{12}O_6$). In contrast, anoxygenic photosynthetic bacteria do not use water as the electron-donor molecule. Some use hydrogen sulfide, thiosulfate, or hydrogen gas instead. For example, the sulfur bacteria use hydrogen sulfide in which $nH_2S + nCO_2$ in the presence of light and chlorophyll yields $(CH_2O)n + H_2O + 2S$. As a consequence, the sulfur bacteria produce globules of sulfur instead of oxygen as their photosynthetic waste product.

As noted, the evolutionary route by which the cyanobacteria acquired the ability to split water and release oxygen is not yet known. One possibility is that a genetically chimeric ancestor possessing the rudiments of photosystem I and photosystem II acquired the ability to use reductants weaker than compounds like hydrogen sulfide but stronger than water. Such an organism may have used Fe^{2+} as a reductant, much like some species of cyanobacteria that have the capacity to oxidize Fe^{2+} to produce the reduced Fe^{3+} ion. The electrons from ions like Fe^{2+} could have been used by a cyanobacteria-like ancestor to oxidize carbon dioxide, and the electron flow could have been used to pump protons across membranes to establish the energy gradients necessary to make ATP.

Much more needs to be known about the evolutionary origin and development of both anoxygenic and oxygenic photosynthetic prokaryotes. At present there are only tantalizing bits of information relating to their origin. What can be said with assurance is that photosynthesis had dramatic biological and physical effects that shaped the course of subsequent evolutionary history. It is clear from geochemical and paleontological data that, after their initial appearance, oxygen-releasing prokaryotes very similar to current cyanobacteria prospered, multiplied, and diversified during the Archean. As a consequence, the primary global productivity of these organisms increased, and, by the end of the Archean and during the early Proterozoic the rate of oxygen production eventually exceeded the rate at which oxygen was consumed by the oxidation of reducing metals in soils and rocks. Somewhere between 2.4 and 2.8 billion years ago, the surface of Earth's ocean and Earth's atmosphere attained 1%–2% of the present-day atmospheric level of oxygen (Knoll 1992). The physical result was the emergence of a stable, oxygen-enriched atmosphere that undoubtedly restricted the habitats available to strict anaerobes to deeper anoxic oceanic regions and that favored the radiation of aerobic prokaryotes near oceanic surfaces. Clearly, aerobic metabolic pathways identical to and differing from that of classical respiration (for example, the flavin oxidase-based respiration of *Thermoplasma*) have evolved independently in a great many different lineages. Thus oxygen, the waste product of cyanobacteria-like Archean organisms, supplied the selection pressure favoring varying degrees of oxygen tolerance in diverse lineages.

The Evolution of the Eukaryotes

The oxygenation of the early Proterozoic atmosphere by early photosynthetic organisms resulted in a physical environment that did not biochemically favor bacteria incapable of tolerating free oxygen. Some aerobic prokaryotes, particularly representatives of the purple non-sulfur bacteria, possess all key features in the respiratory chain found in the mitochondria of eukaryotic cells. These aerobic prokaryotes are physiologically similar to the mitochondrion, which is the site of cellular respiration. Cellular respiration is the most prevalent and energy-efficient catabolic pathway during which oxygen is consumed (as a reactant) and glucose is broken down (to generate ATP). In this sense the mitochondrion uses the waste product of photosynthesis, oxygen, and releases the essential raw materials, carbon dioxide and water, that cyanobacteria require for photosynthesis. Conversely, the chloroplast produces carbohydrates and releases oxygen as a by-product of photosynthesis. Thus, the Earth's first ecosystems may have been comprised of assemblages of oxygenic photosynthetic prokaryotes (chloroplast-like prokaryotes) and oxygen-consuming bacteria (mitochondria-like prokaryotes). These prokaryotic confederations would have derived energy from sunlight and efficiently recycled the chemical compounds essential to photosynthesis and cellular respiration. In many ways, these confederations were organismic analogs to abiotic hypercycles postulated by Eigen and Schuster.

Such a biofilm ecosystem may have prefaced the evolutionary origin and development of the first eukaryotic cells as envisioned by the endosymbiotic hypothesis. First proposed in the late nineteenth and early twentieth centuries (Schimper 1883; Mereschkowsky 1905, 1920) and subsequently championed by Lynn Margulis (1938–2011), the endosymbiotic hypothesis states that eukaryotic cells are the evolutionary consequence of ancient confederations of different kinds of prokaryotes that merged to produce the eukaryotic cell (fig. 1.13). Specifically, the hypothesis argues that mitochondria evolved from once autonomous aerobic bacteria and that chloroplasts evolved from once autonomous cyanobacteria-like bacteria. One of the benefits of a host cell gaining a partnership with a cyanobacterium is gaining a nitrogen-fixing endosymbiont. Once proto-mitochondria and proto-chloroplasts gained entry to a host cell, they continued to function

Endosymbiotic Hypothesis   **Autogenous Hypothesis**

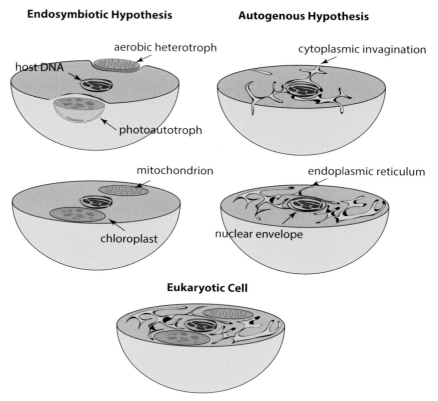

Figure 1.13. Schematic of key stages in the evolution of eukaryotic cells as proposed by
the endosymbiotic hypothesis (left) and the autogenous hypothesis (right). The endosymbiotic hypothesis proposes that chloroplasts and mitochondria evolved as a result of
the incorporation of a photoautotrophic prokaryote (similar to modern-day cyanobacteria)
and an aerobic heterotrophic prokaryote (similar to modern-day proteobacteria) by a host
prokaryotic cell (see fig. 1.2). The autogenous proposes that all organelles evolved by means
of cytoplasmic invaginations that partitioned metabolically specialized cytoplasmic units.
Both hypotheses may reflect the actual course of evolutionary events, since components of
each are not mutually exclusive.

within the cell. It is reasonable to assume that the host was either
an anaerobic, albeit aerotolerant, prokaryote or an inefficient aerobic
prokaryote. Some of the evidence for the endosymbiotic hypothesis
is summarized in table 1.1.

A number of ways a host cell could garner endosymbionts have
been proposed. One way is endocytosis, the ability to engulf an ex-

Table 1.1. Thirteen lines of evidence supporting the endosymbiotic theory

(1) Organelle-specific DNA is non-histonal as in the cytoplasm of prokaryotes.

(2) DNA sequences indicate that extant free-living cyanobacteria and
     α-proteobacteria are the closest relatives of plastids and mitochondria, respectively.

(3) Organelle ribosomes are similar to those of prokaryotes but differ from those
     found in the cytoplasm of eukaryotic cells (70 S- versus 80 S-type, respectively).

(4) The 70 S-type ribosomes of prokaryotes and organelles are sensitive to the
     antibiotic chloramphenicol, whereas 80 S-type ribosomes are not.

(5) The initiation of messenger RNA translation in prokaryotes/organelles occurs by
     means of a similar mechanism.

(6) Organelles and prokaryotes lack a typical (cytoplasmic) actin/tubulin system.

(7) Fatty acid biosynthesis in plastids occurs via acyl carrier proteins (as in certain
     bacteria).

(8) Plastids and mitochondria are surrounded by a double membrane. In the inner
     mitochondrial membrane the bacterial membrane lipid cardiolipin is abundant.

(9) The cardiolipin content of eukaryotic biomembranes is close to zero.

(10) Genome sequences reveal that plastids and mitochondria, which have retained
     large fractions of their prokaryotic biochemistry, contain only remnants of the
     protein-coding genes that their ancestors possessed. Experimental studies have
     shown that DNA has been transferred from organelles to the nucleus of the host
     cell.

(11) Although plastids diverged from their cyanobacterial ancestors at least 1,000
     Mya, the chlorophyll *a/b*–arrangements in land plant chloroplasts and the
     cyanobacterium *Synechococcus* are essentially the same (multi-subunit membrane pigment–protein complexes called photosystems I and II).

(12) Crystallographic analysis of cytochrome b6f (which is a major protein complex
     that mediates the flow of electrons between PS II and PS I) indicates that the
     cyt b6f complexes of cyanobacteria and chloroplasts have almost the same
     molecular structure.

(13) Extensive gene sequence homologies for cellulose synthase are reported for
     cyanobacteria, proteobacteria, algal lineages with cellulosic cell wall constituents, and the land plants, indicating extensive lateral gene flow throughout
     plant evolution.

tracellular particle by surrounding it with a portion of the external
cell membrane and then invaginating the particle encapsulated within
a microvacuole. The endocytotic method of capturing prokaryotic
proto-mitochondria and proto-chloroplasts gains partial support from
the fact that mitochondria and chloroplasts are double-membrane-

bound organelles whose inner membranes are topologically external to the eukaryotic cell. Perhaps the host prokaryote endocytotically ingested prokaryotes with differing metabolic capabilities, failed to digest them, and acquired the ability to work in partnership with its internal occupants. (In this sense contemporary mitochondria and chloroplasts are the consequences of an undigested dinner.) However, one of the stumbling blocks of this hypothesis is that the ability to flex, fuse, and pinch off membranes to form microvacuoles is poorly developed within the majority of living prokaryotes. In part this is due to the general absence of the sterols and polyunsaturated fatty acids found in the cells of eukaryotes that confer membrane flexibility and fusability, and so the capacity for endocytosis may have been the limiting factor in the evolution of endosymbiosis. This objection to the endosymbiotic hypothesis is tempered significantly by the discovery of a deep sea Archaean microbe called *Lokiarchaeum* that carries actin-like genes similar to those of eukaryotes, which participate in bracing the cytoskeleton, and GTPases, which participate in orchestrating the internal transport of vesicles. This microbe also contains genes for components of the ESCRT protein, which facilitates membrane flexure.

As an alternative to endocytosis, which requires membrane flexure, it is possible that protomitochondria and protochloroplasts evolved from internal prokaryotic parasites. This hypothesis receives some support from modern parasitic prokaryotes, such as the aerobic bacterium *Bdellovibrio,* which enters and takes up residency in the periplasmic space of its host cell. Perhaps in the distant past a similar parasite entered an anaerobic but aerotolerant host cell that provided it with some nutrition. In turn the endoparasite could have benefited its host by consuming oxygen through aerobic respiration, thereby detoxifying its host while simultaneously supplying limited quantities of ATP. (Thus the first eukaryotes may have been the product of a disease cured to mutual benefit.)

Regardless of how they gained entry, the first endosymbionts likely were not entirely benign tenants. Because they replicated themselves within its cytoplasm, the host cell had to gain some genetic control of them. This may account for the loss of portions of the chloroplast and mitochondrion genome in modern eukaryotes. Endosymbionts also released exotic materials that may have initially been toxic to their host cells (for example, modern mitochondria release abundant oxidants that can be harmful to prokaryotic organisms). Consequently many genetic and physiological relationships had to be ritualized before ancient prokaryotic confederations evolved into the first eukaryotic cells. We shall briefly touch on this when we consider the export-of-fitness phase of evolution in the context of the evolution of multicellularity (see chapter 7).

The endosymbiotic hypothesis, which is an old concept, had detractors who argued that organelles evolved through the differentiation and invagination of functionally different membrane-bounded areas that eventually pinched off into the cell. This alternative to the endosymbiosis hypothesis is called the autogenous hypothesis (see fig. 1.13). It is particularly attractive for the evolution of the nucleus and endomembrane system (consisting of the endoplasmic reticulum, Golgi apparatus, lysosomes, and other single-membrane-bound organelles). But the autogenous hypothesis encounters the same difficulty as the endocytotic mechanism for the incorporation of protomitochondria and protochloroplasts within a host cell—that is, the generally poor membrane flexibility and fusibility of prokaryotic membranes. Indeed, for either hypothesis to be correct, the flexibility and fusibility of the membranes of the first cells must have differed from the membranes of present-day cells. But the autogenous hypothesis entirely fails to account for the dramatically different genomes found in current chloroplasts and mitochondria, nor can it account for the remarkably conservative nature of the DNA sequences in chloroplasts.

Clearly, the autogenous and the endosymbiotic hypotheses are not mutually exclusive, which may explain different aspects of the eukaryotic cells that are difficult to derived based on one or the other. Although the mitochondrion and chloroplast may have evolved through the lateral transfer among bacterial lineages envisioned by the endosymbiotic hypothesis, the nuclear envelope and endoplasmic reticulum may have evolved autogenically through membrane invagination. Increased cell size may have favored an amplification of membrane systems as a consequence of well-known area-volume (law of similitude) relations that require disproportionate increases in surface area as cell size increases (see chapter 8).

The endosymbiotic origins of mitochondria and chloroplasts are unquestionably confirmed by a variety of data. As early as the late 1960s, Carl Woese (1928–2012) began to assemble catalogs of oligonucleotide sequences released by the in vivo digestion of rRNA isolated from living prokaryotes and eukaryotes. By analyzing shared and unshared oligonucleotide sequences, Woese and his colleagues constructed a phyletic hypothesis for the origins of the eukaryotes and the phyletic relations among the prokaryotes. Although no single molecule can define an evolutionary lineage once lateral gene transfer is admitted to occur, methodologically rRNA was and is the molecule of choice for an evolutionary chronometer because it occurs in all living cells and is functionally conservative, evolving very slowly. The rRNA data amassed by Woese and others indicate that the chloroplast and the mitochondrion found in extant eukaryotes (plants, animals, etc.) descend from very different groups of free-living prokaryotes. Specifically, the oligonucleotide sequences of the chloroplast closely align with those from living cyanobacteria, while the oligonucleotide sequences from mitochondria align with those of a group of purple non-sulfur bacteria called the proteobacteria. This phyletic hypothesis continues to agree with new information about prokaryotic biochemistry and molecular biology. It is also compatible with the essential features of previous classification schemes for bacteria. Additionally, chloroplasts and mitochondria are the appropriate size to have descended from prokaryotes; their inner membranes have several enzymes and transport systems resembling those found in the external membrane encapsulating modern bacteria; both organelles duplicate during the division of the eukaryotic cell by a splitting process that resembles the binary fission of bacteria; and chloroplasts and mitochondria contain DNA in the form of a circular molecule that is not typically associated with histones or other proteins as it is in eukaryotes.

The modern version of the tree of life is reticulated, and not invariably dichotomously branched as Charles Darwin assumed. All the available evidence indicates that the eukaryotes (formally the Eukarya) are the result of the lateral transfer of separate prokaryotic evolutionary lineages. But the reticulate tree of life is even more curious because some photosynthetic eukaryotic lineages are the result of secondary or tertiary endosymbiotic events. The primary endosymbiotic event was a partnership between cyanobacteria-like and nonphotosynthetic prokaryotic organisms, during which the photosynthetic endosymbiont relinquished some of its autonomy to the host cell by giving up part of its genome. This event ultimately led to the red and the green algal lineages. In each of these lineages, the chloroplast is a double-membrane-bound organelle whose inner membrane is derived from the outer cell membrane of the photosynthetic endosymbiont and whose outer membrane is derived from a vesicle created by the prokaryotic host cell. However, the chloroplasts of other algal groups have more than two bounding membranes (cryptomonad and chlorarachniophyte chloroplasts have four membranes). These supernumerary bounding membranes resulted when a nonphotosynthetic eukaryote engulfed a photosynthetic eukaryote and acquired its chloroplast as a secondary endosymbiotic event. The nucleomorph, a nucleus-like organelle sandwiched between the second and third chloroplast membranes, may be the vestigial remains of the nucleus of the engulfed photosynthetic eukaryote. Thus, even if the double-membrane-bound chloroplast evolved only once as a consequence of primary endosymbiosis, the eukaryotic photosynthetic organisms traditionally called the algae likely are polyphyletic because some lineages have chloroplasts that are drastically reduced remnants of photosynthetic eukaryotes. The current view of algal phylogeny gives new meaning to something old, something new / something borrowed, something blue.

Meiosis and Sexual Reproduction

The evolution of the eukaryotic nucleus presaged a dramatic departure from the previous way cells divided and reproduced. Prokaryotes
reproduce by binary fission, during which the single bacterial chromosome replicates and one copy is allocated to each of the two new cells.
Both copies of the duplicated chromosome are bound to the inner surface of the cell membrane and then are drawn apart by the subsequent
growth of the intervening portion of the membrane. During the final
stages of binary fission the bacterial cell pinches inward when the cell
reaches roughly twice its original size. In contrast, the division of eukaryotic cells is far more complex, although it involves the duplication
of the chromosomes and the production of two new cells, just as in
binary fission (fig. 1.14). Unlike the duplication of the prokaryotic
chromosome, the duplicates of each eukaryotic chromosome (called
sister chromatids) are attached to one another by a structure called the
centromere. Each centromere becomes attached to a spindle fiber composed in part of microtubules and their associated proteins radiating
from two opposing spindle poles. Typically the chromosomes in a cell
align on a single plane, called the metaphase plate, at the equator of
the spindle. Because each chromosome arrives at the metaphase plate
with an apparently random orientation relative to the two spindle
poles, chance determines the segregation of the chromatids of each
chromosome into the two new cells.

The events resulting in the duplication of eukaryotic chromosomes
and the division of the nucleus into two new nuclei constitute mitosis,
the process that, barring mutations, obtains two genetically identical new nuclei. During the later stages of mitosis, at about the stage
called anaphase, cytokinesis begins to occur. In cytokinesis, literally
meaning "movement of the cytoplasm," a parent cell divides into two
new cells. At the stage called telophase, when the chromatids arrive at
the two opposing poles and new nuclear envelopes form around them,
materials are deposited along the equatorial plane of the old metaphase plate. These materials are contained in vesicles amassed along

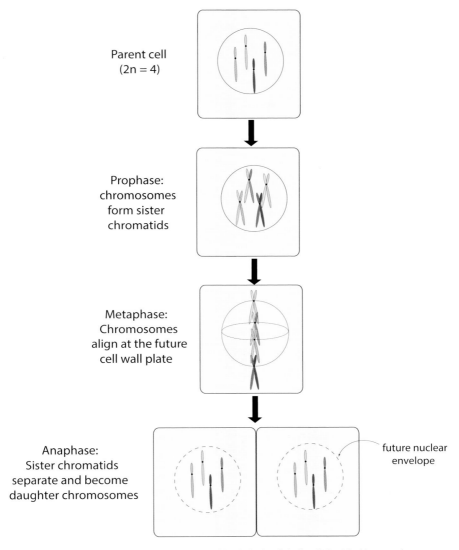

Parent cell
(2n = 4)

Prophase:
chromosomes
form sister
chromatids

Metaphase:
Chromosomes
align at the future
cell wall plate

Anaphase:
Sister chromatids
separate and become
daughter chromosomes

future nuclear
envelope

Figure 1.14. Schematic of mitosis in a typical land plant cell (cell wall depicted by round-edged square; nuclear envelope depicted by circle). This example shows a diploid cell with four chromosomes (2n = 4). During mitosis, chromosomes replicate to form sister chromatids that barring mutations are genetically identical. Sister chromatids are drawn toward opposing poles by spindle fibers that will later be the sites of two nuclei, each of which contains the same number of chromosomes as in the parent cell. Compare this sequence of events and processes to those that occur during meiosis (see fig. 1.15).

the middle of the cell. The coalescence of the vesicles results in the formation of a double membrane that extends from the center of the old metaphase plate and eventually unites with the preexisting cell membrane of the original cell. Once the new membranes unite with the old cell membrane, cytokinesis is complete—each new cell is separated from its twin by its own cell membrane. Among land plants, a new cell wall forms within the double membrane along the old metaphase plate. The formation of the double membrane and the new cell wall is also associated with the phragmoplast, a complex structure composed of microtubules (see fig. 7.11). The distinction between mitosis, the division of the original nucleus into two new nuclei, and cytokinesis, the separation of the original cell's living material into two new cells with their own cell membranes, is important because mitosis is not always followed by cytokinesis. When cytokinesis does not occur, a cell increases in size but doubles the number of its nuclei. Repeated mitosis in the absence of cytokinesis produces a multinucleated cell.

Space does not permit a detailed discussion of the evolution of mitotic cell division. The following references, however, provide provocative reading on this subject: Cavalier-Smith (1978); Carlile (1980); Dodge and Vickerman (1980); Margulis (1981). Nevertheless, attention must be drawn to the suggestion that the terminal replication site on the bacterial chromosome and the centromere of eukaryotic chromosomes are evolutionarily homologous (Maynard Smith and Szathmáry 1995). This suggestion is based on a comparison between bacterial chromosome replication and an atypical form of mitosis called pleuromitosis. Recall that the bacterial chromosome is a single circular strand of DNA. Chromosome replication begins at a single site, proceeds in both directions around the circular DNA strand, and ends at a single point called the terminus. In contrast, chromosome replication in most eukaryotic cells involves a bipolar spindle apparatus that emerges from the two opposing ends of the dividing cell. Each eukaryotic chromosome is attached to the equator of the spindle apparatus by its centromere, the structure involved in pulling a chromosome to one of the two spindle poles (see fig. 1.14).

Although a chromosome's replication involves multiple replication forks (replicons), at first glance it is hard to see how the eukaryotic pattern of chromosome replication could evolve from that of a bacterial antecedent. However, some presumably ancient eukaryotes have an unusual form of mitosis called pleuromitosis. In these organisms the two halves of the spindle apparatus of the dividing cell lie side by side rather than across from one another. During pleuromitosis, the centromeres of chromosomes are attached to the inside of the nuclear membrane very much the way the single chromosome in a bacterial cell remains attached to the rigid bacterial cell membrane. Thus, it is possible that the centromere of the linear eukaryotic chromosome is evolutionarily derived from the terminus of the prokaryotic circular chromosome that somehow split to acquire the familiar eukaryotic linear form. If so, then it is also likely that the site of attachment of the bacterial chromosome to the rigid cell wall and the pole of the eukaryotic spindle apparatus are homologous.

The selective advantages of the transformation from a circular chromosome to a linear chromosome are not immediately obvious. However, if there are limits to the rate at which chromosome replication proceeds, the single site of bacterial chromosome replication may have placed an upper limit on the size of the prokaryotic genome, because genome size tends to correlate positively with the rate of cell division across broad categories of organisms. Thus, one advantage to the linear eukaryotic chromosome is the multiple sites for replication that could have permitted an increase in genome size without a concomitant increase in the duration of cell division.

Naturally, all scenarios for the evolution of mitosis must address the origin of chromosomes, which regardless of their shape or size are collections of linked genes. Why and how this linkage happened remains a puzzle because connected genes would theoretically take longer to replicate than their disconnected counterparts. The molecular mechanisms responsible for linking ancient genetic molecules to form proto-chromosomes are complex and problematic (Szathmáry and Maynard Smith 1993). However, there are a few clear selective advantages to linking genes. First, a complement of unlinked genes can be unequally partitioned during cell division, so that the two derivative cells may have vastly different genetic compositions. Likewise, linked genes are certain to leave copies of themselves in each of the derivatives of division. Second, whereas unlinked genes can potentially replicate at very different rates and thus compete with one another for molecular resources in the same cell, linked genes must replicate synchronously. By means of chromosomes, all linked genes more or less observe the same tempo of cell replication. A third advantage to linked genes is that rapid selfish gene replication can be effectively suppressed by mutations at other loci. In this sense gene linkage serves to democratize originally ancient and very loose confederations of potentially competing genetic molecules. A fourth advantage is that linked genes can be inherited as a functional unit or even trapped in an inversion—for example, the genes involved in many of the sequential steps of biosynthetic pathways often co-occur near one another on the same chromosome. How genes were linked to form the first chromosomes and managed to function in a orderly way remains a matter of conjecture, although it is obvious that this happened well before the advent of the evolution of mitosis and the first eukaryotes.

Barring random mutation or accidental damage to chromosomes, mitotic cell division normally results in no genetic variation among cells tracing their lineage to a single cell (all the cells are chips off the same genetic block). Genetically identical individuals may be advantageous to an organism particularly well adapted to its environment, provided the environment does not change, or to an organism that is capable of migrating and tracking the environment to which it is well adapted, or to an organism whose physiology or morphology is sufficiently plastic to deal with a wide range of environmental conditions. However, over long periods environments change; the ability of many types of organisms to migrate or disperse over great distances to the habitat that suits them best has limits. Thus, genetic variation is critically desirable in a population because it affords the opportunity to produce physiological or morphological variants that may survive when the environment changes or that may colonize new environments. Meiosis and sexual reproduction and the resulting genetic recombination have evolved independently in many lineages, probably because they produce significant genetic variation among the individuals within a biparental population. Sexual and asexual reproduction are not mutually exclusive. Many unicellular eukaryotes reproduce asexually by means of mitosis. These types of organisms benefit from the genetic variation resulting from sexual reproduction when environmental conditions change, and they benefit from asexual reproduction whenever the environment is stable and amenable to growth and survival.

Sexual reproduction (as defined using eukaryotic organisms) has evolved only among eukaryotic organisms because its precondition is an organism that has perfected meiosis and therefore possesses a nucleus (fig. 1.15). To understand meiosis, first appreciate that within the life cycle of all sexually reproducing organisms there exists at least one cell whose nucleus contains homologous pairs of chromosomes, or simply homologues. The cell containing the pair of homologues is called a diploid cell, and the number of chromosomes within the diploid cell is the diploid number, denoted as 2n. During meiosis the diploid number is reduced by half. Typically, four cells are produced when meiosis is complete. These cells are called haploid cells, and the number of chromosomes within each haploid cell is the haploid number, denoted as n. The importance of reducing the chromosome number by half is readily apparent because two haploid cells must fuse to produce a new diploid cell during sexual reproduction, a process called syngamy. If chromosome numbers were not reduced to haploid numbers, the number of chromosomes would double endlessly each time sexual reproduction occurred.

Although each pair of homologues within the nucleus carries the genes that affect the same inherited traits, the genes controlling the same trait are not necessarily identical. They can have different forms, called alleles, each of which may evoke a different phenotypic condition. For example, consider the color of a flower's petals. A gene on one

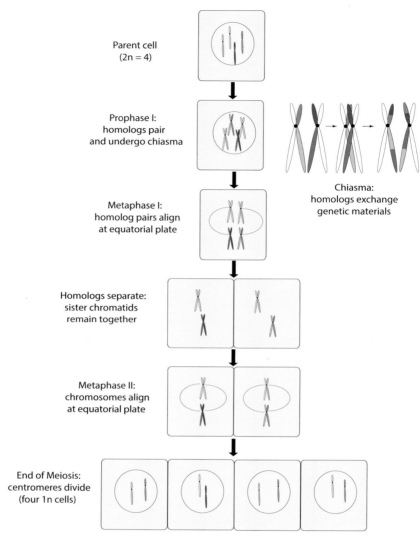

Parent cell
(2n = 4)

Prophase I:
homologs pair
and undergo chiasma

Chiasma:
homologs exchange
genetic materials

Metaphase I:
homolog pairs align
at equatorial plate

Homologs separate:
sister chromatids
remain together

Metaphase II:
chromosomes align
at equatorial plate

End of Meiosis:
centromeres divide
(four 1n cells)

Figure 1.15. Schematic of meiosis in a typical land plant cell (cell wall depicted by round-edged square; nuclear envelope depicted by circle) with four chromosomes (2n = 4). During prophase I, homologous chromosomes pair and exchange genetic material during a process called chiasma (depicted in the diagram to the right of prophase I in which portions of the two homologs are color coded to show that materials have been exchanged between chromatids). During metaphase I, homologs align with the equatorial plate and are drawn to opposing poles. During metaphase II, chromosomes align with the equatorial plate and sister chromatids are drawn to opposing poles. At the end of meiotic cell cycle, four nuclei differing in genetic composition are produced. In this example the nuclei reside in separate cells. However, this is not always the case. For example, multinucleated haploid cells occur during the early formation of angiosperm megagametophytes.

homologous chromosome may code for a plant whose flowers have white petals. In contrast, the gene on the other homologue may result in an individual with blue petals. Since the nuclei of plants with flowers contain both homologues, a plant whose nuclei have the white petal gene and also the blue petal gene may have blue petals (if the white gene is recessive, or pale blue petals if the two genes are codominant), while the flowers of an individual whose nuclei contain homologues with exactly the same form of the gene will have either white or blue petals. These differences among the individuals within a population of flowering plants result from meiosis, because during meiosis the homologous chromosomes within a diploid cell eventually come to reside in four haploid cells. The number of possible permutations of different homologues in haploid cells depends on the number of chromosomes within the diploid nucleus from which they arose. For example, if the diploid nucleus has twenty chromosomes, the possible number of genetically different haploid cells is 2^{20} or 1,048,576 (disregarding the variation resulting from mutation or crossing-over). Each diploid human cell has twenty-three pairs of chromosomes, and so each diploid cell is theoretically capable of producing 2^{23} or 8,388,608 genetically different haploid cells. The genetic variation that can be introduced into a population is compounded because, during sexual reproduction, two haploid cells (called gametes) must fuse to produce a new diploid cell (called the zygote). Even in the absence of mutation or other types of genetic alterations, the number of genetically different permutations of the zygote equals the square of the number of possible genetically different gametes. Thus, the human zygote has $(2^{23})^2$ or 70,368,774,177,664 different genetic combinations, which can be increased by mutation and recombination.

But is the genetic variation obtained from sexual reproduction always evolutionarily advantageous? Are meiosis and sexual reproduction evolutionarily favored because they obtain new, potentially adaptive gene combinations? Consider that whenever an individual that is well adapted to its current environment reproduces sexually it conveys only 50% of its genes to its offspring. Thus, some of the offspring may not be as well adapted to their current environment as the most fit parent. This consequence is called the "50% fitness cost" to sexual reproduction. Clearly, we cannot invoke the argument that sexual reproduction evolves because it confers long-term evolutionary advantages. Because natural selection cannot foresee the future, it cannot have a plan. So how do we resolve the issue of sexual reproduction?

First, recognize that the 50% fitness cost of sexual reproduction holds true only when the relative fitness of mating individuals within the population differs significantly and when asexual reproduction cannot occur. If the former is true, then the precondition that a hypothetical population is particularly well adapted to its environment is not fulfilled—the population may be in the *process* of becoming adapted, but great differences among the fitness of individuals arguably are not the hallmark of a population that has *become* adapted to its particular environment. Second, note that the preclusion of asexual reproduction is biologically unreasonable. There is every reason to assume that the first unicellular eukaryotes could reproduced asexually. Barring high mutation rates, asexual reproduction through mitosis can serve to maintain a well-adapted population in its particular environment. The bonus of sexual reproduction is that it produces virtually endless genetic variants that can supply the raw materials with which to explore new and different habitats or persist in the same habitat whenever environmental conditions change. Thus, the 50% fitness cost of sexual reproduction is illusory in the sense that asexual and sexual reproduction are not mutually exclusive (Williams 1975).

Because sexual reproduction requires diploid cells and the perfection of meiotic cellular division, the critical questions are, What was the initial advantage to diploidy? and How did diploidy first evolve? The most plausible answer to the first question is that diploidy enriches the diversity of a population's gene pool. Each diploid cell contains pairs of homologous chromosomes that can bear different alleles, and so a population of diploid cells consists of heterozygous and homozygous individuals. And in the case of codominance or incomplete dominance, the interactions of alleles in the same cell can establish subtle variations in functional traits. Indeed, a high correlation exists between diploidy and the complexity of developmental interactions (we will explore this issue in greater detail in chapter 4).

Another advantage of diploidy is that undamaged chromosomes can be used as templates to repair damaged DNA of their homologues. Recall that during meiosis prophase I, homologous chromosomes pair. At this juncture, gene conversion can change a defective chromosome back to its original, undamaged state. This phenomenon has inspired the DNA-repair hypothesis arguing that meiosis and sexual reproduction evolved in response to selection pressures for DNA repair. The hypothesis implies that the diploid state need be only transient, existing just long enough to bring homologous chromosomes together to effect DNA repair. The hypothesis gains credibility in that many extant unicellular plants belonging to extremely ancient lineages have a sexual life cycle involving a long-lived haploid cell and a very short-lived diploid cell.

An alternative hypothesis is the selfish DNA or transposon hypothesis. A transposon is a fragment of DNA with the ability to multiply itself and move spontaneously within the genome of an organism. It is called the selfish DNA hypothesis because a transposon's main function seems to be simply to make more copies of itself. In this sense, transposons are genomic parasites. The hypothesis argues that the driving force for the origin of sex was parasitic DNA sequences that enhanced their own fitness by promoting the cycles of chromosome fusion and segregation that occur during meiosis.

How diploidy and meiosis evolved is not known with any assurance, but the first diploid cells resulting from the fusion of two haploid cells (syngamy) most likely benefited from the equivalent of hybrid vigor (heterosis) and the ability to diversify in a changing environment. Most deleterious mutations are partially or wholly recessive and thus negatively affect the heterozygote to a lesser extent than the homozygote. This is referred to as diploid advantage (box 1.3). The first diploid cells may have evolved when haploid cells accidentally

Box 1.3. The Heterozygote Advantage

The most frequently proposed explanation for the evolution and dominance of the diploid phase in many life cycles is called "heterozygote advantage"—when the heterozygote genome has a higher relative fitness than the homozygote dominant or recessive
genome. This explanation revolves around the possibility that heterozygotes carry a
"silenced" mutational load that is free to mutate in ways that can become beneficial
under certain selection regimes. It is true that the diploid condition can be particularly
beneficial in large, multicellular organisms, which can carry recessive somatic mutations. However, diploids do not invariably have an advantage, and the recessive somatic
mutation hypothesis does not explain why multicellularity evolved in both the haploid
and the diploid phase. Consider a hypothetical gene with a bell-shaped expression level
pattern with an optimum at 1.0 and a haploid expression level of 0.5 (fig. B.1.3). Under

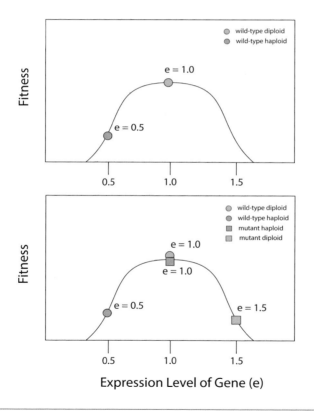

Figure B.1.3. Sche-
matic of how "hetero-
zygote advantage"
can be circumvented.

these circumstances, the wild-type haploid is disadvantaged, because its expression level equals 0.5, whereas the wild-type homozygous diploid has the optimum gene expression level of 1.0. Consider now the effects of a mutation that doubles the expression level in the gene such that the haploid mutant has the optimal expression level of 1.0, whereas the mutant diploid has a suboptimal overexpression level of 1.5. In this scenario, the wild-type homozygous diploid and the mutated haploid have the same expression levels and have an advantage over the mutated heterozygous diploid, provided that mechanisms for non-DNA-sequence modification are unavailable. This scenario is clearly contrived because it rests on a single gene and the opposite effects can be simulated if a mutation confers a slight advantage. It is presented merely as a hypothetical example of heterozygote disadvantage, and not as a blanket denial of the benefits conferred by being a heterozygous diploid, as for example having one allele that could confer an advantage in one environmental setting and another allele that could confer an advantage in another environmental setting. However, changes in gene expression are an important and prevalent mechanism for adaptation, and it is not uncommon for mutations of small effect to be adaptive when they modify expression levels, regardless of whether an organism is haploid or diploid.

fused and remained so for a time as a single, functionally proficient cell. An additional advantage to maintaining the diploid condition is the ability to repair any damage to double-stranded DNA, because repairs can be effected when a second undamaged DNA molecule on a homologous chromosome is available as a template, or when the damaged DNA is repaired using the complementary strand on a nonhomologous chromosome as a template. Assuming that the first eukaryotes were haploid, diploid cells may also have evolved as a consequence of mitotic nondisjunction. That is, the first diploid eukaryotic cells may have been autopolyploids (diploidy resulting from a spontaneous doubling of all chromosomes) and an early form of meiosis may have evolved as a means to revert back to the ancestral haploid condition (this is explored further in chapter 6; see Figs. 6.4–6.6). The phenomenon called endomitosis, in which the chromosome number of haploid cells spontaneously doubles, occurs in a variety of presumably ancient unicellular plants and animals. The immediate adaptive advantage to this form of diploidization may have been the conferral of a smaller ratio of surface area to volume relative to the haploid condition, which may have afforded diploid cells a higher metabolic efficiency than their haploid counterparts. In either case meiosis may have become refined through natural selection as a mode of cell division to reacquire the ancestral haploid condition, which would have been an advantage when nutrients were limited because haploid cells tend to grow faster than diploids. Ancient cell cycles alternating between haploid and diploid phases in asexually reproductive organisms may have promoted the origin of sexual reproduction by providing a preexisting mechanism for a regular reduction in chromosome number.

Multicellularity and Life Cycles

When is an organism multicellular? A generic answer that simultaneously encompasses animal, fungal, and plant life is that multicellularity results when two or more neighboring cells adhere, interact, and physiologically communicate and coordinate their activities. Among multicellular organisms, direct physical contact among neighboring cells is achieved principally in one of four nonexclusive ways: tight junctions, desmosomes, gap junctions, and plasmodesmata (see fig. 7.11 and associated text). Tight junctions occur where proteins built into the external membrane of one cell bond with like proteins in the cell membrane of a neighboring cell. Where they occur, tight protein junctions cause the two neighboring cell membranes to adhere. Desmosomes are regions of intracellular filaments that extend through the external spaces between adjoining cell membranes. The filaments, which cross into the intercellular space, permit substances to pass freely between adjoining cells. Gap junctions are open pores surrounded by transmembrane proteins that pass through the spaces between adjoining cells. Like desmosomes, gap junctions permit materials to move between cells. Plasmodesmata are open channels through the plant cell wall that connect neighboring cells. In multicellular plants these channels are required because each cell is typically surrounded by a cell wall composed of comparatively densely packed materials that may present a physical obstacle to the passage of modestly sized molecules from one adjoining cell to another. Plasmodesmata permit plant cells to interact and intercommunicate physiologically. They are small open channels within the cell walls through which strands of cytoplasm interconnect the living protoplasts of adjoining cells. The cell membranes of neighboring plant cells, which line each of the plasmodesmatal channels, are continuous from one cell to the next. A multicellular plant thus consists of a single living protoplast incompletely subdivided by an infrastructure of cell walls. Throughout this continuous living mass, water and small molecules may pass with comparative ease. One of the exquisite features of the multicellular land plant is that the flow pattern of water and other small molecules can be modified by altering the number and location of plasmodesmata among neighboring cells. Adjoining cells sharing many plasmodesmata may serve as preferred routes of transport for nutrients. Preferential transport of water and nutrients is a requisite for the survival and growth of large aquatic and terrestrial plants.

In contrast to eukaryotes, which include unicellular and multicellular representatives, no prokaryote is known to produce plasmodesmata, tight junctions, desmosomes, or gap junctions. Colonial and filamentous prokaryotes exist (such as *Coelosphaerium* and *Oscillatoria,* respectively), but these growth forms rarely involve cytologically connected cells. Although narrow cytoplasmic channels, called microplasmodesmata, join the cells of some cyanobacteria, it is not known whether these channels serve as cytoplasmic transport conduits. If they do, multicellularity is not restricted to the eukaryotes and finds its equivalent in some prokaryotes. If not, the absence of multicellular prokaryotic organisms indicates that the evolution of nuclei and other organelles was either a prerequisite for multicellularity or a collateral condition.

Phylogenetic analyses indicate that multicellularity has evolved independently in different eukaryotic lineages, which indicates that multicellularity may have directly or indirectly conferred some advantage in survival or reproductive success. One benefit conferred by multicellularity is that cells can divide while organisms move. In contrast, with few exceptions, the dividing cells of unicellular organisms suspended in water are unable to flee predators. Some organisms, like the ciliates, have evolved elaborate nonstandard mechanisms for mitotic cell division. Others have permanently retracted their cilia or flagella and abandoned motility altogether. Still others divide and move in different stages of their cell division cycle. However, one immediate solution is to stick together after cell division and to specialize one cell for locomotion and the other for reproduction (either asexually or sexually). According to this scenario, the evolution of mitosis created a dilemma that was resolved by multicellularity. To be sure, although unicellular organisms with flagella or cilia cannot divide and move at the same time, many of these organisms are eminently successful, and therefore the failure to perform both biological tasks simultaneously has not imposed any obvious "universal" selection pressure. Thus, we are obliged to consider the possibility that the evolution of multicellularity has occurred for different reasons in different lineages (see chapter 7).

Although the origins of multicellularity are shrouded, some of the selective advantages conferred by partitioning the plant, animal, or fungal body into cells are immediately obvious. For example, a multicellular organism continues to survive after it has sexually reproduced. In contrast, a unicellular organisms that sexually reproduces must convert itself into a gamete, or sex cell. From the perspective of its population it ceases to exist—quite literally giving its all to the next generation! The consequences of this self-sacrifice are quickly seen by comparing the growth of populations of sexually reproducing multicellular and unicellular species. In both cases, assume that each adult requires a mate to complete the sexual reproductive cycle; two haploid gametes fuse, and the resulting diploid zygote divides meiotically to produce four functional haploid plants. The only salient difference between these two species is that the multicellular parents continue to exist after sexual reproduction whereas the unicellular parents essentially vanish from the population (see fig. 7.18). Note that each pair of multicellular parents produces six individuals—the two original parents and four offspring. In the more general case, the total number of multicellular organisms in the population increases over successive generations according to the progression N $3N$ 3^2N, 3^3N, 3^4N, 3^5N, and so forth, where N is the original number of parents in the population. For example, starting with two multicellular parents of which only one cell per parent forms a gamete, the population will increase as 2, 6, 18, 54, 162, 486, . . . assuming that no offspring or parents die. In contrast, the number of unicellular organisms merely doubles with each successive generation. That is, starting with any N unicellular adults, the population increases according to the procession N^1, N^2, N^3, N^4, N^5, N^6, . . . such that, when $N = 2$, the population will increase as 2, 4, 8, 16, 32, 64, . . . once again assuming none of the offspring or parents die along the way. By dividing the number of multicellular individuals by the number of unicellular individuals produced in successive generations, one quickly sees that, for the successive generations, N/N, $3N/N^2$, $3^2N/N^3$, $3^3N/N^4$, $3^4N/N^5$, $3^5N/N^6$, . . . Thus, when $N = 2$, these quotients are 1, 6/4, 18/8, 54/16, 162/32, 486/64, . . . etc. By the sixth generation, the population of the multicellular species is more than seven times that of unicellular plant species. Notice that, because we assumed each multicellular organism produces only one gamete at a time, this is the "best case" scenario for the unicellular species. These calculations ignore that reproductive rate, on the average, declines as the 3/4 power of an organism's size. But the difference in the rate of reproduction between a unicellular and a multicellular organism composed of, say, two or three cells is trivial, whereas the slightly larger size of the two-celled organism provides an immediate selective advantage perhaps sufficient to favor an even larger multicellular body size.

Perhaps because of the selective advantage of persistence, multicellularity has arisen in one, two, or even three phases in the life cycles of many plants. A excellent example is the life cycle of some red algae whose sperm cells lack flagella such that fertilization events are potentially rare. In this life cycle, the division of a single fertilized egg

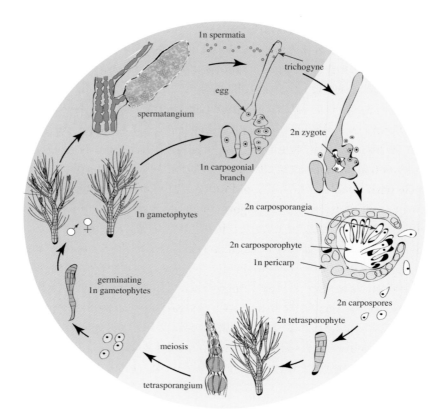

Figure 1.16. Schematic of the tri-phasic life cycle of the marine red alga *Polysiphonia* in which each phase is multicellular. The life cycle consists of three multicellular organisms— the haploid gametophyte, the diploid carposporophyte, and the diploid tetrasporophyte. Among the three, the carposporophyte is the smallest. It is also hemiparasitic within a structure produced by the gametophyte called the pericarp, in which the egg originally resided. One possible explanation for the multiple multicellular organisms in this life cycle is that red algal sperm cells lack flagella, which can reduce the efficiency of fertilization. This explanation is confronted by observations indicating that fertilization has a reasonably high success rate.

is amplified by intercalated mitotic divisions into two morphologically different diploid multicellular organisms (the carposporophyte and the tetrasporophyte) that produce many haploid organisms (gameto-phytes) by virtue of meiosis (fig. 1.16). When both phases in the same life cycle are multicellular, each species consists of two organisms, and both phases have the opportunity to diverge morphologically and anatomically to cope with the particular environmental context facilitating its reproductive function. To understand this, consider a life cycle in which the haploid and the diploid phases are unicellular. In this life cycle, two haploid cells fuse to produce a diploid cell that then divides meiotically to produce haploid cells. A life cycle with a multicellular haploid individual capable of producing many gametes results by intercalating mitotic cell divisions after meiosis (such as the green algae *Spirogyra*, *Chara*, and *Coleochaete*). Conversely, a life cycle with a multicellular diploid individual results by intercalating mitotic cell divisions before zygotic meiosis (such as the brown alga *Fucus*). And intercalating mitotic cell divisions before and after zygotic meiosis produces a life cycle that alternates multicellular haploid and multicellular diploid individuals (such as the moss *Polytrichum*). Although there are significant differences among them, these four life cycles share a very basic feature—an alternation between a diploid cell and a haploid cell. The fact that all plant life cycles are the same and differ only with respect to where or if multicellularity is expressed appears to have gone unnoticed in early debates about the origins of the plant life cycles.

Aside from conferring the potential for morphological and anatomical divergence between or among the life phases in the same life cycle, another advantage of multicellularity is the ability to specialize cells into tissues and tissues into organs. Recall that plant cells are bonded by their cell walls, which among land plants are perforated by plasmodesmata. Small molecules like glucose and water can pass through these perforations, and the more plasmodesmata there are between two adjoining cells, the greater the rate of molecular transport. Modifications in the location and number of plasmodesmata provide a way to direct the flow of metabolites and water along preferred pathways and occur throughout the plant kingdom. For example, some large marine kelps produce "trumpet cells," elongated cells with numerous interconnecting plasmodesmata at their end walls. Very much like the phloem cells in vascular plants and the leptoids found in some mosses, the trumpet cells transport cell sap loaded with sugar from sunlit fronds to more deeply submerged portions of the kelp plant. By the same token, numerous and different plants have evolved functionally analogous preferred routes for water transport (such as the water-conducting cells called hydroids in mosses and xylem elements in vascular plants).

The question remains why different variants of the same basic life cycle dominate different plant lineages and why the diploid and haploid phases in the same life cycle have different durations. The relative duration of the haploid and diploid phases during the sexual life cycle of species varies greatly among organisms, yet good evolutionary explanations for this variation remain elusive. It has been suggested that a prolonged diploid phase evolved because a diploid heterozygous genome can mask the deleterious effects of mutations (the "heterozygous advantage"). Others point out that this advantage may be counterbalanced by the disadvantage that diploids have twice the mutation rate of haploids. Theoretically, diploidy would reduce the mean fitness of the population when the duration of the diploid phase equals or exceeds the duration of the haploid phase. Mathematical models predict that a life cycle dominated by the diploid phase is evolutionarily stable when allelic recombination is entirely free (no linkage), even though prolonging the diploid phase reduces the mean fitness of a population because the mean number of mutations carried per genome is independent of the dominant phase in the life cycle. Models such as this predict that individuals with a prolonged diploid phase have, on average, the same number of mutations per genome as do more haploid individuals. Because reverting to haploidy exposes deleterious mutations that have been previously masked by diploidy, the spread of individuals with long haploid durations is predicted to be very difficult. However, the models predict that the haploid phase will dominate when complete linkage occurs or when mutations are highly deleterious.

In addition, some models predict that life cycles retaining both the haploid and diploid phases are evolutionarily unstable and thus attempt to account for the many different kinds of life cycles in algae even under conditions where selection pressures on the viability of the diploid and haploid phases overwhelm the force of selection against deleterious mutations. For example, some models predict that the phase with the lowest mortality rate will evolve in duration and become the dominant phase in the life cycle provided the differences in the mortality rates of the haploid and the diploid phases are very large. That is, pronounced phase-specific mortality, perhaps resulting from morphological or ecological differences in the two phases, can override the effect of selection against deleterious mutations in the evolution of stable life cycles. A consistent feature of these models is that life cycles containing both the haploid and diploid phases are evolutionarily unstable. This may shed light on the fact that most green algae have life cycles dominated by either the haploid or the diploid phase. This might also account for why, over the course of land plant evolution, the diploid phase in the life cycle has progressively increased in duration as well as size relative to the duration and size of the haploid phase. Technically, all land plants have mixed life cycles in which the multicellular haploid phase alternates with the multicellular diploid phase, but there is nevertheless an evolutionary trend toward reduction of the haploid phase.

A Brief Remark about Self-Organization

This concludes our treatment of four of the eight transformations in the history of plant life. The remaining four transformations are discussed in the next chapter. However, before leaving the topic of life's origins and its early history, it is important to note that early forms of life likely relied on self-organizing abiotic processes that involve the mobilization of physical phenomena. Consider, for example, that land plant cell walls contain cellulosic microfibrils whose orientations in the cell wall influence the extent to which the mechanical properties of the wall are equal in all directions (isotropic) or unequal (anisotropic). If these microfibrils are deposited randomly within the wall, the cytoplasm within the cell wall exerts a uniform pressure over the cell wall that results in isotropic expansion to produce a spherical cell. If however, microfibrils are deposited in a biased way, the uniformly exerted pressure results in anisotropic expansion and a nonspherical cell. Thus, to alter a cell's shape, a unicellular plant must exert an influence on the mechanics of its cell wall, but it can rely on the simple physics of hydrostats to achieve a spherical cell without exerting any additional energy. Other examples of generic physical processes can be relied on. For example, cellulose molecules self-assemble into paracrystalline structures (such that cell walls are intrinsically rigid); mitotic spindles self-assemble in vitro and in vivo; cell plate orientation involves the establishment of the minimum energy (equilibrium) of microtubules tethering the nucleus to the cell membrane; and passive diffusion dynamics account for the self-assembly of cytoplasmic streaming (in the cells of the green alga *Chara*). When viewed collectively, these phenomena present evidence that the earliest life forms need not have evolved genomes that "internalized" the information content conferred by physically generic processes, but that life needed to evolve ways to countermand or control these processes when physics needed to be controlled and manipulated.

Suggested Readings

Bengstrom, S., ed. 1994. *Early life on Earth*. New York: Columbia University Press.

Blankenship, R. E. 1992. Origin and early evolution of photosynthesis. *Photosynth. Res.* 33: 91–111.

Carlile, M. J. 1980. From prokaryote to eukaryote: Gains and losses. In *The eukaryotic microbial cell,* ed. G. W. Gooday, D. Lloyd, and A. P. J. Trinci, 1–40. Thirtieth Symposium of the Society for General Microbiology. London: Cambridge University Press.

Cavalier-Smith, T. 1978. The evolutionary origin and phylogeny of microtubules, mitotic spindles and eukaryotic flagella. *BioSystems* 10: 93–114.

Conway Morris, S. 1998. *The crucible of creation: The Burgess Shale and the rise of animals*. Oxford: Oxford University Press.

Dodge, J. D., and J. T. Vickerman. 1980. Mitosis and meiosis: Nuclear division mechanisms. In *The eukaryotic microbial cell,* ed. G. W. Gooday, D. Lloyd, and A. P. J.

Trinci, 77–102. Thirtieth Symposium of the Society for General Microbiology. London: Cambridge University Press.

Follmann, H., and C. Brownson. 2009. Darwin's warm little pond revisited: From molecules to the origin of life. *Naturwissenschaften* 96: 1265–92.

Gesteland, R. F., and J. F. Atkins, eds. 1993. *The RNA world.* Cold Spring Harbor, NY: Cold Spring Harbor Laboratory Press.

Kasting, J. F. 1993. Earth's early atmosphere. *Science* 259: 920–26.

Keller, M. A., A. V. Turchyn, and M. Ralser. 2014. Non-enzymatic glycolysis and pentose phosphate pathway-like reactions in a plausible Archean ocean. *Mol. Syst. Biol.* 10: 725.

Knoll, A. H. 1992. The early evolution of eukaryotes: A geological perspective. *Science* 256: 622–27.

———. 2003. *Life on a young planet: The first three billion years of evolution on earth.* Princeton, NJ: Princeton University.

Margulis, L. 1970. *Origin of eukaryotic cells.* New Haven: Yale University Press.

———. 1981. *Symbiosis in cell evolution.* San Francisco: Freeman.

Maynard Smith, J., and E. Szathmáry. 1995. *The major transitions in evolution.* Oxford: Freeman.

Mereschkowsky, C. 1905. Über Natur und Ursprung der Chromatophoren im Pflanzen Reiche. *Biol. Centr.* 25: 593–604.

———. 1920. La plante considérée comme un complex symbiotique. *Bull. Soc. Sci. Nat. Ouest Fr.* 6: 17–21.

Miller, S. L. 1953. A production of amino acids under possible primitive Earth conditions. *Science* 117: 527–28.

Nisbet, E. G., J. R. Cann, and C. L. Van Dover. 1995. Origins of photosynthesis. *Nature* 373: 479–80.

Noireaux, V., Y. T. Maeda, and A. Libchaber. 2011. Development of an artificial cell, from self-organization to computation and self-reproduction. *Proc. Natl. Acad. Sci. USA* 108: 3473–80.

Orgel, L. E. 1994. The origin of life on Earth. *Sci. Amer.* 271: 77–83.

Schidlowski, M. 1988. A 3800-million-year isotopic record of life from carbon in sedimentary rocks. *Nature* 333: 313–18.

Schimper, A. F. W. 1883. Über die Entwicklung der Chlorophyllkorner und Färbkornerm, part 1. *Bot. Zeit.* 41: 105–14.

Schirrmeister, B., A. Antonelli, and H. C. Bagheri. 2011. The origin of multicellularity in cyanobacteria. *BMC Evolutionary Biology* 11: 45.

Schopf, J. W., and C. Klein, eds. 1992. *The Proterozoic biosphere: A multidisciplinary study.* Cambridge: Cambridge University Press.

Szathmáry, E., and J. Maynard Smith. 1993. The origin of chromosomes. II. Molecular mechanisms. *J. Theor. Biol.* 164: 447–54.

Taylor, T. N., E. L. Taylor, and M. Krings. 2009. *Paleobotany: The biology and evolution of fossil plants.* New York: Academic.

Williams, G. C. 1975. *Sex and evolution.* Princeton, NJ: Princeton University Press.

Woese, C. R. 1987. Bacterial evolution. *Microbiol. Rev.* 51: 221–71.

2

The Invasion of Land and Air

> A tale, like the universe, they tell us, expands ceaselessly each time
> you examine it, until there's finally no telling exactly where it begins, or
> ends, or where it places you now.
> —CHANG-RAE LEE, *On Such a Full Sea* (2014)

The previous chapter was devoted to evolutionary events occurring in the Precambrian. Here, we turn attention to critical events that occurred in the time interval called the Phanerozoic, although it is just as appropriate to call this time span the Phanerophytic (fig. 2.1). Compared with the early events in life's history, the events in this period may appear less dramatic or important. Yet, four evolutionary events changed the biological world dramatically. These were the invasion of land, the evolution of conducting tissues, the appearance of the first seed plants, and the evolution of the flowering plants. The first organisms to live and reproduce on land were eukaryotic photosynthetic organisms (plants) and lichen-like organisms, and, by providing shelter and food, they paved the way for the invasion of the land by animals. Those of us who study plants can take solace in the fact that the first zoological landfall was contingent on the greening of the terrestrial landscape by plants, which was really more an invasion of air than of

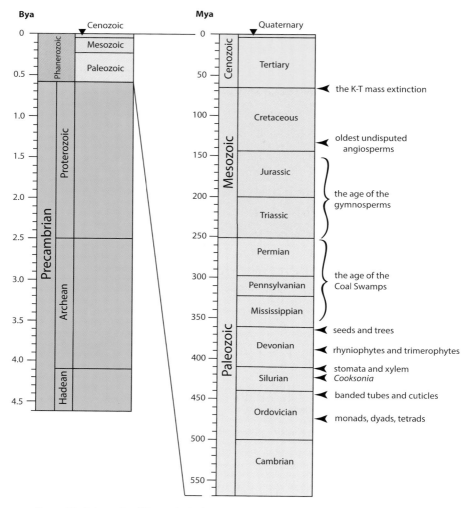

Figure 2.1. Schematic of the geological column emphasizing the time interval called the Phanerozoic (spanning ≈570 Mya to the present day). Some of the key evolutionary events, with an emphasis on land plant evolution, are indicated along with their approximate geological ages.

land. Many organisms can live in the wet interstices of soil and rock, but the first to challenge and later subdue the air were plants. Life's onslaught on air involved the evolution of vascular tissues, which provide low-resistance conduits for the flow of water and food. Without these plant tissues all terrestrial life, animals as well as plants, would remain physiologically anchored to a semiaquatic existence. The final greening of the land required seeds. Seed plants can successfully reproduce as well as survive vegetatively even in the most arid habitats, and so they have radiated into the drier uplands and deserts of the Earth. Eating their way through this green world, animals did not politely follow. With few exceptions, throughout most of life's long history, animals were the mortal enemies of plants. With the evolution of the seed and later the flower, hostilities changed into a truly symbiotic partnership between some plants and some animals. Like their other terrestrial counterparts, flowering plants are sedentary organisms, providing animals with fodder and shelter. But most flowering plants require or benefit from the transfer of pollen from one individual to another. Wind and water can serve in this capacity, but many pollen grains fail to connect with flowers and so are wasted in a reproductive sense. In contrast, animals are effective pollinators because they can recognize flowers belonging to the same species. Flowering plants have evolved intricate rewards and morphological subterfuges to attract animals for this purpose. As noted by Bruce Tiffney, it is not hyperbole to say that the flowering plants have trained animals to become their surrogate brain and muscle. A few gymnosperms have accomplished this as well, as for example the cycads.

The Embryophytes

As noted, when plants invaded the land, they really invaded the air by means of an evolutionary assault on the dehydrating effects of the atmosphere. Plants survive, grow, and reproduce on land because they are able to resist or tolerate desiccation. This ability involves numerous morphological and physiological departures from the ancestral aquatic condition of plants. For example, many land plants, like the mosses and liverworts, hug the moist substrates they grow on. This growth posture maximizes the surface area through which water can be absorbed, minimizes exposure to rapidly moving and drier air, and places them near the source of liquid water. Although many

Figure 2.2. Transverse section through the outer portion of a tomato fruit (*Solanum lycopersicum*). The cuticular membrane, which appears yellow in this unstained section, extends two to four cell layers deep. Mechanical tests show that it is remarkably tough in tension. The tensile modulus of this particular cuticular membrane is equivalent to that of cellophane.

land plants embrace this tactic, hugging a moist substrate is not a dependable method for survival on land. Even short, prostrate plants dehydrate when exposed to high wind speeds. Many moss species can withstand extreme dehydration, but at the expense of suspending growth and reproduction. A far more dependable method of reducing water loss, one that permits the true invasion of air, is to coat external surfaces with a cuticle that resists the loss of water to the external environment (fig. 2.2). The cuticle is a hydrophobic layer that coats the aerial surfaces of the vast majority of land plants, where it serves a wide variety of functions in addition to limiting water loss (for example, it provides a defense against pathogens and ultraviolet light and it prevents organ fusion during organ development). Cuticles are composed of two major components: a mixture of aliphatic and aromatic waxes and a substantial polymeric matrix of interesterified long-chain fatty acids termed cutin that is composed of ω-hydroxy C16 and C18 fatty acids often containing a mid-chain hydroxyl, ketone, or epoxy group. Despite its many important roles, much remains to be learned about cuticle synthesis, and in particular the mechanism of cutin polymerization and assembly at the cell surface.

Unfortunately the cuticle of land plants, which resists the diffusion of water, is also resistant to the diffusion of carbon dioxide and oxygen. The solution was to pierce the cuticle with holes through which these gas molecules can pass. The stomata that perforate the cuticles of vascular plants and some nonvascular plants provide microscopic avenues for gaseous diffusion, and the guard cells flanking stomata adroitly limit the rate at which water is lost through external surface areas while simultaneously ensuring a direct communication between the external atmosphere and the plant's moist interior. Life in the air was further assisted in most land plants by elaborating internal tissues with air-filled labyrinthine intricacy, thereby providing a large moist surface area through which gas can diffuse. Much like the alveoli of the mammalian lung, the aerenchyma of land plants reflects one of the overarching morphological themes of terrestrial life—the topological invagination of external into internal surface area.

Nevertheless, the persistence of plants on land was not ensured by mere vegetative acclimatization to air. The appellation "land plant" must be reserved for photosynthetic eukaryotes capable of successful reproduction on land. The distinction between two fundamental classes of adaptation—those concerned with the survival of the individual and those concerned with genetic continuity among individuals in successive generations—is important. Although reproduction is not essential to the survival of the individual, it is the route by which every species historically persists and evolves. Just like their ancestors, the first land plants undoubtedly were capable of some form of asexual reproduction, multiplying in number by throwing forth genetically identical individuals into the aerial realm. But asexual reproduction cannot account for the persistence of first land plants, which by definition were a significant evolutionary departure from their aquatic ancestors, nor can it account for the dramatic evolutionary changes that followed the first vegetative landfall. Seen in its entirety, the history of land plants could not conceivably have happened in the absence of sexual reproduction. An essential evolutionary innovation was the capacity to surround spore cells with a material that is resistant to both water loss and mechanical damage—the chemical moiety called sporopollenin fulfilled these functional obligations.

If it is true that colonization of the land required reproductive and vegetative adaptations, then a comparatively straightforward ecological definition of "land plant" emerges: *a land plant is any photosynthetic eukaryote that has the capacity to survive and reproduce on land.* This definition combines the two fundamental classes of adaptation, those that permit the survival and growth of the individual and those that ensure the genetic continuity of its species through time, either by means of sexual reproduction or its equivalent, such as prokaryotic pseudo-sexuality. It also emphasizes that two lines of evidence are required to prove an organism is a land plant—the organism must be capable of surviving on land, and its species must be shown to persist evolutionarily on the terrestrial landscape by virtue of sustained reproductive success. By the same token, this definition excludes any type of organism that attempted to colonize the land but ultimately failed. In this regard the fossil record indicates that the terrestrial landscape may initially have been occupied by very different phyletic groups of plants, but it also shows that these forays onto land were largely unsuccessful. In this respect the fossils of these organisms provide a dim glimpse into eclectic but ultimately unsuccessful ways different types of photoautotrophs coped with the desiccating effects of air. In the final analysis, these biological experiments are curiosities that tend to distract from two important conclusions: first, the only bona fide land plants are the embryophytes, and second, the embryophytes are a monophyletic group of plants. That is, all embryophytes ultimately descend from a last common ancestor. The juxtaposition of these two conclusions reveals that the ecological definition of "land plant" and the taxonomic definition are synonymous—the land plants are the embryophytes.

Archegonia and Antheridia

Surprisingly, the term embryophyte may be unfamiliar even to some professional biologists. Yet, all modern land-dwelling plants are embryophytes. Specifically, the embryophytes include the mosses, liverworts, hornworts, ferns, horsetails, lycopods, gymnosperms, and angiosperms. Despite the tremendous diversity in size and appearance among these plants, all modern embryophytes share three important life cycle features: (1) an alternation of generations involving a multicellular haploid generation and a multicellular diploid generation, (2) the retention of eggs and their diploid embryos within a structure called the archegonium (or its evolutionarily modified homolog), and (3) the formation of sperm cells within a structure called the antheridium (or its evolutionarily modified homolog) (fig. 2.3). The haploid generation is called the gametophyte because this multicellular plant produces gametes—sperm cells or egg cells or both. The diploid generation is called the sporophyte because it produces spores by meiosis, each encased in a spore wall composed of the highly chemically resistant biopolymer sporopollenin. The term embryophyte literally means "embryo-bearing plant," alluding to the fact that the gametophyte generation of every embryophyte retains its eggs, zygotes, and the resulting juvenile diploid embryos within archegonia.

Unlike the egg- and sperm-bearing reproductive structures of most algae, embryophytes enclose their gamete-producing cells within a sterile jacket of cells. The egg is enclosed within the archegonium; the sperm cells are enclosed within the antheridium (fig. 2.4). Although the evolution of the antheridium is important, the evolutionary significance of the archegonium is our focus because this structure serves

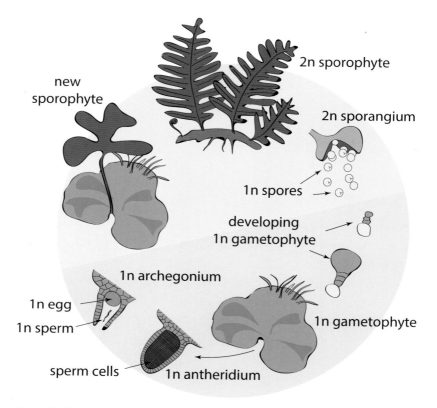

2n sporophyte

new
sporophyte

2n sporangium

1n spores

developing
1n gametophyte

1n archegonium

1n egg
1n sperm

1n gametophyte

sperm cells   1n antheridium

Figure 2.3. Schematic of the life cycle of the land plants (embryophytes), illustrated here for a fern. The life cycle is called the "alternation of generations" (also called the diplobiontic life cycle) because it involves two multicellular individuals, a diploid (2n) sporophyte and a haploid (1n) gametophyte. The sporophyte is the generation that produces haploid spores by meiotic cellular divisions in a structure called the sporangium. Among ferns and other non-seed plants, the spores are released from the sporangium and develop into a gameto- phyte. The gametophyte of most non-seed plants is typically photosynthetic and capable of producing both eggs and sperm (although generally not at the same time). Sperm are produced in a multicellular structure called the antheridium, whereas the egg is produced in a structure called the archegonium (see fig. 2.4). When an egg is fertilized, the resulting dip- loid zygote and subsequently formed diploid embryo are retained, protected, and nourished for a period of time within the archegonium. Consequently, the young sporophyte develops and initially grows within and upon its gametophyte, which typically dies thereafter. All land plants that reproduce sexually have some version of this life cycle, although the sporophyte and gametophyte have gone through dramatic evolutionary modifications in their appear- ance and functional intricacies.

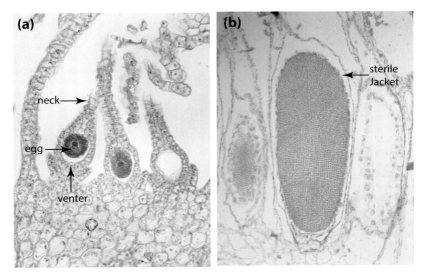

Figure 2.4. Transverse sections through an archegonium (a) and an antheridium (b) on haploid gametophytes. a. The archegonium consists of a neck, venter, and egg. The neck is a tube of cells whose inner core (called the neck tube cells) eventually deliquesces to provide sperm cells access to the egg. The venter consists of the cells immediately surrounding the egg. It provides nutrients to the egg and subsequently developing diploid embryo. b. The antheridium consists of a sterile jacket of cells surrounding a mass of spermatogenous cells that will eventually give rise to sperm cells.

to protect, nourish, and developmentally influence the growth of the juvenile sporophyte embryo.

Technically, the archegonium is a multicellular organ ontogenetically derived from a single haploid cell on the gametophyte. Although superficially simple in appearance, the archegonium is an extremely sophisticated reproductive device. In addition to protecting the egg and the diploid embryo that will eventually develop within it, the archegonium serves as a physiological interface shuttling nutrients produced by the gametophyte to the developing diploid embryo. This interface is physically manifested in the form of specialized transfer cells at the base of the archegonium, called the venter. These cells have highly branched or convoluted walls that dramatically increase the surface area of the cell membranes through which nutrients pass. The cells of the venter are physiologically active, transporting sugars from the gametophyte into the fertilized egg and developing embryo. In this respect the sterile cells of the archegonium function in a way analogous to the "placental transfer cells" occurring at the junction between the female placental mammal and its embryo.

How the embryophyte life cycle and archegonium evolved has been the subject of intensive research and speculation, but it is clear that, despite their complex appearance and physiology, the embryophytes evolved from an algal ancestor. We can say this with absolute, albeit ironic, confidence because by definition the "algae" are all plants lacking archegonia. Thus, the algae represent the only possible types of plants from which the embryophytes could have evolved. Nonetheless, there are about 70,000 known fossil and extant species of algae, and the algae are polyphyletic as a result of primary, secondary, or tertiary endosymbiotic events. Indeed, some comparative studies suggest that the algae constitute at least six distinctly different eukaryotic lineages. Within the past century this broad field of candidates for the embryophyte "cenancestor" has been dramatically narrowed. Only the green algae, the Chlorophyta, share with the embryophytes the same cell wall chemical composition (a variety of polysaccharides including cellulose), the same complement of photosynthetic pigments (chlorophyll *a* and *b*; carotenoids), the same principal carbohydrate food reserve (starch consisting of amylose and amylopectin), and the same number, location, and type of flagella on motile cells (two apically or laterally inserted whiplash flagella of equal length). Further, among the green algae, only the class Charophyceae shares with the embryophytes the same mode of cell division, called phragmoplastic cell division, which involves a persistent mitotic spindle and the disintegration of the nuclear envelope (Pickett-Heaps 1975, 1976; Graham 1993). Additional features shared by the Charophyceae and the embryophytes are the presence of a very particular photorespiratory enzyme called glycolate oxidase, found in organelles called peroxisomes, and the presence of a multilayered cellular structure at the base of the flagella of reproductive cells. Thus, all the evidence indicates that, of the 70,000 known species of algae, the species assigned to the

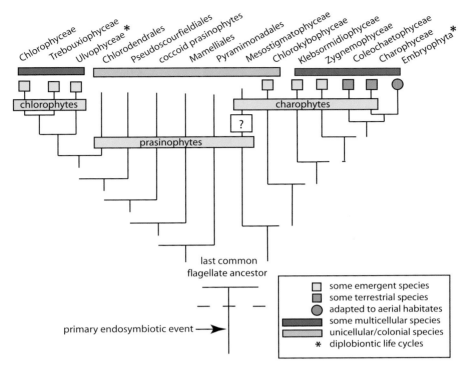

Figure 2.5. Cladistic schematic of the phylogenetic relationships among the extant green algae and the land plants (embryophytes), which are collectively called the Viridiplantae. This schematic also diagrams some of the shared features between the two large groups within this clade, the chlorophytes (Chlorophyta) and the Charophyceae and the Embryophyta. The latter two groups are called the streptophytes. For example, some chlorophytes have a life cycle with a diploid and a haploid multicellular life form (the "alternation of generations" life cycle). Some of the phylogenetic relationships shown in this diagram are problematic (as indicated by the broken lines). For example, some molecular studies place the Zygnemophyceae closer to the embryophytes than either the Charophyceae or the Coleochaetaceae.

order Charophyceae share the largest number of biological features with the over 250,000 living species of embryophytes (fig. 2.5).

Additional comparisons between the Charophyceae and the Coleochaetophyceae on the one hand and the embryophytes on the other help with the reconstruction of the evolutionary steps leading to the first land plants (fig. 2.6). For example, most molecular based phylogenies of the green algae and the land plants identify algal species

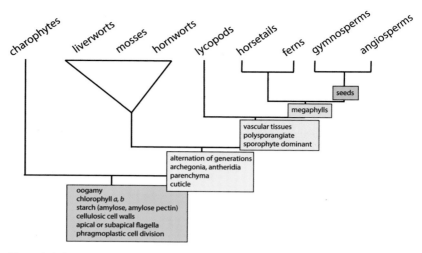

Figure 2.6. Schematic of the phylogenetic relationships among the green plants (Viridiplantae) showing a few of the features shared among the different lineages within this large monophyletic clade. For example, the charophycean algae and the embryophytes share oogamy (large, nonmobile eggs) and a host of physiological and cytological attributes. However, the charophycean algae lack a life cycle with an alternation of generations. The only multicellular organism in their life cycle is the haploid gametophyte. In turn, the nonvascular land plants (the liverworts, mosses, and hornworts, which are collected called "bryophytes") possess an alternation of generations (in which the gametophyte is persistent and the sporophyte short-lived), archegonia, and antheridia, and many have a cuticle, but lack vascular tissues. The phylogenetic relationships among the bryophytes are not resolved in this schematic because cladistic analyses using different data sets yield different phyletic relationships. All of the vascular plants are polysporangiate (that is, their sporophytes produce more than one sporangium as opposed to the monosporangiate bryophytes). The lycopods differ from the rest of the vascular plants in possessing leaves that are called microphylls (and should be called lycophylls) as opposed to the leaves of horsetails, ferns, and the seed plants, which are called megaphylls (also inappropriately called euphylls).

with large, nonmobile large eggs such as *Coleochaete* and *Chara* as the nearest living species to the embryophytes. From these studies, it is reasonable to surmise that the ancestor to the embryophytes was oogamous (that is, the ancestor produced one or more large nonmobile eggs). In addition, all of these algal species retain their eggs and provide them with nutrients as do embryophytes. In the case of *Coleochaete* this is accomplished by the growth of cortical cells around the zygote, and in at least one species, *Coleochaete orbicularis*, these cells develop localized wall ingrowths similar to those produced in the nutrient-supplying (venter) cells of archegonia. There is some evidence that cortical cells may actively transport sugars to fertilized eggs, which is evidence of a physiological condition called matrotrophy (nutrition provided by the mother). For example, *Coleochaete* and *Chara* zygotes continue to accumulate large stores of lipids and starch after the egg is fertilized, and overall vegetative growth is stimulated by adding glucose to the medium in which *Coleochaete* is cultivated (Graham et al. 1994). Thus, although the evidence is circumstantial, there are good reasons to believe that vegetative and reproductive cells have the ability to actively transport sugars. Additionally, distinctive acetolysis-resistant, autofluorescent materials, which bear a striking similarity to those found in the placental cells of the hornwort genus *Anthoceros,* are produced in the cortical cell walls of all *Coleochaete* species examined. All of these similarities may be the result of convergent evolution, but viewed in the broader context of other shared features it seems reasonable to conclude that the common ancestor to the embryophytes and algae like *Coleochaete* and *Chara* had the ability to produce a jacket of sterile cells around its zygotes, and that these cells functioned physiologically in a way analogous to the venter cells of the archegonium.

Yet, it is obvious that the algae typically placed closest to the land plants lack two very important features—the archegonium itself and a multicellular diploid phase in the life cycle. In these algae, the only diploid cell found in the life cycle is the fertilized egg (which subsequently divides meiotically). These differences make it unwise to speculate on whether the multicellular diploid generation of the embryophytes evolved before, after, or concurrently with the evolution of the archegonium. Notice that all extant plants with archegonia have a multicellular diploid individual in their life cycle. But this correlation comes from a curious intellectual hindsight. Just because all *modern* embryophytes have archegonia and a multicellular sporophyte, there is no reason to suppose that these two features evolved concurrently.

Indeed, because there are numerous advantages to each of the biological innovations that collectively define an embryophyte, there is no reason to assume that "archegoniate plants" and "embryophytes" are synonymous. Likewise there is no a priori reason to assume that all past embryophytes were land plants. Consider a hypothetical sequence of evolutionary events beginning with an organism very like *Coleochaete* or *Chara* that lived in shallow bodies of freshwater or in moist microhabitats, perhaps near streams, rivers, or lakes (fig. 2.7). For the sake of argument, we will assume that this hypothetical organism had large, nonmobile eggs protected by sterile cells produced either before or after fertilization. The sterile cells would have reduced the probability of microbial attack on eggs and zygotes, in addition to reducing the chances that zygotes would be washed away or dehydrate when the habitat flooded or the water level dropped precipitously. We will assume further that this organism already had some type of active sugar transport system, because the large, nonmobile eggs of most green algae, even those produced by algae unrelated to the charophycean, continue to accumulate nutrients after the egg cell severs its cytoplasmic connections with other adjoining cells. However modestly developed, an active nutrient transport system would confer an immediate benefit to flagellated cells resulting from the meiotic divisions of the zygote—greater vigor and possibly larger size, and therefore greater success at establishing the next generation of gametophytes.

The next stage in the hypothetical sequence is an organism whose eggs were retained within a true archegonium capable of transporting nutrients to its egg (the evolution of matrotrophy). The advantages are obvious. The egg would be supplied with nutrients and protected from inimical physical or biological environmental conditions *before* fertilization, thereby fostering a healthy population of eggs and subsequently assisting the establishment of a healthy population of zygotes. The next step in the hypothetical sequence is the evolution of delayed zygotic meiosis, which would result in a multicellular diploid phase in the life cycle. The advantage of this innovation is that even a few mitotic cellular divisions would amplify the reproductive benefits of each fertilization event because every cell of the diploid phase would

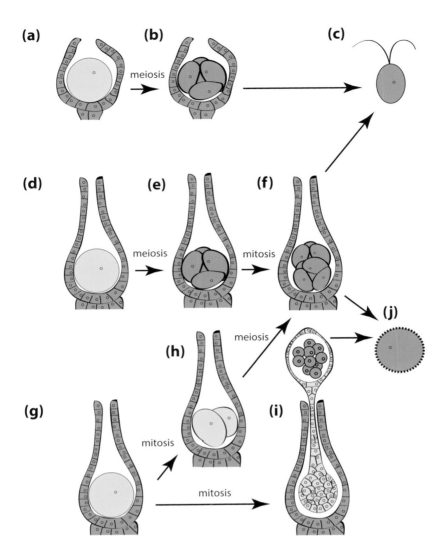

Figure 2.7. Scenarios for the evolution of the embryophytic alternation of generation life cycle from an ancestor possessing archegonia. Diploid cells shown in yellow; haploid cells shown in green. a–c. The life cycle of the ancestor involves an archegonium containing a fertilized egg (a) than undergoes meiosis to yield cells (b) that are subsequently released as flagellated cells (c). d–f. The fertilized egg (d) undergoes meiosis (e) and the haploid cells undergo mitosis (f) to yield cells that develop either into flagellated cells (c) or spores with spore walls (j). g–j. The fertilized egg (g) undergoes mitosis to produce either a few diploid cells (h) that subsequently undergo meiosis (f) or that develop into a multicellular sporophyte (i). Note that each of these life cycles involves a gametophyte with archegonia (a, d, g), yet two of these life cycles lacks an alternation of generations (a–c and d–f). Note also that the presence of spores with spore walls (j) does not provide unequivocal evidence for the existence of an alternation of generations (or for a terrestrial organism).

be functionally equivalent to a single fertilized egg. Thus even very rare fertilization events could potentially yield a population of new adults. The last step in the hypothetical sequence is the evolution of a multicellular diploid sporophyte composed of vegetative (somatic) and reproductive (germ line) cells. The growth and division of somatic cells could have elevated the reproductive portion of the sporophyte, thereby enhancing the probability of long-distance dispersal of spores and the establishment of new populations of gametophytes far away from the parent population (box 2.1). Even for an aquatic organism, the ability to form spores with walls that resist the loss of water (and that resist mechanical damage) would be advantageous in habitats prone to seasonal drying (and herbivory). Many freshwater algae encase their zygotes with walls composed of sporopollenin. It is possible that the genetic modifications attending the evolution of delayed zygotic meiosis in the first embryophytes also delayed the initiation of the biosynthetic pathway leading to the formation of sporopollenin-encased spores.

There is no good reason to believe that this hypothetical "gradualistic" scenario reflects the actual course of events leading to the appearance of the first embryophytes. Conceivably the archegonium and delayed zygotic meiosis could have simultaneously and suddenly appeared as a consequence of some type of macromutation (a "hopeful monster"). Nonetheless, the gradualist scenario serves a pedagogical function, showing that we should not assume the first archegoniate plants were embryophytes, or that all past embryophytes were land plants.

Earlier, we discussed features that permitted plants to survive on land, such as cuticles and stomata, and features that ensure successful reproduction on land, such as spores with walls composed of sporopollenin. Although the sequence of historical events that procured these functional traits may never be known with absolute assurance, the fossil record provides evidence that land plants most likely evolved in the Ordovician period and certainly no later than the Silurian (see fig. 2.1). Spores with sporopollenin walls, cuticle-like sheets bearing

Box 2.1. Gametophyte Dispersal and Sexuality

The fossil record provides evidence showing that the most ancient land plants dispersed their haploid spores by means of wind. Elevating spores above ground in sporangia borne at the tips of sporophytes increased the range of dispersal because ambient wind speeds generally increase (to a limit) in the vertical direction away from the surfaces of substrates. However, on the basis of the study of modern-day nonvascular land plants (mosses, liverworts, and hornworts, collectively the bryophytes), there are insufficient grounds to argue as to whether the most ancient land plant gametophytes produced antheridia and archegonia on the same plant, or whether they produced antheridia and archegonia on separate plants.

Unisexual (*dioicous*) gametophytes cannot self-fertilize themselves and therefore require a gametophyte of the opposite sex to successfully produce sporophytes, spores, and a new generation of gametophytes. One advantage of this reproductive syndrome is that sexual reproduction will invariably result in genetic diversity within a population. In contrast, bisexual (*monoicous*) gametophytes can self-fertilize themselves (fig. B.2.1) and thereby colonize a new location by means of a single viable spore. Arguably, this reproductive syndrome provides an advantage in terms of establishing new colonies. However, one disadvantage is that all of the sporophytes resulting from self-fertilization are homozygous. This minimizes the genetic variation within the next generation of gametophytes and abrogates the presumed benefits of sexual reproduction. In this sense, the apparatus for sexual reproduction (antheridia and archegonia) has been hijacked for cloning.

The potentially negative consequences of having bisexual gametophytes can be avoided in a number of ways: (1) the evolution of self-incompatibility, (2) the maturation of antheridia and archegonia at different times on the same gametophyte, and (3) the physical separation of antheridia and archegonia on the same plant.

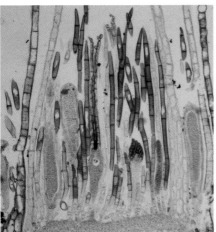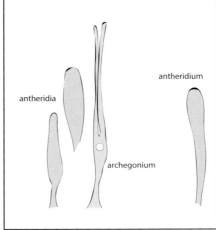

Figure B.2.1. The tip of the reproductive shoot (called the gametophore) of the moss *Mnium* showing an egg-bearing archegonium surrounded by sperm-bearing antheridia and sterile filaments (called paraphyses). Serial sections through this gametophore revealed additional archegonia. Some species have evolved self-incompatibility, which prevents plants from fertilizing their own eggs. Other species are self-compatible and can highjack the apparatus of sexual reproduction to reproduce asexually.

the impressions of cells, and microfossils called banded tubes appear in Ordovician sediments. Morphologically similar microfossils are also found in early Silurian rocks (see Pratt, Phillips, and Dennison 1978). Although spores with sporopollenin walls are not prima facie evidence for the existence of multicellular terrestrial sporophytes (Banks 1975), spores with resistant walls provide evidence of plants with reproductive cells that could survive limited exposure to dry conditions. That some of the most ancient spores are the products of meiosis, and therefore derived from diploid sporophytes, is indicated by the presence of a trilete marking on spore walls. The trilete mark is a Y-shaped ridged surface on the spore wall that occurs when a diploid cell divides to produce four pyramidially arranged, closely abutting spores (fig. 2.8). The co-occurrence of cuticle-like sheets bearing cellular impressions is also suggestive of terrestrial plants. A true cuticle is totally useless (even detrimental) to an aquatic plant because it impedes the passage of carbon dioxide and oxygen from the water into living plant cells (Raven 1984). And finally, the microfossils called banded tubes could represent extremely ancient water-conducting cell types, signifying a level of anatomical sophistication expected for a bona fide land plant.

Unfortunately, the biological affinities of cuticular sheets and banded tubes are problematic. Although some paleontologists contend these microfossils are the remains of ancient plants, others find troubling morphological parallels with the remains of aquatic and semiaquatic animals (Banks 1975). The arguments marshaled on either side of this debate are cogent. Those who believe that some Ordovician cuticular sheets and banded tubes belong to plants rather than animals point out that these microfossils are found alongside plant spores and therefore most likely all belong to the same kind of organism. Perhaps in frustration, these experts also point out that the botanical affinities of the Ordovician remains would not be contentious if "cuticles" and "banded tubes" had been found in Silurian rocks. In turn, skeptics argue that the burden of proving the botanical affinities of these remains lies with those who would push the origin

Figure 2.8. A clay model of a spore with a trilete mark (left) removed from what was originally a tetrad of spores (right) produced when a diploid cell (called a sporocyte) undergoes
meiosis to produce four spores. The spores of non-seed plants and the pollen of seed plants
posses a rigid spore wall composed of sporopollenin, which is deposited around the spore
membrane after meiosis. The tetrad of spores remains aggregated until the spore walls are
fully formed. The aggregation results in each spore interfacing with each of the other three
spores. The three contacting surfaces produce the Y-shaped trilete marking.

of land plants into the Ordovician. Few doubt that the first land plants
evolved during early Silurian times and that the vascular land plants
had evolved by the end of the Silurian, but the Ordovician microfossils
are more dubious in terms of what may be said about their biology.
Nonetheless, we can accept the perspective that, although a skeptical
attitude is warranted, particularly when the implications of a hypothesis might chafe against our conventional thinking, land plants most
likely evolved in the Ordovician. Whether all these Ordovician "land
plants" were embryophytes or whether some were only transient ex-
plorers of the terrestrial domain, derived from algal lineages unrelated
to the charophycean algae, remains to be seen.

The Sporophyte

Before moving on to the next major evolutionary transition (from nonvascular to vascular land plants), we need to explore how and why the alternation of generations evolved—more specifically we need to explore how and why the multicellular sporophyte evolved. Clearly, all plant life cycles have a diploid and a haploid cell in them. Among the charophycean algae that are phylogenetically nearest to the land plants, the diploid cell in the life cycle is the zygote. Thus, the evolutionary innovation called the land plant sporophyte can be thought of as a zygote that has achieved multicellularity. Why should this have happened?

Life cycles that have an alternation between a multicellular haploid and a multicellular diploid generation occur in the brown algae, the red algae, the green algae, and the land plants. In each case, these life cycles are posited to have evolved from a life cycle similar to that of *Chlamydomonas* (fig. 2.9). Comparative phenotypic and molecular studies also indicate that multicellularity most likely appeared first in the haploid generation and subsequently in the diploid generation within each of these three lineages. This parallel sequence of events raises a simple question—why did multicellularity evolve in the haploid phase first and in the diploid phase second?

Whether multicellularity evolves in the haploid or diploid phase, or in both phases, depends on a number of factors including an organism's population genetics and its environmental context. Here, we focus on the former, since the frequency of convergent evolution in the life cycles of marine and freshwater algae and in the land plants is very high. If the haploid phase is multicellular, the gametes produced by an individual will be genetically homogenous, and, if self-compatible, many of its progeny resulting from sexual reproduction will be clones. Meiosis will recombine and re-sort the haploid genome after syngamy, but the fitness of a *population* of haploid multicellular organisms (with a *Chara*-like life cycle) is nevertheless limited because all zygotes contain at least one-half of the haploid's genome. In contrast, if the dip-

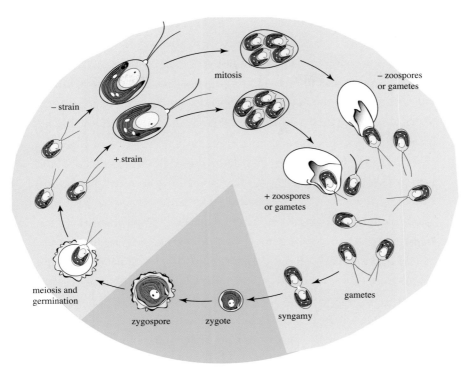

Figure 2.9. Schematic of the life cycle of the unicellular green alga *Chlamydomonas*. In this life cycle, meiosis produces the unicellular haploid phase, which has two different mating types (− and + strains). When stressed, the haploid cells encyst and undergo mitotic division, which increases the number of cells that can subsequently develop into asexual zoospores or gametes, which join and fuse together (a process called syngamy) to produce diploid zygotes. Note that the only diploid cell in this life cycle is the zygote (and its dormant equivalent called the zygospore).

loid is multicellular, its gametes will be genetically diverse as a result of meiosis to produce gametes (gametic meiosis). Thus, an organism with zygotic meiosis will likely produce progeny with less genetic variance than the progeny produced by an organism with gametic meiosis. Naturally, the extent to which this holds true depends on other factors. For example, if genetic variation is predominantly maintained by selection acting on gene frequency, gametic meiosis will be more effective at maintaining genetic variation than zygotic meiosis. Likewise, as long as the expression of deleterious alleles is sufficiently masked by the wild-type allele in the heterozygous condition, genetic models show that diploidy is favored over haploidy. However, when the rate of recombination is low, diploidy is much less likely to be favored over haploidy, regardless of environmental context. In fact, according to one model, the evolution of diploidy is impossible without significant levels of recombination, even when the masking of deleterious alleles is strong.

It is interesting to consider the consequences of selection acting primarily on gamete function with or without its effect on diploid fitness, a phenomenon called meiotic drive. In life cycles with only a multicellular diploid phase (such as human beings and the brown alga *Fucus*), the alleles for gamete function will segregate during meiosis to subsequently create a population of diploids that can maintain these alleles under constant selection for alleles that enhance gamete-function. In contrast, zygotic meiosis (such as in the *Chara*-like life cycle) will result in haploids containing alleles that are either less or more fit for gamete function, such that multicellular individuals bearing the less fit alleles will be at a disadvantage, thereby reducing the effective population size of reproductively viable individuals. The effect of gametic meiosis on the genetic variance of alleles affecting gamete function will also depend on the intensity of competition among gametes. If competition is intense among gametes, only those gametes bearing superior alleles for gamete function will succeed at producing zygotes, such that multicellular diploids with gametic meiosis will have an advantage over their haploid counterparts. If, however, competition among gametes is low, the majority of gametes will be successful at forming zygotes such that multicellular haploids with zygotic meiosis will have the advantage; this may explain the parallel evolution of zygotic meiosis in so many algal lineages.

Finally, it is worth noting that achieving an alternation of generations likely did not require extensive genomic reorganization, since almost all unicellular eukaryotes are capable of asexual reproduction. Nevertheless, post-fertilization mitosis necessitated modifications in the onset of the cell cycle. If the ancestral haploid was multicellular, the

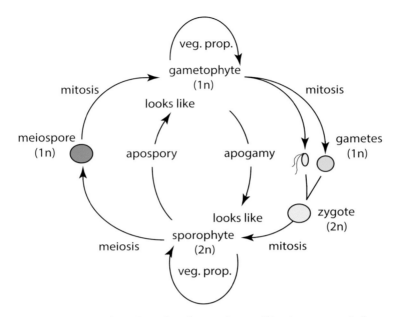

Figure 2.10. Schematic of the alternation of generations and the phenomena called apospory and apogamy. Among the extant land plants, the gametophyte and the sporophyte differ in their appearance (they are dimorphic). However, under certain conditions, the haploid cells of a gametophyte can be induced to grow into a plant that has the morphological appearance of a sporophyte (apogamy, literally *without the benefit of gametes*). Likewise, the diploid cells of a sporophyte can be induced to grow into a plant that has the morphological appearance of a gametophyte (apospory, literally *without the benefit of spores*). Note that each of the two multicellular generations is capable of vegetative propagation (veg. prop.) and can thus clone itself.

expression of multicellularity in the diploid phase arguably presented no great challenge because the genomic potential for multicellularity already existed. Two reasonable deductions, therefore, are (1) the ancestral diplobiontic life cycle was isomorphic (as in the green alga *Ulva*) and (2) the evolution of a dimorphic life cycle had to confer some advantage as for example niche-partitioning as a counter-measure perhaps against predation. In passing, it is worth noting that isomorphy is not an unknown even among modern-day plants. The phenomena called apospory and apogamy occur in many bryophytes and ferns (fig. 2.10). It is even possible to make a pollen grain, which is haploid, develop into a plant that looks like a diploid sporophyte (fig. 2. 11).

Figure 2.11. An illustration of experimentally induced apogamy in a diploid species of *Brassica*. The plant on the right is a normal diploid sporophyte. The plant on the left was generated by heat-shocking a haploid pollen grain, which was subsequently grown in culture. Both plants produced flowers. The haploid plant failed to produce seeds.

This transformation is particularly easy to perform for species that are polyploids. In this context, it is interesting to note that the deletion of the *KNOX2* gene in the moss *Physcomitrella patens* results in the development of the gametophyte morphology by diploid embryos without meiosis (apospory). The *KNOX* genes are transcription factors that occur in the green plants. There are two subfamilies in land plants, *KNOX1* and *KNOX2*, which resulted from gene duplication in the lineage leading to land plants. *KNOX1* genes regulate sporophytic (diploid) meristematic genes, which indicates that the evolution of *KNOX* genes was important in establishing the alternation of generations in land plants.

The Evolution of Vascular Tissues

When juxtaposed, the evolutionary histories of the nonvascular embryophytes (the bryophytes) and the vascular plants (the tracheophytes) indicate that the embryophytes as a group evolutionarily explored one of two largely mutually exclusive routes, either the amplification of the size of the gametophyte generation or the amplification of the size of the sporophyte generation. The bryophytes followed the former route, and the vascular plants followed the latter.

As noted before (see fig. 2.6), the bryophytes include the mosses, liverworts, and hornworts (see table 0.1). As a group, the bryophytes are paraphyletic. A group of organisms is said to be paraphyletic when it does not include all of the descendants of a last common ancestor. Yet, each bryophyte lineage is characterized by having a persistent gametophyte and small ephemeral sporophyte. The tracheophytes include the ferns, horsetails, lycopods, the various lineages of seed plants (gymnosperms and angiosperms), and a host of extinct plant groups. Each of these tracheophyte lineages is characterized by having sporophytes producing multiple sporangia and comparatively small, short-lived gametophytes. The either/or choice in embryophyte evolutionary history reflects the fact that the sporophyte generation is physically attached to the gametophyte generation and is incapable of an independent existence during its early ontogeny. Because the sporophyte is physically attached to its egg-bearing gametophyte and, at least initially, physiologically dependent on it, indeterminate growth of the sporophyte generation necessarily has destructive consequences for the continued survival and growth of the gametophyte. Conversely, the elaboration of the size of the gametophyte generation can be entertained only when the sporophyte generation is comparatively small or short-lived.

The elaboration in size of the sporophyte or the gametophyte generation literally takes two very different "directions," or more precisely orientations, because the two multicellular generations of every embryophyte fulfill very different reproductive roles that in turn have different biological and physical requirements. The reproductive functions of the gametophyte are to produce gametes, to facilitate the fertilization of eggs, and to maintain at least for a time the developing sporophyte. These functions require continued access to liquid water for the survival of the gametophyte, the transfer of sperm to egg, and the initial growth of the sporophyte embryo. For this reason the free-living gametophytes of most embryophytes tend to be short, prostrate organisms that hug their moist substrates, thereby reducing the distance motile sperm cells must travel to reach eggs and minimizing the rate of water loss from growing tissues. In this sense, indeterminate vertical growth of the gametophyte generation is not favored, whereas continued horizontal growth against a moist substrate is possible and even beneficial. In contrast, the reproductive roles of the sporophyte generation are to produce and eventually disseminate meiospores. In the terrestrial landscape, wind provides a cheap and more or less dependable abiotic vector for spore dispersal. Because wind velocity diminishes toward ground level and because the time it takes for spores to settle out of the moving air increases as spores are released higher into the air, the reproductive roles of the sporophyte generation of land plants are favored by vertical rather than horizontal growth.

The elaboration in the size of a more or less prostrate gametophyte or a vertical sporophyte requires water- and food-conducting tissues

once either organism reaches a critical size. The functional significance of conducting tissues can be explained in terms of a basic physical principle: the time for even small molecules like water and glucose
to passively diffuse through living cells increases with the square of
the distance to be traversed. This principle follows directly from Fick's
second law of diffusion. In the absence of specialized conducting tissues, height is physiologically limited by the time it takes molecules to
passively diffuse through cells and reach aerial portions of the plant.
Simple calculations show that it takes about one second for water
molecules to passively diffuse through a living cell measuring 50 μm
in length but about one year for the same number of molecules to diffuse through a string of five hundred of these cells placed end to end
(a total distance of roughly 2.5 cm). Clearly, passive diffusion of water
and glucose molecules is rapid enough to maintain the metabolic requirements of plants over short distances but woefully insufficient to
sustain even modestly elevated aerial portions of plants. Thus, simple
physics shows that some kind of conducting tissue is a physiological
necessity for the continued vertical growth of land plants.

Physical laws and engineering principles further dictate the nature of conducting tissues. The efficient and rapid passage of water
requires tubular conduits offering low resistance to fluid flow along
their length, and the volume of water passing through a conduit is
proportional to the fourth power of the radius of the conduit. Therefore, broad hollow tubes are the best water-conducting devices. Water
conduction through plant tissues is accomplished by removal of the
protoplasts of living cells whose principal (longitudinal) axes are
aligned parallel to the direction of preferred flow as for example in the
water-conducting cells of some mosses, called hydroids, and in the
xylem tracheids and vessel members of vascular plants (fig. 2.12). In
some plants, the end walls of these cells remain after the death of
the protoplasts (hydroids and tracheids). In other plants, where the
demand for water is high, the end walls of water-conducting cells are
partially or entirely removed by protoplasts before their death (as in
vessel members), thereby minimizing resistance to water flow from

Figure 2.12. Longitudinal section through the permineralized xylem stand of *Rhynia gwynne-vaughnii*, a Devonian age plant. Note the secondary wall thickenings and the tapered end of the tracheid indicated by the arrow. Polarized light indicates that some native cellulose is still present in some of these cells. Although organic materials are preserved in the Rhynie Chert, the coloring in the figure is largely the result of minerals that have perfused cell walls.

one cell to another and increasing the rate of flow. A characteristic of many water-conducting cells is a cell wall that is differentially thickened internally (tracheids and vessel members). These thickenings mechanically reinforce the cell wall from mechanical implosion as a consequence of the negative internal pressure created when a fluid moves rapidly through a tube. Thus, these thickenings are important when the demand for water is great and high flow rates through conducting tissues are required. They are of less importance when the demand for water is small. It is hardly surprising, therefore, that the water- and food-conducting cells of embryophytes—bryophytes and tracheophytes alike—tend to be tubular, stacked end to end along the principal dimension of organs like stems, leaves, and roots. Nor is it surprising that the water-conducting cells of mosses lack internally thickened cell walls and that taller plants possess them.

Physical principles also reveal the magnitude of the critical size mandating water- and food-conducting tissues. To be sure, the idea of "critical size" is vague because it depends on environmental conditions (ambient wind speed, temperature, and so forth), the extent to which the plant is elevated or reclines against a moist substrate (and therefore is exposed to the drier atmosphere), and the extent to which the surface of the plant is protected from water loss through evaporation (presence or absence of a cuticle and stomata). However, calculations based on Fick's second law of diffusion indicate that in the absence of a cuticle a cylindrically shaped plant 2 cm tall growing in a still atmosphere at 70% relative humidity requires a water-conducting tissue even if the plant has free and unlimited access to liquid water at its base. These calculations are necessarily crude, but they help show the order of magnitude of "critical size"—the threshold of vertical height expected to impose selection pressures favoring conducting tissues.

It should not escape attention that the size threshold mandating a conducting tissue is totally indifferent to whether a plant is a sporophyte or a gametophyte. Any selection pressure favoring tall gametophytes or tall sporophytes is expected to also favor organisms with conducting tissues. In this regard it is hardly surprising that tall moss gametophytes have water- and food-conducting tissues—the hydrome and the leptome, respectively—that are strikingly similar to the xylem and phloem tissues found in the sporophytes of vascular plants. By the same token, any selection pressure favoring a reduction in the vertical size of either the gametophyte or the sporophyte would likely reduce anatomical complexity to a parallel degree. Conducting tissues are highly reduced or wholly absent in short and small moss species phylogenetically related to tall and large species with conducting tissues (Hébant 1977).

The relation between critical size and anatomical complexity for land plants is embedded within the duality of the embryophyte life cycle. Because the embryophyte life history presents an either/or choice of whether to elaborate the size of its sporophyte or that of its gametophyte generation, an evolutionary amplification of the size and anatomy of one generation will be mirrored by an evolutionary reduction in the size and anatomical complexity of the other. Crudely put, the embryophyte life cycle permits one "degree of freedom," although the "choice" is not irreversible. A lineage that has aggrandized its sporophyte generation in the past could conceivably reverse this trend. The only constancy is in the consequences of the choice— beyond a critical size threshold, increasing the size of one generation mandates the presence of conducting tissues and a diminution in the relative size and anatomy of the other generation.

This either/or choice and its anatomical consequences make the "bryophytes" and the "tracheophytes" a foregone conclusion in the sense that they are the only two broad categories of biological organization available to the earliest terrestrial embryophytes. By virtue of the presence and absence of vascular tissues, the tracheophytes and the bryophytes were also considered two distinctly different levels of anatomical organization. Largely for these reasons, the bryophytes and the tracheophytes were traditionally viewed as two distinct lineages, though they trace their evolutionary history back to a common ancestor. Thus, the bryophytes and the tracheophytes were simultaneously thought of as clades (monophyletic groups of plants) and grades (different levels of biological organization). These phyletic and anatomical commonalties were formally codified by placing all the bryophytes into the division Bryophyta and all the vascular plants into the Tracheophyta. Additionally, Bryophyta was positioned between Thallophyta (collectively, all the algae) and Tracheophyta, thereby conveying the notion that the nonvascular embryophytes were ancestral to the vascular land plants.

This classical view of the bryophytes and tracheophytes is now known to be wrong. As noted, comparative phylogenetic studies indicate that the bryophytes are a paraphyletic group of plants. However, it remains uncertain which of the bryophyte lineages is the most closely related to the vascular plants. The mosses, liverworts, and hornworts most likely had separate evolutionary origins from a last common ancestor. Thus, the "bryophytes" cannot be considered a clade, but rather a set of lineages sharing the same grade of biological organization (nonvascular embryophytes). An interesting line of speculation revolves around the possible independent evolution of the sporophyte in each of the three bryophytic lineages. It is not unreasonable to suggest that the gametophyte phase in the life cycles of the ancestor to each of the current moss, liverwort, and hornwort lineages evolved first from a common plexus of charophycean algae and that the sporophyte phase in the life cycle of each of the three bryophyte lineages evolved later. This possibility would help to explain why phyletic analyses have had difficulty resolving the relationships among the three byrophytic lineages (and it may help to explain why the sporophytes of some early Devonian plants such as *Aglaophyton major* morphologically mirror those of vascularized sporophytes, yet lack tracheary elements).

Curiously, the monophyletic nature of the tracheophytes is not secure either, at least not in terms of the origin of vascular tissues. The fossil record provides evidence that the lycopods are a derived group that evolved from an extinct lineage called the Zosterophyllophytina that in turn diverged from the other vascular plant lineages well before the organographic distinctions among leaves, stems, and roots evolved (and possibly before the land plants evolved vascular tissues). Consequently, the things botanists call "leaves," "stems," or "roots" are not necessarily anatomically and morphologically the same things. For example, the leaves of lycopods (called microphylls, which is inappropriate because some of the leaves of the tree lycopods were roughly one meter long) differ significantly from the leaves of all other vascular plants (called euphylls or megaphylls, which is inappropriate because they are neither "true" nor invariably "large"). Accordingly, whether we advocate tracheophyte monophyly in a strict sense depends on whether certain biological features shared among different lineages are homologous features or are simply analogous. As we shall see, the determination of what constitutes *homology* is far from easy (see chapter 4). Clearly, all the evidence indicates that the land plants are monophyletic, but the relationships among the vascular and non-vascular lineages are intensely complex, particularly since phylogenies based exclusively on molecular data cannot incorporate data from the fossil record and because phyletic hypotheses based on cladistic analyses are extremely sensitive to the data matrices used (as well as how character states are scored by a researcher).

Finally, it is worth mentioning that the most ancient vascular tissue may have been phloem-like rather than xylem-like. Whether the first hydraulic tissues transported water through dead space like xylem or through cytoplasm like phloem remains conjectural, since the function of such a delicate tissue as the phloem would not be easily deduced from fossils. However, it is noteworthy that phloem, and not xylem, is the principal venue for water transport in developing leaves, and it transports significant quantities of water as a result of phloem-loading in mature organs (fig. 2.13 a). It is conceivable, therefore, that the first specialized tissues for transport in the earliest land plants conveyed both water and photosynthates through living cytoplasm (that is, symplastically) (fig. 2.13 b), just as it is possible that the phloem-like tissue (called the leptome) of some extant mosses transports water as an indirect consequence of the basipetal transport of photosynthates in the xylem-like tissue called the hy-

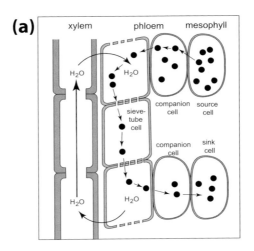

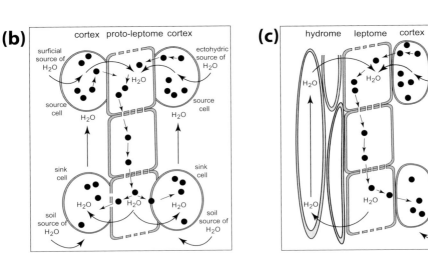

Figure 2.13. Schematic of phloem loading in an angiosperm such as a willow (*Salix*) (a), a hypothetical symplastic (= through cytoplasm) water transport system driven by external water sources (ectohydric) and a solute concentration gradient (b), and a hypothetical hydrome-leptome (xylem- and phloem-like tissue, respectively) tissue system with a phloem-loading component (c). Solutes denoted by black dots; concentration gradients indicated by numbers of black dots. The phrase *phloem-loading* refers to the process whereby carbohydrates produced in photosynthetic cells enter the conducting cells in the phloem tissue system and are transported to other cells within the plant body. The process involves the generation of a concentration gradient of photosynthates from a source (photosynthetic cells) to a sink (as for example cells in the root).

drome (fig. 2.13 c). If so, a plausible scenario is that, because of their small size, the most ancient land plants could rely on passive diffusion for the acropetal symplastic transport of water using basipetally decreasing solute concentrations in a leptome. Plant stature could have increased with the evolution of a xylem-like hadrome tissue capable of apoplastically transporting water (supplied in part by the external, ectohydric wicking of water) driven by a simple hadrome solute loading system. The transition from this condition to a true vascular system with xylem and phloem-loading required the evolution of cell wall patterning and differential thickening and programmed cell death, both of which are prefigured in nonvascular plants. The differential thickening and lignification of xylem cells empowered tracheids and vessel members to resist implosion by virtue of the mechanical reinforcement of cell walls and by virtue of lignin's hydrophobic ability to prevent the hydration of cellulose, which is weaker when wet than when dry.

Heterospory and the Seed Habit

Most Devonian vascular land plants had sporophytes that freely shed their spores into the air. The gametophytes of these organisms grew detached from the sporophytes that produced them. Vascular plants that freely shed their spores are called pteridophytes. The term literally means "fernlike plant" and alludes to the fact that the sporophytes of plants like ferns, horsetails, and lycopods shed their spores and therefore have "free-living" gametophytes. The independent lifestyle of the free-living gametophytes in part dictates the ecological distribution of pteridophyte species. Although the sporophyte may be well adapted to survival and growth in a dry habitat, the gametophyte generation of the same species typically requires moist conditions for vegetative growth, the transference of sperm cells to archegonia, the fertilization of eggs, and the survival of the sporophyte embryo. In a very crude sense, pteridophytes may be thought of as the amphibians of the plant kingdom in the sense that the completion of their sexual life cycle requires a "return to water." Although this simile is somewhat whimsical, pteridophyte sexual reproduction is ecologically confined either to comparatively moist habitats or to those that have a wet season during which free-living gametophytes can grow and develop eggs and new sporophyte embryos.

The colonization of new sites by early vascular land plants was likely aided by their comparatively small spores that developed into bisexual gametophytes. Small spores have a greater probability of being carried long distances by wind currents than larger spores because, other things being equal, the rate at which any particulate settles out of the air is proportional to the size and density of the particles. This simple physical relation also helps to explain one of the major trends in early vascular land plant evolution—the comparatively rapid evolutionary increase in the size and height of sporophytes. If it is true that most early vascular land plants relied on wind currents to transport and deposit spores in microhabitats conducive to the growth and sexual function of gametophytes, then the elevation of sporangia far above ground level undoubtedly enhanced the long-distance dispersal of spores regardless of their size. This elevation likely increased the probability that at least some spores would reach a moist refuge where they could grow and develop into sexually mature gametophytes. It seems reasonable to conclude, therefore, that species with tall sporophytes would have had a selective advantage over their shorter counterparts, and that with time the size and height of sporophytes would tend to increase.

Regardless of the size and density of spores or the size and height of sporophytes, spores developing into bisexual gametophytes would have initially conferred an advantage in colonizing new sites. Provided a gametophyte can produce both egg and sperm and that the gametes produced by the same gametophyte are sexually compatible and can form viable zygotes, only one gametophyte is required to colonize a new site by establishing a sporophyte. On the other side of the life cycle coin, the sporophytes of many modern pteridophytes are capable of asexual reproduction and may live for many years. Assuming that the sporophytes of ancient pteridophytes were also capable of asexual reproduction, once a sporophyte was established in a dry habitat (by virtue of the felicitous survival of a bisexual gametophyte in a temporarily moist microhabitat) it could establish a "population" of genetically identical individuals (a clone) and simultaneously colonize distant sites by discharging literally millions of spores.

But what is the basis for assuming that the earliest vascular land plants had bisexual gametophytes? The answer lies in two characteristics of pteridophytes that produce bisexual gametophytes—their spores tend to be comparatively small, and all the spores produced by a particular species have the same general external appearance (Chaloner 1967). Plants that produce only one kind of spore are called homosporous, meaning "same spores." Because the spores of the most ancient vascular land plants were comparatively small and because the spores found in the fossil sporangia of these plants tend to all look alike, these plants are generally assumed to have produced bisexual gametophytes. This assumption is somewhat problematic, however, because some modern mosses produce spores that look alike but develop into sexually dimorphic gametophytes (as for example those of *Macromitrium comatum*). This condition is called anisospory, literally "unequal spores." Here we see the distinction between "spores of unequal size" and "spores that develop into unisexual gametophytes." The former reflect morphological heterospory; the latter presage sexual dimorphism.

Many modern species of ferns, lycopods, and horsetails have retained the presumably ancient homosporous condition and therefore produce bisexual gametophytes. However, a significant number of living and extinct pteridophyte species are heterosporous; that is, they produce two kinds of spores: small spores that develop into sperm-producing gametophytes and comparatively larger spores that develop into egg-bearing gametophytes. In recognition of their relative size, the small spores of heterosporous species are called microspores and the sperm-bearing gametophytes they develop into are called microgametophytes. The larger spores produced by hetero-

sporous species are called megaspores, and the egg-bearing gametophytes they develop into are called megagametophytes. However,
even though "micro" and "mega" imply that size is the salient feature
of heterosporous plants, this is not the case. The essential feature of
heterosporous plants is that they produce unisexual gametophytes.

Heterospory appears to confer a selective advantage, because it
evolved independently in at least four very different plant groups:
ferns, lycopods, horsetails, and seed plants. Likewise, the first heterosporous vascular plants make their appearance fairly early in the
fossil record. By plotting the frequency distribution of fossil spore
sizes during the Devonian, William G. Chaloner (1967) showed that
lower Devonian spores tend to be uniformly small, typically less than
100 μm in diameter, but that toward the middle of the Devonian the
frequency distribution of spore size begins to expand toward larger
spores, some much more than 200 μm in diameter. By the end of the
Devonian, some spores measure 2,200 μm in diameter (as for example
Cystosporites devonicus). This trend in the distribution of spore size,
which incidentally reinforces the view that the earliest Devonian vascular plants were homosporous, supports the hypothesis that heterosporous plants evolved about 386 million years ago.

Another line of evidence supporting the hypothesis that heterospory evolved early in the history of plant life are fossil sporangia containing both small and large spores ("mixed" sporangia) and sporangia attached to the same plant containing either small or large spores
(micro- and megasporangia). Both situations occur by Devonian times,
and both fall in line with the logical expectation that heterosporous
plants would eventually evolve sporangia dedicated to the exclusive
production of microspores or megaspores. For example, the Devonian plant *Barinophyton citrulliforme* had fairly large mixed sporangia
containing several thousand small spores (each measuring 50 μm in
diameter) and approximately thirty large spores (each measuring up
to 900 μm in diameter). In addition to bearing mixed sporangia, the
Devonian plant *Chaleuria cirrosa* produced microsporangia (containing only small spores measuring from 30 to 48 μm in diameter) and

megasporangia (containing spores measuring 60 to 156 μm in diameter). The presence of mixed sporangia on *Barinophyton* and *Chaleuria*
plants indicates that the sporophyte generation had gained some
physiological control over the sexuality of the spores it produced. The
presence of micro- and megasporangia on *Chaleuria* plants suggests
that this control was tighter, perhaps as a consequence of regulating
the flow of nutrients to cells destined to undergo meiosis and produce
spores. In contrast, heterospory among bryophytes seems to be a consequence of sex chromosomes (Sussex 1966; Bell 1979). However, it
is interesting to consider the possibility that the sex chromosomes of
heterosporous bryophytes might be involved with the regulation of
nutrient flow to developing spores.

Although the "when and how" of vascular plant heterospory can
be traced with some precision by virtue of the fossil record, no one is
quite sure why heterospory evolved. If the colonization of new sites
was important and favored by producing bisexual gametophytes,
then it seems counterintuitive for plants to abandon homospory in
favor of heterospory. Nonetheless some did so, and rather quickly
in terms of the evolutionary time scale. The reason may lie in the benefits that come from reapportioning metabolic resources in favor of
the spores that will develop into megagametophytes. Consider that
heterosporous plants tend to produce many more microspores than
megaspores and that the microspores of each species are generally
much smaller than megaspores. Assuming that each sporangium is
invested by its sporophyte with roughly the same amount of metabolic resources, then partitioning this resource into "many small
microspores" reduces the amount of metabolites allocated to each
microspore, which in turn may reduce the vigor of the microgametophyte. Although this may appear "biologically foolish," recognize
that once sperm are released, the microgametophyte has fulfilled its
reproductive function and is expendable. By producing many small
microspores, heterosporous plants increase the chances that many
sperm-bearing plants will be available for sexual reproduction even
if the gametophytes that produce them die shortly thereafter. In con-

trast, consider the other side of the reproductive coin—producing a "few large" megaspores means that each is invested with a comparatively large amount of metabolic reserve that the megagametophyte can use to initially establish itself and to provide to developing sporophyte embryos. The megagametophyte must continue to survive after it produces eggs if the developing embryo within it is to survive and eventually take up an independent existence. In this sense many small microgametophytes and a few very large (and presumably healthy) megagametophytes ensure a high probability of fertilization and the recruitment of at least a few new sporophytes in the next generation. By loose analogy, the heterosporous condition is a biological extension of the logic underwriting oogamy: eggs (and egg-bearing plants) are metabolically favored over sperm (and sperm-producing plants).

One consequence of this reproductive allocation is that large spores tend to settle out of air currents faster than small spores, and so the megagametophytes and microgametophytes produced by an individual sporophyte are likely to part ways aerodynamically when wind carries them a great distance from the parent plant. Curiously, aerodynamic sorting based on spore size may confer a reproductive advantage in terms of outbreeding and heterosis (hybrid vigor). Because they share the same genetic background, eggs fertilized by sperm produced by the same microgametophyte invariably produce homozygous sporophytes, diploid plants whose homologous chromosomes have genes encoding for exactly the same character states. Homozygous sporophytes can be at a disadvantage, particularly when deleterious mutations occur. Thus, inbreeding among gametophytes produced by the same sporophyte can have long-term deleterious genetic effects on a population. In contrast, the production of very small microspores, some of which can be carried a long distance, has the positive effect of ensuring the exchange of genetic information among genetically different gametophytes, thereby increasing the frequency of heterozygous sporophytes in the populations of heterosporous species (see chapter 3).

In light of the resource allocation model for the evolution of heterospory, we would expect an inverse relation between the number and the size of the megaspores produced within a megasporangium, ultimately resulting in some species with a single megaspore in each megasporangium. These predictions appear to be borne out. For example, among the extant species of the lycopod genus *Selaginella*, some produce sixteen megaspores within each megasporangium, others produce eight megaspores per megasporangium, and in one species only four megaspores are found in a megasporangium. Among these species, megaspore size proportionally increases as the number of megaspores per sporangium decreases, showing that the contents of sporangia are partitioned more or less evenly among the cells destined to divide meiotically to form megaspores. Much more dramatically, the Devonian fossil *Cystosporites devonicus* consists of a single quartet of spores; one measures 2.2 mm in diameter while the other three are aborted megaspores only 100 μm in diameter. All four spores are surrounded by a delicate membrane, indicating that the megaspore tetrad was the meiotic product of a single diploid cell that occupied virtually the entire contents of its megasporangium.

The partitioning of resources among the spores produced within a sporangium may have set the stage for the evolution of the seed habit. In contrast to heterosporous pteridophytes, which freely shed their microspores and megaspores and have free-living gametophytes, seed plants retain megaspores within their megasporangia. Importantly, the number of functional megaspores per megasporangium has been reduced during the evolution of seed plants. Among the most derived lineages, the number of functional megaspores has been reduced generally to one. The retention of the megaspore (and subsequently the megagametophyte) within sporophytic tissues confers considerable benefits. It eliminates the requirement for an external supply of water to fertilize eggs, and it affords the opportunity to protect and nourish megagametophytes, their eggs, and eventually the diploid embryos that develop from fertilized eggs. Seed plants are sometimes spoken of as the "amniotes of the plant kingdom" because their sporophyte generation retains the egg and embryo just as the female amniotic animal does. Of course this fanciful simile ignores the fact that the egg of the seed plant is produced by a multicellular haploid organism, the megagametophyte, which is a parasitic organism living within the host sporophyte and deriving its nourishment from it. It also ignores the fact that the sporophyte embryo is contained within the megagametophyte and not directly attached to the diploid parent as it is in amniotic animals. Nevertheless, we are seeing precisely the same biologic in the sense that seed plants and amniotic animals have evolved reproductive syndromes that eliminate the dependency on freestanding water for fertilization and enhance the survival of the zygote by a reliance on the terrestrial adaptations of the diploid parent. Naturally, plants did this first.

The evolution of seed plants involved the acquisition of a *syndrome* of reproductive features that goes well beyond the evolution of the structure called the ovule/seed. To understand this syndrome, it is necessary to briefly examine how the ovule/seed develops and to define a few technical botanical terms along the way.

Strictly speaking, the seed is a mature ovule that typically contains at least one viable sporophyte embryo. The ovule consists of a single megasporangium (whose wall, among seed plants, is called the nucellus) invested with one or two ensheathing structures called integuments (thus, the ovule is an integumented megasporangium). Among modern seed plants, the integument has a clearly defined opening at its tip called the micropyle, through which the microgametophyte gains entry to ultimately deliver its sperm. In contrast, the integuments of the oldest known seed plants either were deeply lobed or consisted of a girdling truss of sterile branchlike axes (fig. 2.14). Strictly speaking, many of these ovules lacked micropyles and integuments. The integument of modern seed plants is believed to have evolved by the fusion of these lobes (Long 1977). Another hypothesis is that the integument is an evolutionary modification of the nucellus (Andrews 1961). Regardless of the origins of the integument, among extant seed plants meiosis occurs in the nucellus (or what remains of it after evolutionary reduction) and results in a single functional

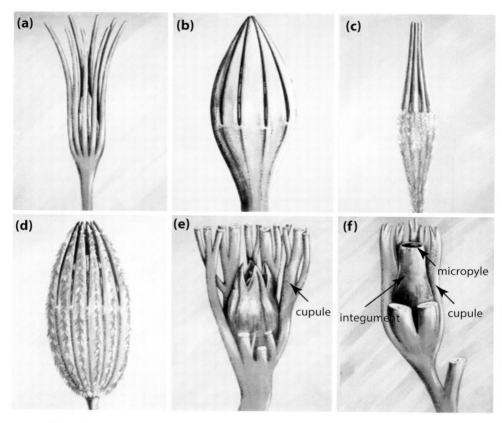

Figure 2.14. Reconstructions of six early Paleozoic seeds and seed-like structures arranged in a series purporting to show stages in the evolution of the micropyle and the integument. According to this hypothesis, the sterile axes surrounding the megasporangium (a) became reduced, basally fused, and recurved around the megasporangium (b–d). Continued reduction and basal fusion resulted in an integument with a micropyle-like opening (e) that subsequently evolved into an integument with a micropyle similar to that of modern day seed plants (f). (The micropyle is the opening formed by the integument that permits the delivery of sperm cells. The megasporangium is the sporangium that produces megaspores that will develop into egg-bearing gametophytes.) Additional trusses of sterile axes are posited to have undergone similar morphological and anatomical modifications to produce a structure beneath one or more ovules, called the cupule. This hypothesis ignores the fact that all of the fossils illustrated here are contemporaneous in the fossil record. None can be viewed as the ancestor to the others, and no legitimate *series* of morphological transformation can be proposed. It is far more likely that selection acting on reproductive success eliminated plants possessing seed-like structures lacking well-defined micropyles and integuments.

megaspore that subsequently develops into a megagametophyte that bears at least one egg. (Among some seed plants eggs are borne in archegonia, just as they are in pteridophytes and bryophytes. However, among flowering plants, the distinguishable features of the archegonium are absent.) The death and breakdown of nucellar cells provides sperm access to the megagametophyte and its egg(s), which are contained within the nucellus. In some species, notably among some gymnosperms, such as *Ephedra* and *Pinus*, the disintegration of cells near the tip of the nucellus simultaneously provides a droplet of fluid, called the pollination droplet, that fosters the entry of sperm cells into the ovule (fig. 2.15). Under any conditions, the fertilization of the egg requires the development of microgametophytes, which produce sperm cells. The microgametophytes of seed plants are called pollen grains. The microgametophyte initially develops within the wall of the pollen grain, which opens to release sperm cells in some manner. For example, among flowering plants, the microgametophyte develops a pollen tube that grows outside of the spore wall and delivers sperm cells to the megagametophyte. Among some gymnosperms, such as the cycads and *Ginkgo*, sperm cells bear flagella and swim to the archegonia of megagametophytes, which are submerged in a liquid pool. After fertilization, the ovule continues to develop. Its integument(s) mature(s) to form the seed coat, and the sporophyte embryo grows and matures within the megagametophyte.

This brief review of the ovule and seed shows that the reproductive syndrome of all seed plants involves at least five features: (1) the retention of megaspores within megasporangia, (2) the reduction of the number of megaspores per megasporangium to yield a single functional megagametophyte, (3) the enclosure of the megasporangium by an integument, (4) the modification of the megasporangium to receive microspores and their sperm-bearing microgametophytes, and (5) the elaboration of microspores, enabling them to deliver sperm cells to eggs.

Clearly, a number of important features had to evolve before the seed habit evolved. Some of these features appear comparatively early

Figure 2.15. Two ovules of the gymnosperm *Ephedra* showing pollination droplets emerging from micropyles. Pollen grains landing on the surface of a pollination droplet are drawn into the ovule (when the pollination droplet is reabsorbed by the nucellus) where they deliver sperm cells. The nucellus is a special term reserved to describe the megasporangium of a seed plant.

in the history of vascular plants. The oldest known fossil seed plant evolved in the Late Devonian, roughly 350 million years ago. It is called *Elkinsia polymorpha* (Rothwell, Scheckler, and Gillespie 1989). Its ovules were located at the tips of slender branching stems. Each was surrounded by sterile truss of branches collectively called a cupule (fig. 2.16). Each ovule had a single integument consisting of four or

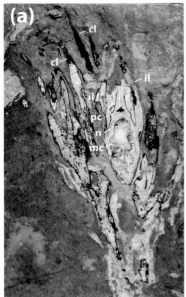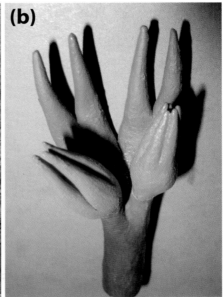

Figure 2.16. A fossil and a reconstruction of the currently oldest known seed plant, *Elkinsia polymorpha*. a. Lateral view of a permineralized cupulate ovule. cl. = cupulate lobes. il. = integumentary lobes. n. = nucellus. pc. = pollen chamber. mc. = megaspore chamber. Courtesy of Gar W. Rothwell (Ohio University, Athens, Ohio, and Oregon State University, Corvallis, Oregon). b. A three-dimensional reconstruction of two quadrants of an *Elkinsia* cupule. Ovules shown in red; cupule shown in green.

five vascularized lobes that were fused together at their base. The tip of the nucellus was extended outward to receive microspores. Each nucellus contained a single large megaspore as well as three smaller aborted megaspores. A slightly younger Devonian plant, *Archaeosperma arnoldii* (Pettitt and Beck 1968), had a similar morphology, but the cupules of this plant, which were borne in pairs, had extremely long vascularized extensions fused at the base. Two ovules were attached within each cupule, and each ovule had an integument serrated into a number of lobes that formed a rudimentary micropyle. Because the ovules are poorly preserved, it is not possible to determine how the tip of the nucellus may have been modified to receive pollen grains, but the nucellus contained one large megaspore and three aborted spores toward the nucellar tip, just like *Elkinsia*. Curiously, unlike the megaspores of modern seed plants, which have membranous megaspore walls, the megaspores of *Archaeosperma* had recognizable spore walls. This distinction likely was important from a functional point of view. Thick spore walls can be an obstacle to the transfer of nutrients from the nucellus to the developing megaspore and subsequently to the megagametophyte. The megaspores of modern seed plants have reduced spore walls that are very thin and membranous, permitting the passage of nutrients. Some early seed plants, like *Archaeosperma*, may not have been able to supply nutrients to their developing megagametophytes as efficiently as modern seed plants, particularly after their megaspore walls were completely developed. The large size of *Archaeosperma* megaspores indicates that the rudiments of differential nutrient allocation existed early in the history of seed plants.

The earliest seed plants, like *Elkinsia* and *Archaeosperma*, were gymnosperms ("naked seed" plants). Just like the ovules of modern species of pine and spruce, the ovules of these fossil plants were directly exposed to the air when they were receptive to pollination rather than enclosed within the tissues of their parental sporophytes as they are in flowering plants. This is not to say that gymnosperm ovules and seeds are totally unprotected at all stages in their development. The term gymnosperm refers to the *topology* of ovules. The ovules of modern pine and spruce species are borne on the upper surfaces of scales attached to a stem. These scales flex backward when ovules are receptive to pollination, reflex toward the stem after pollination has occurred, and subsequently flex backward once again to permit the release of mature seeds. The choreography of the scale ensures that (1) airborne pollen grains can reach receptive ovules, (2) developing seeds are later provided a modicum of protection from animals and microbes, and (3) seeds are eventually released when mature.

A good deal of evolutionary innovation and "experimentation" occurred among the gymnosperms of the Late Devonian, roughly 358 million years ago. One overriding theme in this experimentation, at least during the Devonian period, was the extension of the nucellar tip into a funnel-like tube called the salpinx. This tubular extension appears to have directly received pollen grains. In this sense the salpinx was functionally analogous to the micropyle of modern seed plants. As noted, the integuments of the most ancient seed plants were variously lobed, and a distinct micropyle was not present. It has been suggested that the deeply lobed integument may have functioned aerodynamically to direct airborne pollen grains to enter the salpinx, thereby increasing the efficiency of pollination. If so, then lobed integuments functioned much like a snow fence that directs windswept snowflakes to accumulate in a preferred site. This hypothesis does not discount the role of the integument in protecting the nucellus and the megagametophyte within it from predatory animals seeking nutrition, nor does it exclude the possibility that the integument of some fossil species may have attracted ancient pollinating animals. Indeed, the structure of some fossil ovules indicates that the integument may have served in a variety of ways. What can be said with assurance is that the seed plants whose ovules had a salpinx eventually became extinct, and species whose ovules had a true micropyle gained ascendancy.

By Carboniferous times, some now extinct gymnosperms evolved microgametophytes that formed pollen tubes that grew within and through the nucellus (Rothwell 1972). Among living seed plants, pollen tubes have been experimentally shown to draw nutrients from the nucellus so as to grow and develop. In this sense, microgametophytes with pollen tubes are parasitic plants that subsist for a time by drawing metabolites from the nucellus of their host sporophyte. The nourishment of the microgametophyte by the nucellus may have been an important mechanism by which the sporophyte generation biochemically and immunologically recognized pollen produced by its own species. It also would have provided a trigger to kill or retard the growth of the microgametophytes from other species. The presence of fossil microgametophytes with pollen tubes indicates that some Carboniferous gymnosperms may have evolved extremely important and sophisticated pollen recognition systems, perhaps prefiguring the pollination biology of flowering plants, many of which either retard the growth of "exotic" pollen grains that accidentally land on their stigmata, or kill them outright.

The Flower

Horticulturists have modified flowers into such beatific essays in color, shape, and size that it is easy to forget the flower's raison d'être—seed production. From a functional perspective, the flower differs not a jot from the less affected pinecone. Likewise, it is easy to forget that the flower is nothing more than a determinate shoot bearing sterile and reproductive appendages. Molecular and genomic analyses of floral development confirm what the great botanist Johann Wolfgang von Goethe (1749–1832) deduced—the flower is a determinate shoot bearing highly modified leaves (fig. 2.17). Some of these leaves, like the sepals, may be scale-like in appearance and function to protect the inner parts of the flower as they mature. In some species, however, the sepals can be large and brightly colored, supplementing or entirely substituting for the petals that attract pollinators in most flowering plants. The sepals and petals collectively compose the perianth, which literally means "around the flower." Within the perianth, the stamens and carpels develop. The stamen bears microsporangia (the anthers) in which pollen grains develop from microspores. The carpel is a modified angiosperm leaf bearing one or more integumented megasporangia—that is, one or more ovules. Just as in gymnosperms, the megagametophyte typically develops from a single functional megaspore. Unlike gymnosperms, whose ovules typically have one integument, the ovules of the flowering plants are typically bitegmatic— each ovule is surrounded by two integuments.

Because the carpel surrounds the ovules within it and therefore hides seeds from direct view, the flowering plants are called angiosperms, literally "seeds are contained" (*angio* is the diminutive of the Greek word for container, referring in this case to the carpel). More important, the eggs within the ovules of flowering plants are secluded

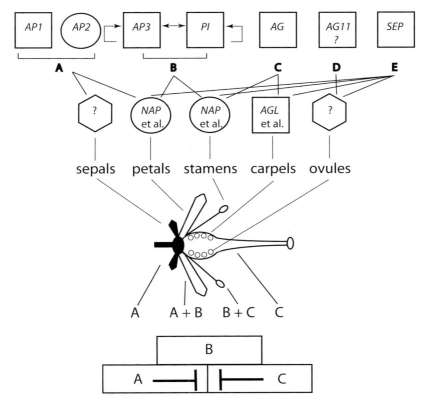

Figure 2.17. Highly simplified schematic of the ABCDE model for floral development. A series of genes, such as *APETALA1* and *APETALA2* (*API* and *AP2*), interact with other genes to regulate the development of each of the floral parts, such as the sepals. The compartmentalization of the effects and interactions among the various series of genes control the morphological and anatomical characteristics of each of the floral parts. For example, the interactions between the proteins AP1 and AP2 and between the genes *APETALA3* (*AP3*) and *PISTILLATA* (*PL*) participate in the formation of petals as opposed to sepals. Mutations in this gene regulatory network can result in significant departures from the normal appearance of a flower and can result in a change in the normal biotic pollination syndrome of a species. Such mutations can result in the evolution of a new species.

from pollen grains in the sense that once a pollen grain attaches to the stigma of a carpel it must grow through the tissues of the carpel before it can deliver its sperm cells to an egg. The pollen grains of all living angiosperms produce pollen tubes that grow through the style and deliver two sperm cells to the megagametophytes within an ovule. One sperm cell eventually fertilizes the egg, and another sperm cell enters the central cell wherein its nucleus fuses with the nuclei in the central cell to initiate the formation of endosperm. Elements of this "double fertilization" have evolved in some gymnosperms, perhaps because it afforded a mechanism whereby the sporophyte could immunologically identify pollen from its own species and conversely reject or kill foreign pollen grains. For example, the two sperm cells of some *Abies* species enter the ventral canal cell of the archegonium before one of them fertilizes an egg.

The passage of the pollen grain from one flower to another is a painstakingly investigated affair among angiosperms and accounts in part for the stunning diversity in flower size, shape, and number of parts (we shall take up this topic in greater detail in chapter 6). Some flowers, like the common duckweed *Lemna minor*, measure less than 1 mm in length and have waterborne pollen; other flowers, like those of the parasitic Sumatran plant *Rafflesia arnoldii*, approach 1 meter in diameter and are pollinated by carrion-loving beetles or flies. Some are bilaterally symmetrical with fused parts that, like many orchids, mimic the appearance of a pollinator's mate or prey, thereby attracting an enterprising animal and dusting it with pollen grains. Other flowers, like *Magnolia*, have spirally arranged parts so numerous and fleshy that portions of the flower may be gratuitously tendered as food to attract pollinators (fig. 2.18). Although most flowers are "perfect" in the sense that they have both stamens and carpels, some are self-sterile or configured to release their pollen before their own carpels mature, and many species have evolved "imperfect" flowers bearing exclusively either stamens or carpels. Many plants are wind pollinated and have evolved floral traits that reduce the chances of capturing pollen from the same plant (as for example, the grasses). These and other floral contrivances can decrease the probability that a flower will pollinate itself and so suffer the deleterious genetic effects of inbreeding.

The intricate relation between floral size and shape and the biology of pollinators offers a potential explanation for the extraordinarily large number of angiosperm species and the apparent rapid rise of the angiosperms compared with other plant groups (Grant 1963, 1971).

Figure 2.18. A flower of *Magnolia macrophylla.* Cornell University Plantations.

Any mutation altering the pollinator attracted to the flowers of a species could invoke a reproductive barrier between the mutant and its parent population. Subsequent genetic and phenotypic divergence could lead to speciation. Likewise, species hybrids that attract a novel pollination syndrome would be reproductively isolated from both parent species (Grant 1949). Nevertheless, it is highly problematic that the success of the flowering plants is entirely due to the evolution of the flower per se. A constellation of morphological, anatomical, developmental, and ecological features coevolved with the flower and likely contributes to the great success of the angiosperms (Tiffney 1981). The taxonomic separation of the flowering plants from the gymnosperms is technically vouchsafed not merely by the enclosure of ovules within carpels. (The term "angiosperm" focuses on only one of the biological features that truly set this group of seed plants apart from all other groups.) The angiosperms differ from the gymnosperms not because they have "hidden seeds" but because their biology entails a large number of additional features such as the ability to propagate asexually, grow rapidly, and produce specialized xylem elements called vessel members and phloem cells called sieve tube members in addition to the features collectively defining the flower. When this constellation of vegetative and reproductive features is viewed in its entirety, the adaptive breakthrough of the flowering plants is seen to be not the evolution of the flower but the embedding of this reproductive organ within a backdrop of many other adaptations.

"Biological success" is not easy to measure in an unambiguous way, but by almost any yardstick the angiosperms appear to be the most successful seed plant group. Consider the number of living angiosperm species and families compared with the number of surviving bryophyte, pteridophyte, and gymnosperm species. Depending on the taxonomic authority consulted, there are anywhere between 300 and 400 distinct angiosperm families containing over 250,000 species. In contrast, there are currently only 22,400 bryophyte, 9,000 pteridophyte, and roughly 750 gymnosperm species. Thus there are about 7 flowering plant species for every extant nonflowering embryophyte. In fact, the number of species within a single flowering plant family can far outnumber all extant pteridophyte and gymnosperm species combined! The bean family alone has 14,000 species, compared with the roughly 10,000 surviving pteridophyte and gymnosperm species. Yet another yardstick of biological success is ecological diversity. Here too the flowering plants are the most successful. Aside from occupying polar regions, deserts, and bodies of freshwater, the angiosperms have also returned to the sea (as for example the "sea grass," *Zostera*), a habitat totally devoid of any bryophyte, pteridophyte, or gymnosperm species. Yet another gauge of evolutionary success is how rapidly one plant group assumes dominance over previously successful ones. The fossil record shows that the angiosperms rose to taxonomic (but not necessarily ecological) dominance over their gymnosperm contemporaries within 40 million years or so. It is not clear whether this was a consequence of competitive displacement or of the radiation of angiosperms into niches vacated when earlier gymnosperm and pteridophyte species became extinct just before the Cretaceous. Nevertheless, in terms of their current species richness, their ability to survive and prosper in many different habitats, and their historical rise as the dominant element in the flora, the flowering plants are the most successful type of land plant.

Attempts to lay the success of the angiosperms squarely on the reproductive virtuosity of the flower are easily thwarted. Some of the most successful flowering species have totally abandoned sexual reproduction. For example, although it has all the accoutrements for sexual reproduction, the common dandelion produces seeds with viable embryos not from the fusion of sperm and egg but from the asexual division of cells within the nucellus. The flowers of the lemon and mango equally eschew the zygote as the means of reproduction. By the same token, duckweed crowds the surfaces of lakes and the water hyacinth clogs lowland waterways with genetically identical copies produced by asexual reproduction through fragmentation. On land the kudzu and poplar, like the weedy dandelion, duckweed, and water hyacinth, illustrate the same theme—rapid colonization and occupancy of a particular site by means of vegetative growth and propagation. Even among those species that normally reproduce sexually, many rely on rapid vegetative growth to establish individuals in new sites and to hold on to a previously occupied habitat when populations are disturbed either physically or biologically. The rapidity with which some angiosperm species grow vegetatively relates to their annual or herbaceous growth habits, which are totally absent in other extant seed plants. The seeds of the mouse-ear cress *Arabidopsis* grow into a flowering adult within twenty-four days. Even the most rapidly growing gymnosperms, like *Ephedra,* typically take years to reach sexual maturity, and the seeds of most gymnosperms take months or years to develop and mature from fertilized ovules.

Even though the evolutionary success of the angiosperms cannot be ascribed solely to benefits conferred by possessing flowers, the features of the flower are typically used to trace the evolutionary origin of the angiosperms to some type of gymnosperm ancestor. Tantalizingly, many of the morphological and developmental features that characterize the "flower" overlap with those found in the reproductive organs of fossil gymnosperms. For example, the ovules of the Jurassic gymnosperm *Caytonia* were surrounded by carpel-like cupules, and the cupules of another group of Jurassic gymnosperms, the Czekanowskiales, had papillate flanges that superficially resembled the stigmatic surfaces of modern flowers. Although the pollen grains of these fossil gymnosperms appear to have landed directly on the micropyles of the ovule, many ancient extant angiosperms have carpels that are not completely closed around their ovules and thus lack styles sensu stricto (such as *Drimys*, *Degeneria*, and *Exospermum*). Clearly, a number of ancient gymnosperm lineages appear to have explored the benefits of enclosing their ovules with carpel-like structures.

Another morphological feature, the bitegmatic ovules of angiosperms, may have been explored well before the angiosperms made their first appearance. The seeds of the Pentoxylales, a group of Jurassic and Cretaceous gymnosperms, have two integuments. Likewise the seeds of the Bennettitales, another gymnosperm group extending from the Triassic into the Cretaceous period, are interpreted to be bitegmatic. The Bennettitales are of interest for another reason. Some of these organisms had bisporangiate reproductive organs; that is, pollen and ovules were produced in the same structure. This is unusual among gymnosperms, which typically produce their pollen and ovules in separate organs. In sum, familiarity with the diversity of fossil and living gymnosperms quickly shows that some of the most salient "floral" features appeared in ancient gymnosperms before the first angiosperms made their evolutionary entrance.

Yet, one feature of the flowering plants remains unique—double fertilization and the formation of endosperm, the storage tissue within the seed. The endosperm is a consequence of the biological phenomenon called double fertilization. As noted, after the angiosperm pollen tube grows through the style of a flower, it delivers two sperm cells to the megagametophyte. The haploid nucleus of one sperm cell fuses with the haploid nucleus of the egg cell to produce a diploid zygote that subsequently develops into a sporophyte embryo. The nucleus of the second sperm cell enters the central cell. Initially, the central cell is binucleate. Its two nuclei are called the polar nuclei; in the simplest case (monosporic megagametogenesis), each of the polar nuclei is haploid. As a consequence of double fertilization, a triploid nucleus is produced in the central cell. The triploid central cell subsequently divides mitotically and produces a triploid endosperm tissue. As it develops, the endosperm absorbs nutrients from the parent sporophyte and later relinquishes these storage materials to the embryo.

A form of double fertilization was reported in the living gymnosperm *Ephedra* (Friedman 1994). However, this form of double fertilization is not homologous with the double fertilization seen in the angiosperms. Among other differences, no polyploid endosperm is formed. The *Ephedra* megagametophyte provides the developing embryos' nutrition. In contrast, the concept of double fertilization posits the interaction of two sperm cells with two specialized cells within the megagametophyte—the egg cell and the central cell—whose post-fertilization development differ profoundly (the fertilized egg develops into the embryo; the polyploid central cell develops into endosperm).

The absence of a polyploid endosperm or its analog distances the reproductive biology of *Ephedra* from that of the flowering plants. But the pattern of fertilization observed in *Ephedra* is not unique. The pollen grains of many conifers produce two sperm cells, and one of them "left with something to do." In some conifers, the two sperm cells differ in size, and the larger one carries out fertilization while the smaller sperm cell disintegrates (or in some cases as for example *Abies* enters the ventral canal cell to produce a diploid nucleus that can divide but subsequently aborts). In *Ephedra*, the second sperm cell can produce supernumerary embryos within the same developing seed. It is very likely that the gnetalean and angiosperm lineages evolved along a variety of nearly parallel reproductive avenues that in some cases produced analogous features.

Charles Darwin called the origin of the angiosperms an "abominable mystery." Things have not changed appreciably. By its nature, evidence of angiosperm double fertilization has a low probability of being preserved in a fossil seed. And even when preserved, the endosperm can be distinguished only based on the ploidy of cells, a cytological feature that is also not likely to be preserved.

A Little Chemistry

A detailed and comprehensive discussion of the complex history of land plants is not possible, but one aspect deserves special mention here because it provides additional insights into the phylogenetic relationships among the algae and the land plants as well as the importance of chemical evolution to plant survival. This is the apparently unique ability of some green algae and all their terrestrial descendants to divert large amounts of carbon from aromatic amino acid metabolism into the biosynthesis of phenylpropanoid compounds. These compounds include the flavonoids, lignins, and a host of small phenolic molecules that are used by plants as pigments, UV screens, and chemical defenses. Among the tracheophytes, the biosynthesis of phenylpropanoid compounds is developmentally activated to construct specific cells or tissues (as for example xylem and phloem fibers). These compounds are also mobilized in response to environmental stress such as UV irradiation or localized wounding or infection. Thus, these compounds serve in proactive and reactive capacities. In terms of the insights they provide, consider the simplest UV-absorbing compounds, coumaric acid and its chemical derivatives (as for example, the ubiquinones). The formation of these substances is governed by a key enzyme, phenylalanine ammonia lyase, which has been detected in the chlorophycean alga *Dunaliella*. The presence of this enzyme in this alga and others supports the hypothesis that the land plants evolved from a green algal ancestor. The flavonoids are other UV-screening chemicals. These compounds, which pigment the flowers of many angiosperms, are produced by the charophycean algae *Chara* and *Nitella* (Markham and Porter 1978), which further supports the view that the embryophytes descended from a charophycean-like ancestor. Also, the presence of flavonoids in the green algae and the embryophytes suggests that the first land plants could synthesize a UV screen, which was undoubtedly important to the successful colonization of land (Stebbins and Hill 1980; Swain and Cooper-Driver 1981; Chapman 1985). The capacity to synthesize lignins was also important to land plant evolution because these chemicals stiffen the secondary cell walls in vascular tissues and make these walls impermeable to water. The lignins, which are complex three-dimensional polymers composed of *p*-hydroxyphenyl, guaiacyl, or syringyl monomers, typically contribute up to 25% of the dry weight of wood, making them among the most abundant organic molecules on earth. Plants also use lignins to seal off microbial infection or wounding sites. Thus, the ability to synthesize lignins provided chemical defenses against animal attack in the most ancient land plants and fostered the subsequent evolution of the arborescent growth habit.

A Word about the Fungi

The fungi also played a critical role in land plant evolution just as they continue to play important roles today. Unfortunately, space precludes a detailed treatment of fungal evolution and plant-fungal ecological relationships (for excellent reviews of the fossil record of fungi, see Taylor et al. 2009; Taylor et al. 2014; for a provocative thesis about the role of fungi and the evolution of early life, see Moore 2013). However, in the context of this chapter, a few features stand out as particularly important. Although the fungi are extremely diverse in their morphology, reproduction, and ecology, all fungi are eukaryotic and heterotrophic life forms that acquire their nutrients by absorption (rather than by ingestion, as animals do). This mode of acquiring nutrients has permitted fungi to evolve into highly specialized parasites, saprophytes, or mutualistic symbionts. The role of parasitic fungi in the present-day ecosystem is well known, but the importance of saprophytic and mutualistic symbiotic fungi is often underappreciated, especially as it relates to early Earth's ecology, at a time when exposed rock surfaces experienced extensive fluctuations in temperature and hydration and likely were low in nutrients other than those released by decaying bacteria or ancient eukaryotes. During Earth's early history, saprophytic fungi probably were very important because of their ability to decompose the organic remains of other ancient life forms and return available nutrients to the ecosystem. Fossil fungi have been reported from Precambrian and Cambrian rocks, by which time a variety of saprophytic and mutualistic symbiotic forms of fungi had evolved. That some of the latter invaded the land as lichens, formed by a mutually beneficial relation between a fungus and a cyanobacterium or a green alga, cannot be doubted.

Equally mutually beneficial relationships in the form of mycorrhizal associations between fungi and photosynthetic multicellular eukaryotes appear to have played a critical role in the survival of nutrient-starved early land plants lacking roots. Mycorrhizae are mutualistic associations between plant roots and a fungus. These associations require specific kinds of bacteria that aid the recognition and receptivity of the fungus and its host plant. When fully established, the fungus increases the ability of roots to absorb water and nutrients from the substrate while the roots provide the fungus with metabolites and a safe haven. Mycorrhizal associations appear to be very ancient. The axes of vascular plants preserved in the Rhynie Chert are known to have contained an arbuscular mycorrhiza that probably increased the uptake of phosphorus, nitrogen, and other nutrients from rocks and soil as it does in modern-day plants with true roots. There is no reason to believe that these ancient plants were unique in this regard. Similar mycorrhizal associations, possibly assisted by "helper" bacteria, will probably be found for other early land plants, and so the greening of the continents was the product of very ancient, albeit complex, relationships among bacteria, fungi, and plants.

Suggested Readings

Andrews, H. H., Jr. 1961. *Studies in paleobotany.* New York: John Wiley.

Banks, H. P. 1975. Early vascular land plants: Proof and conjecture. *BioScience* 25: 730–37.

Bell, P. R. 1979. The contribution of the ferns to the understanding of life cycles of vascular plants. In *The experimental biology of ferns,* ed. A. F. Dyer, 57–85. London: Academic.

Chaloner, W. G. 1967. Spores and land-plant evolution. *Rev. Palaeobot. Palynol.* 1: 83–93.

Chapman, D. J. 1985. Geological factors and biochemical aspects of the origin of land plants. In *Geological factors and the evolution of plants,* ed. B. H. Tiffney, 23–45. New Haven: Yale University Press.

Friedman, W. E. 1994. The evolution of embryogeny in seed plants and the developmental origin and early history of endosperm. *Amer. J. Bot.* 81: 1468–86.

Graham, L. E. 1993. *Origin of land plants.* New York: John Wiley.

Graham, L. E., J. M. Graham, W. A. Russin, and J. M. Chesnick. 1994. Occurrence and phylogenetic significance of glucose utilization by charophycean algae: Glucose enhancement of growth in *Coleochaeteorbicularis. Amer. J. Bot.* 81: 423–32.

Grant, V. 1949. Pollination systems as isolating mechanisms in Angiosperms. *Evolution* 3: 82–97.

_____. 1963. *The origin of adaptations.* New York: Columbia University Press.

———. 1971. *Plant speciation.* 2d ed. New York: Columbia University Press.

Gray, J., and A. J. Boucot. 1977. Early vascular land plants: Proof and conjecture. *Lethaia* 10: 145–74.

Gray, J., D. Massa, and A. J. Boucot. 1982. Cardocian land plant microfossils from Libya. *Geology* 10: 197–201.

Gray, J., and W. Shear. 1992. Early life on land. *Amer. Sci.* 80: 444–56.

Hébant, C. 1977. *The conducting tissues of bryophytes.* A. R. Gantner Verlag Kommanditgesellschaft. Vaduz, Ger.: J. Cramer.

Hutchinson, A. H. 1915. Fertilization in *Abies balsamea. Bot. Gaz.* (currently called *Intl. J. Plant Sci.*) 60: 457–72.

Long, A. G. 1977. Lower Carboniferous pteridosperm cupules and the origin of angiosperms. *Trans. Roy. Soc. Edinburgh* 70: 13–35.

Markham, K. R., and L. J. Porter. 1978. Chemical constituents of the bryophytes. *Prog. Phytochem.* 5: 181–272.

Moore, D. 2013. *Fungal biology in the origin and emergence of life.* New York: Cambridge University Press.

Pettitt, J. M. 1970. Heterospory and the origin of the seed habit. *Biol. Rev.* 45: 401–15.

Pettitt, J. M., and C. B. Beck. 1968. *Archaeosperma arnoldii:* A cupulate seed from the Upper Devonian of North America. *Contrib. Mus. Paleont. Univ. Michigan* 22: 139–54.

Pickett-Heaps, J. D. 1975. *Green algae: Structure, reproduction and evolution in selected genera.* Sunderland, MA: Sinauer.

———. 1976. Cell division in eukaryotic algae. *BioScience* 26: 445–50.

Pratt, L. M., T. L. Phillips, and J. M. Dennison. 1978. Evidence of non-vascular land plants from the early Silurian (Llandoverian) of Virginia, U.S.A. *Rev. Palaeobot. Palynol.* 25: 121–49.

Raven, P. H. 1984. Onagraceae as a model of plant evolution. In *Plant evolutionary biology,* ed. L. D. Gottlieb and S. K. Jain, 85–107. London: Chapman and Hall.

Rothwell, G. W. 1972. Evidence of pollen tubes in Paleozoic pteridosperms. *Science* 175: 772–74.

Rothwell, G. W., S. E. Scheckler, and W. H. Gillespie. 1989. *Elkinsia* gen. nov., a late Devonian gymnosperm with cupulate ovules. *Bot. Gaz.* 150: 170–89.

Stebbins, G. L., and G. J. C. Hill. 1980. Did multicellular plants invade the land? *Amer. Nat.* 115: 342–53.

Sussex, I. M. 1966. The origin and development of heterospory in vascular plants. In *Trends in plant morphogenesis,* ed. E. G. Cutter, 141–52. London: Longmans.

Swain, T., and G. Cooper-Driver. 1981. Biochemical evolution in early land plants. In *Paleobotany, paleoecology, and evolution,* ed. K. J. Niklas, 103–34. New York: Praeger.

Taylor, T. N., E. L. Taylor, and M. Krings. 2009. *Paleobotany: the biology and evolution of fossil plants.* New York: Academic.

Taylor, T. N., M. Krings, and E. L. Taylor. 2014. *Fossil fungi.* Amsterdam: Elsevier.

Tiffney, B. H. 1981. Diversity and major events in the evolution of land plants. In *Paleobotany, paleoecology, and evolution,* ed. K. J. Niklas, 1: 193–230. New York: Praeger.

3

Population Genetics, Adaptation, and Evolution

> Evolution may lay claim to be considered the most central and the most important of the problems of biology. For an attack upon it we need facts and methods from every branch of the science—ecology, genetics, paleontology, geographical distribution, embryology, systematics, comparative anatomy—not to mention reinforcements from other disciplines such as geology, geography and mathematics.
>
> −JULIAN HUXLEY, *Evolution: the Modern Synthesis* (1942)

Thus far, we have examined some of the major evolutionary transformations in Earth's history as revealed by juxtaposing data drawn from the living and from the fossil record. However, we have not talked about the mechanisms of evolution—we have not delved into *how* things evolve. In this chapter, we will address this omission by reviewing the sources of genetic variation and the consequences of variation by making use of mathematics to understand inheritance, fitness, adaptation, and other aspects of evolution that are described statistically.

To understand why evolution is a statistical process, consider that physicists and chemists investigate the properties and interactions of elementary particles, which are physically uniform and invariant in each of their characteristic traits and behavior. A single experiment adducing the properties of a single entity such as an electron or proton can be used to productively extrapolate the properties of all comparable entities in the universe. Nevertheless, to understand how a gas produces pressure in a contained vessel, the physicist must deal with the statistical behavior of billions of molecules, each moving randomly with endless variations. The behavior of a single molecule in this case is irrelevant because it is the behavior of the population of molecules that matters. In biology, the situation is somewhat similar. The organisms that biologists randomly withdraw from a population are typically different from all of the rest. With the exception of identical twins or cloned individuals, no two members of the same species look exactly alike (even identical twins may differ physically as a result of their individual histories or epigenetic events). Thus, the strict concept of "types" or "individual" is not relevant in the context of evolution. Clearly, there are limits to biological variation and these have literally shaped evolutionary history. No population is ever capable of generating all possible theoretical genomic variants, in part because the existence of any particular population is finite. Therefore, biological variation, which provides the "raw material" for evolutionary change, is confined by random as well as nonrandom events. Nonrandom processes certainly shape evolution. The "struggle for existence" among the offspring of each generation eliminates variants that are less well fitted to their particular environment, whereas chance occurrences eliminate what could have been considered the best adapted had they lived. (This duality eliminates the circular argument that fitness equals survival.) Those that survive pass their genetic information on to the next generation. In this way, evolution is the statistical summation of random events (mutation and sexual recombination) and selection, which is nonrandom.

It is also important to bear in mind that the use of statistics requires large numbers and with this requirement posits that an individual organism does not evolve—it lives, it may reproduce, and it dies, but it does not evolve into another organism. By definition, the statistical perspective offered by population genetics argues that only populations evolve, and so we must look at how populations change over successive generations, and this requires learning the mathematics that has emerged from studying the genetic compositions of populations. This explains why "population genetics" is in the title of this chapter. Fortunately, learning the mathematics of population genetics is not onerous because much of it is intuitive, at least at an elementary level.

Early History

Before we explore some of the principles of population genetics, it is instructive to examine its history during the development of evolutionary theory, particularly in the context of the Modern Synthesis, which was mentioned in the Introduction. Here, we will delve a little deeper into this historical period and see how population genetics emerged and in general took over the early development of evolutionary theory.

To begin, recall that the fundamental concept called the "principle of natural selection" was conceived independently by two nineteenth-century British naturalists, Charles Darwin and Alfred Russel Wallace, and was substantially elaborated upon in the early part of the twentieth century with the rediscovery of Mendelian genetics and subsequent advances in population genetics. Importantly, this "modern synthesis" continues to the present day, as insights are gained from diverse fields of study, particularly molecular biology, which is rapidly detailing the precise mechanisms whereby genomes (and the phenotypes they engender) are altered. Here, our aim is to review the historical development and the progress made in evolutionary theory from the time of Darwin and Wallace to the present. Clearly, no such summary can ever be complete, because the literature dealing with evolutionary biology is vast and complex. Here, we can only sketch the broad outlines of the basic history of evolutionary theory and enquiry.

In August 1858, two of the most influential publications in the

Table 3.1. The five assertions of Darwinian evolution

(1) All organisms produce more offspring than their environments can support.
(2) Intraspecific variability of most characters exists in abundance.
(3) Competition for limited resources leads to a struggle for life (Darwin) or exis-
     tence (Wallace).
(4) Descent with heritable modification occurs.
(5) As a result, new species evolve into being.

history of biology were published. These concurrent papers by Darwin and Wallace contained a "very ingenious theory to account for
the appearance and perpetuation of varieties and of specific forms on
our planet" (foreword by C. Lyell and J. Hooker). Therein, Darwin and
Wallace presented for the first time the hypothesis of descent with
modification by means of natural selection. This hypothesis makes
five fundamental assertions (table 3.1).

Darwin supported his arguments with a large body of facts, drawn
mostly from breeding experiments and the fossil record. He also provided detailed direct observations of organisms existing in their natural habitats. Thirty years later, natural selection's codiscoverer published a series of lectures under the title *Darwinism* (Wallace 1889),
which treated the same subjects as Darwin but in light of facts and
data that were unknown to Darwin (who died in 1882). A detailed
comparative analysis of the Darwin/Wallace publications reveals that
Wallace's contributions were more significant than is usually acknowledged, so much so that the phrase "the Darwin/Wallace mechanism
of natural selection" more properly acknowledges the importance of
the "second Darwin." Although Darwin is usually credited as the "principal author" of evolutionary theory, Ernst Mayr points out that it is
not correct to refer to "Darwin's theory of descent with modification"
(the word "evolution" does not appear in the original 1858 papers of
Darwin and Wallace; only in later editions of *On the Origin of Species*
and in Wallace's Darwinism). If we equate the word Darwinism with
the content of the book *On the Origin of Species*, we can distinguish

five separate concepts: (1) evolution as such, (2) theory of common descent, (3) gradualism, (4) speciation, and (5) natural selection.

The concept that all organisms on Earth have evolved from a common ancestral life form by means of genomic and morphological transformations (evolution as such) was not "invented" by Darwin or Wallace. Historians of science have shown that the idea of organismic evolution can be traced back to several Greek philosophers. Likewise, the hypothesis of continuous transformations was proposed by numerous eighteenth- and nineteenth-century authors who are credited by Darwin in the first chapter of his book. However, Darwin was the first to summarize a coherent body of observations that solidified the concept of organismic evolution into a true scientific theory (defined here as a system of hypotheses that can be experimentally refuted). When Darwin (1859) proposed his theory of descent by means of slight and successive modifications (gradualistic evolution), the available fossil record was still very fragmentary. Indeed, the early fossil record (Precambrian) was entirely unknown and thus unexplored. Nevertheless, Darwin concluded that if his theory of evolution was valid, aquatic creatures must have existed before the evolutionary appearance of the first hard-shelled organisms (such as trilobites) in the Cambrian.

Darwin's dilemma—the apparent missing pre-Cambrian fossil record—was used as a major argument against his proposal (see table 3.1). However, as discussed in chapter 1, this dilemma no longer exists. Scientists have explored the Precambrian in detail (see Schopf 1999, Knoll 2003 for summaries). We now know that life is far more ancient than believed in Darwin's time. We also know that these ancient forms of life were the ancestors to all subsequent organisms on this planet (see fig. 1.2). However, Darwin had other worries, as for example the apparent absence of intermediate forms (connecting links) in the fossil record of life, which challenged his gradualistic view of speciation and evolution. With the exception of the famous Urvogel *Archaeopteryx*, which displays a mixture of reptile- and bird-like characteristics, virtually no intermediate forms were known during Darwin's lifetime. This dilemma has also been resolved by more recent discoveries of intermediate forms in the evolutionary history of many plant and animal lineages, as for example the Early Cretaceous four-legged snake *Tetrapodophis* and the seed fern-like angiosperm *Archaefructus* (box 3.1).

Darwin's second postulate (the common ancestor, alluded to by the phrase "a few forms or one") has been verified by a large body of molecular data that has altered our perspectives in many important ways, as for example the "RNA world" hypothesis discussed in chapter 1. The principle of common descent is documented in the

---

Box 3.1. Missing Links and *Archaefructus*

Darwin's critics frequently pointed out that few if any fossil intermediates linking major groups of animals or plants could be found in the fossil record. If Darwin's hypothesis concerning gradualistic evolution was true, why didn't the fossil record provide evidence for "missing links"? Darwin responded to this criticism by saying that the fossil record was largely incomplete owing to the erosion of sediments containing fossils and the low probability that organisms will be preserved at all. Although it is true that the fossil record is incomplete, the extent to which it had been explored in Darwin's day was minimal, just as it is also true that evolution is not invariably gradualistic—macromutations such as those postulated by Richard Goldschmidt have been well documented.

It is somewhat ironic that one of the more recently found missing links is a fossil plant named *Archaefructus*, which appears to be a missing link between an ancient gymnosperm lineage and the flowering plants. The irony revolves around one of Darwin's great dilemmas—the rapid radiation of the angiosperms in the absence of any early fossil record revealing their origins. *Archaefructus* is described as an herbaceous aquatic angiosperm found in the Yixian Formation in northeastern China (Sun et al. 2002). Specimens consist of reproductive axes bearing highly dissected leaves, each of which is subtended by what appears to be a bract-like stipule (fig. B.3.1). The reproductive axis is interpreted to consist of helically arranged carpels (each containing two to four ovules) that are elevated above a cluster of stamens, each consisting of a short filament and four pollen sacs. There is no evidence of sepals or petals (there is no perianth). The entire reproductive axis is interpreted to be a flower with extended internodes. An alternative interpretation posits that the reproductive axis is an inflorescence.

The taxonomic position of this genus is problematic, but its characteristics position it between the seed ferns (in which ovules are subtended by cupule-like sterile structures) and the earliest flowering plants (in which ovules are retained in conduplicate carpels).

well-supported universal phylogenetic tree of life, which identifies a "last universal common ancestor" of all Earth's organisms as a diverse community of prokaryotic protocells that subsequently evolved into true prokaryotic organisms attended by the emergence of the genetic code and subcellular constituents (Woese 1987). When these diverse lines of evidence are taken together, there is no question that all life on Earth arose in a manner not unlike what was proposed by Darwin (propositions 1 and 2).

Finally, it is clear that Darwin didn't get everything right. For example, his concept of inheritance (called pangenesis) never benefited from Mendel's seminal discoveries, which excluded the inheritance of

Figure B.3.1. Schematic of a reproductive axis of *Archaefructus eoflora* (a) with a growth habit similar to *Myriophyllum* or *Utricularia* (b). a. The genus *Archaefructus* is interpreted to be a species of an aquatic herbaceous flowering plant found in ≈125 million year old sediments in the Yixian Formation of northeastern China. Each of its reproductive axes is interpreted to be a flower consisting of extended internodes bearing a terminal cluster of carpels subtended by a cluster of stamens. b. The genus *Utricularia* is an herbaceous aquatic angiosperm that produces emergent inflorescences. The comparison between *Archaefructus* and *Utricularia* is made solely for the purpose of showing comparable growth habits and is not intended to imply any phylogenetic relationship.

acquired characteristics at least in the sense of the Lamarckian theory of inheritance. The inheritance of epigenetic features is more Lamarckian than Darwinian, although it would be historical revisionism to suggest that Lamarck or Darwin anticipated this mode of inheritance. It is also clear that Darwin's theory was incomplete. Discoveries continue to show that gradualistic evolution is not the only mode of evolution. Although it stands at the heart of evolutionary theory, Darwin's conception of evolution is limited owing to continued advances in our understanding of heredity, genetics, and molecular biology.

The Modern Synthesis

The five proposals in Darwin's *Origin of Species* concentrate on two separate aspects of biological evolution (see table 3.1): the evolutionary process as such, and the mechanisms that bring about evolutionary change. Whereas biologists no longer debate the existence of evolution as a fact of life (literally), the mechanisms that account for the transformation and diversification of species are still very much under investigation. Pertinent studies are usually carried out on populations of living organisms (neontology) in contrast to historical reconstructions of evolution based on the fossil record (paleontology). The theory of evolution continues to grow as scientists strive to account in precise detail for the mechanisms of evolutionary change.

Historically, the early development of a more modern theory of evolution can be divided into three stages: (1) Darwinism as represented by the 1859 publication of *On the Origin of Species*, (2) Neo-Darwinism as represented by the work of the German zoologist/cytologist August Weismann (1834–1914) who provided experimental evidence against "soft (Lamarckian)" inheritance and who postulated that sexual reproduction (recombination) generates variation in every generation, and lastly (3) the modern synthetic theory, which ushered in the field of population genetics and was deduced primarily from observations and quantitative genetic data obtained with eukaryotic, bisexual model organisms such as the fruit fly and the hamster.

Table 3.2. The basic tenets of the Modern Synthesis

(1) The units of evolution are populations of organisms.
(2) Genetic and phenotypic variability in eukaryotes is brought about by genetic recombination resulting from sexual reproduction and random mutations.
(3) Natural selection, particularly directional selection, shapes the course of phenotypic evolution.
(4) Speciation can be defined as the stage in evolutionary process at which members of the same species can no longer interbreed.
(5) New species evolve from preexisting varieties by slow processes and maintain at each stage their specific adaptations.
(6) Macroevolution (evolution above the species level or the occurrence of higher taxa) is a gradual step-by-step-process that is extrapolated from microevolution processes.

The modern synthetic theory was based largely on the contents of books authored by the Russian American naturalist/geneticist Theodosius Dobzhansky (1900–1975), the German American naturalist/systematist Ernst Mayr (1904–2005), the British zoologist Julian Huxley (1887–1975), the American paleontologist George G. Simpson (1902–1984), the German zoologist Bernhard Rensch (1900–1990), and the American botanist G. Ledyard Stebbins (1906–2000). These six scientists are often referred to as the "architects" of the Modern Synthesis, which offered a reductionist explanation for the *how of evolution* based on non-Mendelian genetics. The basic postulate of the Modern Synthesis is that evolution is literally "irreversible changes in the genetic composition of populations." Indeed, conceptually, evolution was viewed (if not defined) as *changes in allele frequencies*. This approach concentrated on the genotypic level of organismic organization and posited that (1) evolution can be explained in terms of small genetic changes ("mutations") and recombination, and the ordering of this genetic variation by natural selection, and that (2) evolutionary phenomena, particularly macroevolutionary processes and speciation, can be explained in a manner that is consistent with known genetic mechanisms (table 3.2).

The post-Modern Synthesis has revealed that some of the propositions of the Modern Synthesis are not universal. As we will see in chapters 4 and 5, speciation is not invariably a slow process, and macroevolution is not invariably the result of small genetic changes. Phenomena such as linkage, epistasis, lateral gene flow, and trans posons also challenge the notion that a phenotype can be mapped directly onto its genotype. Nevertheless, the Modern Synthesis has worked brilliantly in the majority of cases, and it ushered in the era of population genetics, which emerged largely as a reconciliation of Mendelian genetics and statistical genetic models addressing allele frequencies in natural populations. Its biometric architects were undeniably Ronald A. Fisher (1890–1962), John B. S. Haldane (1892–1964), and Sewall Wright (1889–1988). Fisher, who studied physics at the time when wave-particle duality was actively investigated, showed that continuous (nonparticulate) variations in a population (as opposed to the particulate variations revealed by Mendelian genetics) can emerge as a consequence of the combined action of many discrete genes and selection acting on the allele frequency of those genes. Haldane provided a mathematical theory for natural selection that provided insights into the direction and the rate of gene frequency changes. Wright introduced the concept of genetic drift and the "adaptive landscape;" and, along with Fisher, developed methods of computing the frequency distributions in natural populations.

Simple Rules of Probability and the Binomial Distribution

Population genetics is predicated on the rules of probability and models that stipulate, in the simplest cases, populations consisting of individuals carrying genes with generally only two allelic forms. Although much more complex models exist, it is important to know the basic rules of probability that apply to these simple models. It is also important to understand the structure of two-class populations, which is described by the Binomial Theorem.

The study of probability probably began the moment humans started gambling, when it became profitable to know the likelihood of a certain event happening in a game of chance. In more formal terminology, each event such as the role of the die or the flip of a coin is called a *trial* that can have a certain number of different outcomes. The flip of a two-headed coin is a trial that have two different outcomes (heads or tails). The role of a six-headed die has six different equally probable outcomes (1, 2, 3, 4, 5, or 6), just as a gene with six allelic forms has six different equally probable chances of being drawn. Likewise, the probability of drawing an *e* out of the sequence of letters *a, b, c, d, e, f, g* is one out of seven. These assertions can be stated as a general rule: *If a trial (a drawing from a population) has n equally probable outcomes, of which only one will happen, the probability P of any individual outcome O equals the inverse of n*:

Rule 1: $P(O) = 1/n$.

To illustrate, suppose we want to know the probability of drawing a vowel out of the sequence of letters *a, b, c, d, e, f, g*. There are two vowels among the seven letters, and therefore the probability of drawing a vowel is 2/7. Similarly, the probability of drawing one letter from among the first four letters is 4/7. This result emerges from the second rule of probability: *If an event E is satisfied by any one group of mutually exclusive outcomes O, the probability of event E happening P(E) is the sum of the probabilities of the outcomes of the group of mutually exclusive outcomes*:

Rule 2: $P(E) = P(O_1) + P(O_2) + P(O_3) + P(O_4) \ldots + P(O_i)$,

where $P(O_i)$ is the probability of the *i*-event. This rule is broken if the events are not mutually exclusive. Consider what happens if we what to know the probability of drawing a letter from *a, b, c, d, e, f, g* that is *either* a vowel *or* from among the first three letters. Because only three letters satisfy the second condition (*a, b, c*), the probability of this event happening is 4/7. Notice that in this case "or" equates to "and." However, we get the wrong answer if we reason that there are three letters, but only one vowel among them, and conclude that the probability of drawing one of the first three letters is 3/7 and that the probability of drawing a vowel is 2/7 such that the probabilities are additive and thus 5/7.

Now consider the probabilities of successive and *independent* trials: *The probability that a specified series of events will happen is the product of the probabilities of the individual events happening*:

Rule 3: $P(E_1 \text{ and } E_2 \text{ and } \ldots \text{ and } E_i) = P(E_1) \times P(E_2) \times \ldots \times P(E_i)$.

For example, the probability of flipping a coin and getting a head is ½ regardless of how many times the same coin is flipped. Thus, the probability of getting two heads successively is ½ × ½ = ¼ and the probability of getting three heads successively is ½ × ½ × ½ = ⅛.

We can now apply these rules to populations consisting of individuals carrying genes, each of which has only two alleles. Such a two-class population is described by the proportion p of the members of the population that fall into one class and by the proportion q of the members that fall into the other class. Since there are only two classes in the population, we see that $p + q = 1$. Now consider a single gene in this population with allelic forms A and a. In a single drawing of individuals from this population, the probability of drawing an individual with A is p and the probability of drawing an individual with a is $q = 1 - p$. Therefore, the probability of drawing two successive As is pp or p^2 and the probability of drawing two successive as is qq or q^2 (see Rule 3). Similarly, the probability of drawing A and a successively is pq and the probability of drawing a and A successively is qp. Therefore, the probability of getting the sequence a and A or the sequence A and a is $2pq$, and

$$1p^2 + 2pq + 1q^2 = (p + q)^2 = (p + 1 - p)^2 = 1^2 = 1.$$

This general structure conforms to the binomial formula $(p + q)^n$, where n is the number of successive drawings from the population. Note again that the sum of the probabilities in any binomial expansion always equals unity because $(p + q)^n = (p + 1 - p)^n = 1^n = 1$.

It is useful to know that the binomial coefficients (such as 1, 2, 1 in $1p^2 + 2pq + 1q^2$) conform to well defined progressions that are summarized in what is known as Pascal's Triangle wherein each row gives

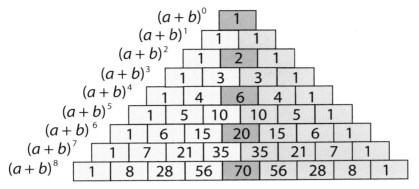

Figure 3.1. Pascal's Triangle is used to visualize the relationships among the coefficients in binomial distributions in which there are only two possible outcomes such as heads or tails. The rows give the sequence and the numerical values of the coefficients of the binomial expansions shown to the left. The sum of the values in a row gives the value of 2^n, and the powers of 11 can be read off directly from each row.

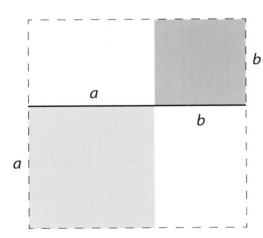

Figure 3.2. The Binomial Theorem dates back to Euclid (Elements II:4), who wrote, *"If a straight line be cut at random* [as for example into *a* and *b*], *the square on the whole is equal to the squares on the segments* [a^2 and b^2] *and twice the rectangle of the segments* [$2ab$]." Mathematically, this translates into $1a^2 + 2ab + 1b^2$.

the coefficients of successively larger binomial expansions (fig. 3.1). It is also fun to know that the Binomial Theorem dates back to Euclid (fig. 3.2) and earlier.

The Hardy-Weinberg Equilibrium Equation

Our previous discussion of probability and the Binomial Theorem has prepared us to consider what to expect from a non-evolving

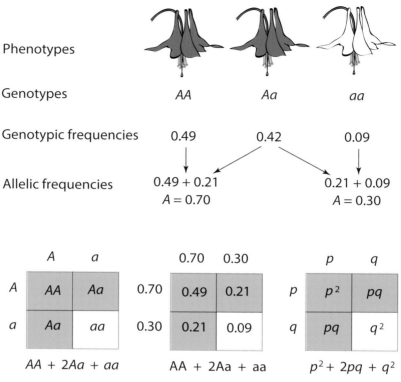

Figure 3.3. A derivation of the Hardy-Weinberg equilibrium equation showing that the gene pool of a non-evolving population remains constant over successive generations barring mutations. The illustration is for a population of flowering plants where flower color is dictated by one gene with two alleles (the dominant allele is denoted by *A* and the recessive allele is denoted by *a*). Three diploid genotypes are possible (*AA*, *Aa*, and *aa*). Two genotypes result in red flowers (*AA* and *Aa*), and one produces white flowers (*aa*). The allelic frequencies in the first generation (*A* = 0.7 and *a* = 0.3) remain the same in subsequent generations, provided all the conditions for a non-evolving population hold true (see text).

population—that is, a population in which the proportions of *A* and *a* do not change over successive generations. By way of illustration, consider a case in which reproductive success is governed by flower color and that flower color is dictated by a single gene (fig. 3.3). Assume that this gene has only two alleles: a dominant allele, denoted by *A*, that results in an individual bearing red flowers, and a recessive allele, denoted by *a*, that produces a white-flowered individual. Because every plant in the hypothetical population is diploid, there are only three possible genetic combinations, or genotypes, in terms of flower color—*AA, Aa,* and *aa.* Because *A* is dominant and *a* is recessive, there are only two phenotypes—red-flowered individuals (*AA* or *Aa*) and white-flowered individuals (*aa*). Simple particulate Mendelian genetics shows that in the absence of selection against flower color, the frequencies of the three genotypes in the population will not change from one generation to the next. That is, if the reproductive success of individuals bearing white flowers equals the reproductive success of individuals bearing red flowers, the frequencies of the *A* and *a* alleles will not change generation by generation in the absence of mutation.

Imagine now a population consisting of 700 individual plants in which *AA, Aa,* and *aa* are represented by 343, 294, and 63 individuals. The frequencies of the three genotypes are $AA = 343/700 = 0.49$, $Aa = 294/700 = 0.42$, and $aa = 63/700 = 0.09$. Because each individual is diploid, there are 1,400 genes for flower color in the population. The *A* and *a* alleles account for 70% and 30% of the entire gene population, respectively, because each of the 343 *AA* genotypes has two "doses" of the *A* allele, while each of the 294 *Aa* genotypes has one dose of the *A* allele and one dose of the *a* allele. Thus, the total number of *A* alleles equals $(343 \times 2) + (294 \times 1) = 686 + 294 = 980$, which is 70% of 1,400. Since there are only two alleles for the gene governing flower color, the remaining 30% of the gene population must consist of the *a* allele.

Now, during sexual reproduction each parent can transmit only one of its two alleles for flower color. Assuming that the union of sperm and egg is completely random, the probability of drawing the *A* or *a* allele is equal to the original frequency of occurrence of either allele in the entire population. That is, the probability of randomly drawing the *A* allele is 70% (or, expressed as a decimal fraction, 0.7), while that of randomly drawing the *a* allele is 30% (or 0.3). Consequently, in the absence of mutation or some phenomenon that biases the random draw of alleles during reproduction, the frequencies of the three genotypes remain the same from generation to generation. For example, the probability of drawing the *A* allele from the total population of sperm cells and the probability of drawing the *A* allele from the total population of egg cells to produce an individual with the *AA* genotype is 0.7 × 0.7 = 0.49. Likewise, the probability of drawing two *a* alleles from the pool of sperm and egg to produce the *aa* genotype is 0.3 × 0.3 = 0.09.

Following standard practice, we designate the probability of drawing the *A* allele as *p* and the probability of drawing the *a* allele as *q*, such that $p + q = 1$. Thus, the probability of producing the *AA* genotype is p^2, the probability of producing the *Aa* genotype is *2pq*, the probability of producing the *aa* genotype is q^2, and $p^2 + 2pq + q^2 = 1$. This "sum of probabilities" equation is called the Hardy-Weinberg equilibrium equation, in honor of Godfrey H. Hardy (1877–1947) and Wilhelm Weinberg (1862–1937) who independently derived it from Gregor Mendel's laws of inheritance.

Unless they are absolutely linked, genes, even those lying on the same chromosome, are expected to behave in populations as if they are independent. Here "independent" means that, in the absence of selection, the frequencies resulting from the fusion of sperm and egg for various combinations of alleles at any two loci equal the product of the frequencies of the two separate loci (that is, alleles, whether linked or not, are expected to be associated at random within populations of randomly cross-fertilizing individuals). However, as we will discover, the independent assortment of genes can be thwarted by absolute linkage or by differential selection preventing the attainment of equilibrium frequencies. Under these conditions, selection against a particular allele conferring low fitness reduces the allelic frequencies of other linked genes.

The Hardy-Weinberg equilibrium equation is a formal mathematical statement showing that the frequency of occurrence of genotypes within a population will remain constant provided six conditions hold true:

(1) The population consists of sexually reproductive organisms.
(2) The population is and remains very large.

(3) Individuals do not migrate into or out of the population.

(4) Mating is completely random.

(5) No mutation occurs.

(6) All possible genotypes have an equal probability of reproductive success.

These conditions describe what will happen in the absence of mutation or selection, and as long as segregation and independent assortment occur during meiosis. However, collectively and individually these conditions never hold true for real populations of organisms, and so they proscribe the genotypic frequencies expected in a non-evolving population (the Hardy-Weinberg equilibrium equation provides a null hypothesis—the condition obtained when evolution does not occur).

Some of the implications of the Hardy-Weinberg equation may be less obvious, especially to those who do not think of individuals as transient beings. For example, we often speak of a gene as *dominant* or of a gene's *penetrance* with respect to an individual's phenotype. However, as the Hardy-Weinberg equation shows, these two terms apply to an individual's phenotype and have little or nothing to do with the allelic frequency of the gene in a population—a dominant gene does not translate into a disproportionate number of individuals carrying the gene. Far more important, the Hardy-Weinberg equation highlights two important phases in the life cycles of biparental organisms—the *gametes-to-zygote* phase and the *zygote-to-adult* phase. The *gametes-to-zygote phase* is dealt with by assuming random mating and random union of sperm and egg. The *zygote-to-adult* phase deals with the selection factors that change allele frequencies.

Finally, it must be noted that many of the conventional statistical tests for Hardy-Weinberg equilibria (such as the chi-square test) are not especially sensitive to deviations from the genotype frequencies predicted by the formula. Thus, we cannot assume that all of the assumptions of the Hardy-Weinberg equation hold true simply because observed genotype frequencies happen to fit those predicted by the equation. In this sense, the Hardy-Weinberg equation has more pedagogical than practical value.

Sources of Variation: Mutation and Recombination

When viewed through the perspective of population genetics, evolution involves three components. The first is genotypic-phenotypic variation in a population. The second is that the variation is heritable. The third is that the variation must result in differences in fitness such that selection can operate. Excluding transgenerational epigenetic mutations (as for example DNA methylation), among eukaryotic organisms there are five primary sources of genetic variation: (1) mutation, (2) recombination during meiosis, (3) transposable elements (DNA sequences capable of replication that can insert into different sites in the genome), (4) lateral gene exchange among species, and (5) the migration of individuals carrying unique gene combinations among populations of the same species. Two additional sources of genetic variation, which are far more common among plants than among animals, are hybridization and autopolyploidy with subsequent gene divergence and specialization. Nevertheless, mutation is perhaps more important because it is the ultimate source of genetic novelty. Unlike recombination, transposons, lateral gene exchanges, and migration, mutations generate new genes. Recombination (see fig. 1.15) is also important because it provides a mechanism in sexually reproducing species to bring beneficial mutations together and increases the combinatorial diversity of genotypes.

Among eukaryotes, mutations typically occur on the order of 10^{-4} to 10^{-6} new mutations per gene per generation. These rates may seem excruciatingly slow and therefore of little effect. However, in large diploid populations, mutations can produce many mutated alleles because each genome has two copies of the same gene and because there are many copies of each gene in a population. For example, if the mutation rate is 10^{-6} per nucleotide pair per generation and if there are 10^6 nucleotide pairs in a gamete, three mutations will occur in each generation such that, on average, each fertilized egg could have six mutations. In a population numbering on the order of one thousand, this would result in 6,000 new mutations per generation.

However, most mutations are lethal or neutral. Theoretically, those that are neutral have an initial frequency of occurrence equal to $1/(2N)$, where N is the size of the diploid population. (Note that we have $2N$ because we are dealing with diploid organisms and it is the alleles and not the individuals that are being counted.) Recurrent mutations may occur, but they increase the mutated allele very slowly. Mutations can be divided into two broad categories—those that are irreversible and those that are reversible. Using the previous example of gene A, an irreversible mutation means A can mutate into a, but a cannot mutate into A. Let us consider the case of irreversible mutation (a condition that is sometimes referred to as *mutation pressure*). Assume that A has the probability μ of mutating into a, and designate the allele frequency of A as p and the allele frequency of a as q. Next we obtain a formula that allows us to calculate the allele frequency of A in terms of the allele frequency of A in the previous generation. A simple formula for this purpose is

$$p_g = p_{g-1} (1 - \mu),$$

where g designates the current generation and $g - 1$ designates the previous generation. The same mathematical reasoning shows that $p_{g-1} = p_{g-2} (1 - \mu)$ such that $p_g = p_{g-2} (1 - \mu)^2$. If this form of reiteration is followed to its logical conclusion, we see that

$$p_g = p_o (1 - \mu)^n,$$

where p_o is the initial frequency of A and p_g is its frequency in generation g after n generations. Note that in the first generation $p_o = 1.0$ because A has not yet mutated. We can now calculate the number of generations it will take for the frequency of A to decrease to zero using biologically established upper and lower mutation rates ($\mu = 10^{-4}$ and $\mu = 10^{-6}$) and plotting the allele frequency of A as a function of

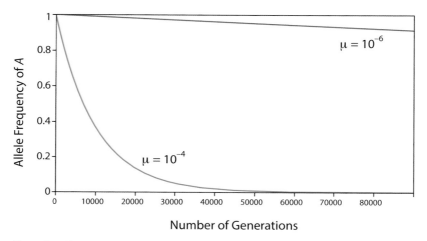

Figure 3.4. The change in the allele frequency of A when the mutation rate of A to a equals 10⁻⁴ and the mutation rate of a to A equals 10⁻⁶. At these biologically realistic mutation rates, the equilibrium between A and a requires tens of thousands of generations. In this example, the equilibrium allele frequency of A is 0.091.

the number of generations n (fig. 3.4). Such a plot shows that at low mutation rates ($\mu = 10^{-6}$) the allele frequency of A changes very little over tens of thousands of generations. Even at significantly higher mutation rates ($\mu = 10^{-4}$), A's frequency drops to 55% only after 6,000 generations, which is over 700 years for a fruit fly and over 12,000 years for a biennial plant. This simple example illustrates why the architects of the Modern Synthesis argued that speciation is a slow process, although a paleontologist might point out that the average time interval of a geological stage is 5,000,000 years.

Population genetics also shows that in theory A will never be eliminated from a population if reversible mutation occurs, that is, if A can mutate into a and if a can mutate back into A because at some point an equilibrium will be reached such that the allele frequencies of A and a in a population (p and q) will undergo no statistically significant changes. This can be shown by returning to our previous notation and noting that an A allele in generation g can occur in two ways: A may not have mutated into a (with a probability of $1 - \mu$), or A might have been an a allele that mutated into A (with a probability of ν). These possi-

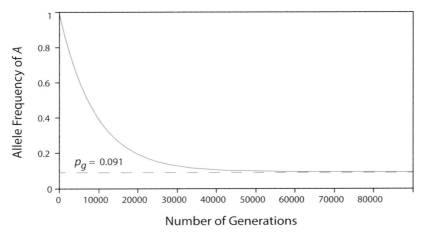

Figure 3.5. The change in the frequency of allele A in a population over successive generations assuming that allele A is subject to reversible mutation. Over successive generations,
A reaches its equilibrium condition (denoted by p_g) in the population. Assuming comparatively high mutation rates for A to a and a back to A (specifically, $\mu = 10^{-4}$ and $v = 10^{-5}$), the
equilibrium condition is $p_g = 0.091$ (depicted by a dashed line).

bilities can be expressed mathematically as $p_g = p_{g-1}(1 - \mu) + (1 - p_{g-1})v$.
Manipulating this formula yields

$$p_g = [p_o - v/(\mu + v)](1 - \mu - v)^n + v/(\mu + v).$$

The equation above can be simplified to find the equilibrium condition
because $(1 - \mu - v)^n$ approaches zero as n gets larger and larger. Therefore, the equilibrium condition is given by the formula

$$p_g = v/(\mu + v).$$

Plotting the allele frequency of A in a hypothetical population over
successive generations shows that thousands of generations are required to reach allelic equilibrium under comparatively high mutation
rates ($\mu = 10^{-4}$ and $v = 10^{-5}$) (fig. 3.5).

Many assumptions are hidden in the preceding mathematics. In
addition to assuming that mutations are neutral and thus not subject
to selection, it was also assumed that gene A has only two allelic forms
(A and a). Things get much more interesting if multiple alleles for each
gene are considered. For example, a small protein composed of 100

amino acids has a coding sequence of 300 nucleotide pairs. Since each nucleotide can be occupied by A, T, G, or C, the total number of theoretically possible alleles is 4^{300}, or roughly 4×10^{180}. Thus in theory, every mutation could create an allele that is unique. Some of these mutations may have no effect (synonymous code substitutions), whereas others may (non-synonymous code substitutions). The mathematical model dealing with this condition is called *the infinite-alleles model*. Because each allele is unique, this model assumes that identical alleles in a diploid organism are identical because of common descent. That is, homozygous alleles are copies of the same ancestral gene as a result of a mating between related individuals. In the parlance of genetics, this kind of homozygous genotype is called *autozygous*. Under these circumstances, a population can be randomly sampled to determine the frequency of autozygous individuals (and in this way determine a population's homozygosity) to evaluate the effects of mutation on allele frequencies. This can be achieved by determining the probability that two alleles chosen from a population are homozygous. Notice that "identical by descent" implies that neither allele has mutated in the preceding generation. If an allele had mutated in the previous generation, its locus would be heterozygous.

To implement the infinite-alleles model, note that a gamete carrying any particular allele in a population has a frequency of $1/(2N)$, where once again N is the size of the population. Thus, the probability of a heterozygous pair of alleles equals $1 - 1/(2N)$ from which it follows that the probability that two randomly chosen alleles shared descent (the probability of autozygosity) is given by the formula

$$p_g = (1 - \mu)^2/(2N) + [1 - 1/(2N)]((1 - \mu)^2 p_{g-1},$$

and the equilibrium condition is given by $p_g = (1 + 4N\mu)^{-1}$. Since the proportion of homozygosity in a population is inversely proportional to the proportion of heterozygosity, it follows that the proportion of heterozygosity equals $1 - (1 + 4N\mu)^{-1} = 4N\mu (1 + 4N\mu)^{-1}$. Plotting these two expressions against $4N\mu$ reveals a very narrow range of $4N\mu$ within which intermediate levels of heterozygosity exist (fig. 3.6).

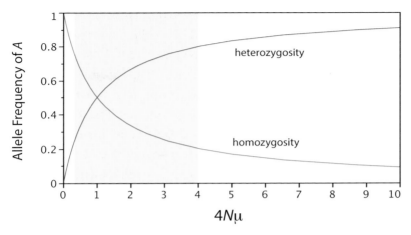

Figure 3.6. The proportion of homozygosity and heterozygosity in a hypothetical population of a biparental species plotted as a function of 4Nμ. The proportion of heterozygosity
increases as either the population size (N) or mutation rate (μ) increases, and only a very
narrow range of 4Nμ exists within which intermediate levels of genetic variation occurs
(shaded in blue).

Sources of Variation: Recombination

Although mutation is considered the ultimate source of genetic variation, recombination during sexual reproduction allows linked genes
to recombine in manifold potentially adaptive combinations (fig. 3.7).
This variation occurs during the first division of meiosis when homologous chromosomes pair and portions of homologues intertwine,
break, and reconnect in different ways. The process is called crossing-
over. It creates new combinations of linked alleles. In randomly mating diploid populations, linked alleles become randomly associated
and eventually reach a condition called *linkage equilibrium*. The rate at
which this happens obviously depends on the frequency of recombination. But it also depends on the distance between alleles. For purely
physical reasons, very closely linked alleles are less likely to cross-
over and become separated than those further apart. In this sense,
some alleles move and thus evolve together. Population size is also
important. The advantage of recombination decreases with decreasing
population size because, as already noted, the probability of mutation

Recombination   **No Recombination**

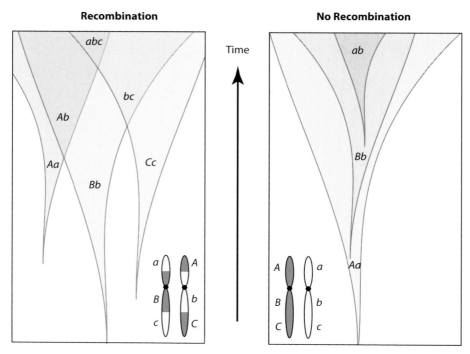

Figure 3.7. Schematic of the effects of recombination on the association of beneficial muta-tions *a, b, c* of three genes *A, B, C* in a large diploid population. Recombination resulting from crossing-over and the exchange of genetic material in homologous chromosomes (see lower right insert) increases the combinatorial diversity of the three mutations and al-lows combinations to occur in a nonlinear manner. In the absence of recombination (shown to the left), beneficial combinations of *a, b, c* must be achieved linearly because there is no mechanism to bring the three mutations together in the same genome. Each mutation must also achieve a higher frequency in the population to have a sufficient probability that next mutation will occur in its favorable genetic background.

decreases as a function $1/(2N)$. Thus, the fate of a favorable mutation can depend on the alleles with which it is closely related, the genetic background it finds itself in, and the size of a population.

Nucleotide polymorphisms give insights into the extent to which neighboring alleles are shuffled between chromosome homologues (fig. 3.8). For example, assuming that mutation rates are uniform along the length of a chromosome, regions of low nucleotide poly-morphism, such as the tip of a chromosome or the region near the cen-tromere, characterize regions that have a low probability of crossing-

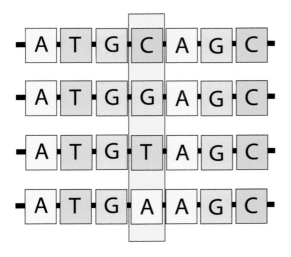

Figure 3.8. Schematic of four single nucleotide polymorphisms (occurring in the nucleotide sequence site shown in yellow). More extensive polymorphisms involving two or more substitutions at neighboring or adjoining sites can occur.

over. Tight linkage can also explain reduced levels of polymorphism either as a result of the rapid fixation of beneficial mutations in a population (called *selective sweep*), or as the genomic elimination of harmful mutations (called *mutation purging*). During a selective sweep, the beneficial allele "takes its neighbors for a ride" and its frequency and that of the hitchhiking region in a population increase, resulting in a monomorphic nucleotide region.

A Nod to Polyploidy

As mentioned earlier, an important driver for variation in plant populations is gene duplication and divergence resulting from polyploidy. A polyploid individual contains more than one complete set of the normal complement of chromosomes for its species. Polyploidy appears to have occurred frequently in plants, as illustrated by the frequency distribution of haploid chromosome numbers among seedless vascular plants and the angiosperms. The dominance of even numbers over odd numbers for haploid chromosomes among species within the *same lineage* is expected when polyploidy occurs frequently. The extremely high haploid chromosome numbers reported for pteridophytes, particularly ferns, indicate that these species are particularly prone to polyploidization. In terms of their time of origin, polyploids have been classified into "neopolyploids," species with haploid chromosome numbers that are multiples of the basic diploid chromosome numbers of their respective genera, and "paleo-polyploids," evolutionarily ancient species that have become repeatedly diploidized and therefore have exceptionally large basic chromosome numbers. As the "paleo" and "neo" indicate, the salient difference between neo- and paleo-polypoidy is time. Polyploid animals also exist, but most of these are insects, flatworms, leeches, or brine shrimp, some of which reproduce parthenocarpically—that is, offspring are produced by the mitotic cell divisions of parent cells. Among vertebrate species, polyploidy is rare, perhaps because the sex-determination mechanisms in late divergent metazoans depend on a delicate balance of chromosome numbers that is easily disrupted by polyploidy (Mable 2004).

Of course the inference that polyploidy underpins the dominance of even over odd chromosome numbers must be approached cautiously because of the confounding effects of dysploidy (a change in chromosome number resulting from the breakage and fusion of chromosomes without net gene loss or gain). In a polyploid sequence, lower chromosome numbers are ancestral to higher ones in the sequence. In a dysploid sequence, the chromosome numbers may run in either direction, from low to high (as in polyploidy) or from high to low. If all the vital genetic material of a chromosome is transferred to one or more other chromosomes, the remnant of the original chromosome is no longer essential and can be lost, and so the original haploid chromosome number is reduced by one. This type of dysploidy occurs in the flowering plant genus *Echeveria*, which contains 143 species whose haploid chromosome numbers include every number from 12 to 34. Unlike polyploidy, dysploidy does not lead to gene duplication and subsequent divergence in gene function. Nevertheless, the gene rearrangements resulting from the consolidation of the genetic material of many chromosomes into fewer chromosomes may dramatically affect phenotypic appearance because the influence of a gene may be altered by its position in the genome (a phenomenon called *positional effects* that results from both the physical interactions among portions of chromosomes and genetic interactions among adjoining genes). Gene duplication by means of polyploidy followed by gene divergence and position effects caused by dysploidy could have profound evolutionary significance. The role of polyploidy in plant speciation is treated in greater detail in chapter 5.

In summary, the basic message of this section is threefold: (1) evolution, adaptive or otherwise, would essentially come to a standstill unless new variation is introduced into a population; (2) all new variation essentially comes from gene and chromosome mutations; and (3) the rate of mutation, which is the maximum rate of evolution, is held in check by selection.

Darwinian Fitness

With this background, we can now explore the concept of Darwinian fitness—the measure of an individual's relative contribution to the gene pool of the next generation. The theory of natural selection posits that some naturally occurring factor or force biases the draw of different alleles, because some genotypes have lower rates of reproductive success than others. As a consequence of a biased draw, the frequencies of different genotypes within successive generations of a population will change. Suppose that, in our hypothetical population, a red-flowered plant is always reproductively successful—that is, each individual invariably passes its genes on to the next generation. Because Darwinian fitness is a relative term, each red-flowered individual is said to have 100% reproductive success. Denoting Darwinian fitness as w, the fitness of all red-flowered individuals will be $w = 1$. Now suppose that each white-flowered plant passes its genes on to the next generation at a lower rate because of some disadvantage. White flowers may fail to attract pollinators as well as pigmented flowers because they offer less desirable nectar rewards. White-flowered individuals may have lower survival rates or set fewer seeds than those bearing pigmented flowers. In these circumstances, white-flowered individuals have a fitness less than 100% or, in notation, $w < 1$. For the sake of argument, suppose each white-flowered plant produces 20% fewer offspring than a red-flowered plant does. If so, its fitness relative to that of a red-flowered individual is $w = 1 - 0.2$, or 0.8. The fitness of the red- and the white-flowered plants differs by 0.2 owing to some form of selection. This difference in fitness is called the selection coefficient; it is symbolized by s. The selection coefficient is a *relative measure* of selection *against* reproductively less successful genotypes. Thus, s ranges in value from 0 (for a genotype experiencing no selection) to 1 (for a genotype that is lethal to its carriers).

In the absence of selection or some other equivalent phenomenon, the Hardy-Weinberg equilibrium equation shows that sexual reproduction will keep the amount of genetic variation within a population constant from generation to generation. Conversely, when selection occurs, the genotypic variant with the highest Darwinian fitness is expected to gradually prevail at the expense of other genotypes. This can be seen by comparing the allelic frequencies in our hypothetical population of red and white flowered plants before and after selection against white-flowered individuals. For example, suppose that the starting frequencies of AA, Aa, and aa are initially p^2, $2pq$, and q^2. Suppose that between the formation of zygotes and the attainment of reproductive maturity, only $(1 - s)$ aa individual plant survives for each AA or Aa plant:

Genotype	AA	Aa	aa
Darwinian fitness (w)	1	1	$(1 - s)$
Pre-selection frequencies	p^2	$2pq$	q^2
Post-selection frequencies	p^2	$2pq$	$q^2 - sq^2$
Average fitness	$(\bar{w})$		$1 - sq^2$

The frequency of a after selection is $(q^2 - sq^2)/(1 - sq^2)$ and the change in the frequency of a is $\Delta a = -sq^2(1 - q)/(1 - sq^2)$. If q is very small, q^2 becomes negligible and $\Delta a \approx -sq^2$.

These relationships reflect the passage of one generation. If selection persists, the difference in the frequencies of the two alleles is amplified generation by generation. Note that the *rate* at which the genetic composition of a population changes in absolute time depends on how long it takes a particular type of organism to complete its reproductive cycle. The generation time of small organisms is, on the average, much shorter than that of large organisms. Thus, changes in genotype frequencies are expected to occur more rapidly for small organisms, such as unicellular plants or bacteria. We will see in chapter 9 that this is a characteristic of what has been called r-selection.

The Difficulty of Measuring Fitness

Although relative individual fitness poses little conceptual difficulty, Darwinian fitness is virtually impossible to measure directly. Many biological properties influence an individual's ability to contribute to the gene pool of the next generation relative to that of other individuals. For flowering plants, the individual seed must germinate, grow quickly, and establish a healthy, reproductively mature plant. Flowers must be successfully pollinated and then produce and support viable seeds. All other things being equal, seed number is also important. Consequently biologists have measured Darwinian fitness by comparing individuals in terms of their relative viability, longevity, developmental speed, fertility, or fecundity. Clearly each of these biological properties reflects a separate component of Darwinian fitness, but for practical reasons all of these are rarely measured for the same individual.

It is generally assumed that all the components of Darwinian fitness are correlated with the one measurement actually taken. This assumption may be true, especially in terms of the relation between survival and reproductive success. We may expect that the longer a plant survives, the larger it becomes and the more reproductive organs it can bear and metabolically foster. The relation between the ability to survive and the ability to successfully reproduce can be remarkably intimate. Every individual invests some physical measure of itself in the production of its offspring. The formation of spores, seeds, and fruits requires energy supplied by the plant bearing them. Nevertheless, the assumption that different measurements of the components of an organism's biology give equivalent measures of Darwinian fitness is problematic at best. A sterile plant that lives for hundreds or perhaps thousands of years is very fit in terms of survival but evidently is not equipped to perpetuate its own kind. The converse is true for a plant that lives for a few days or weeks but leaves behind thousands of offspring during its brief lifetime. This comparison between a "sterile but essentially deathless plant" and a "fecund but ephemeral plant" is purely hypothetical, but it suffices to show that there is no a priori reason to assume that survival and reproductive success are correlated. The only practical justification for assuming that Darwinian fitness can be measured in terms of one among many important components contributing to an individual's fitness is that it is largely impractical to simultaneously measure all the components.

Another consideration is how to measure and compare the fitness of different populations. How do we calculate the *adaptiveness* of one population compared with that of another? Perhaps the most obvious measure is the average population fitness. But what does the *average* value of relative fitness mean? Suppose we have two populations of the same kind of flowering plant. If in one population 90% of the plants produce 70 seeds each, while the remaining 10% produce 35 seeds each, the average population fitness, $\bar{W}_1$, equals $[(0.9 \times 70) + (0.1 \times 35)]/70 = 0.95$. If, in the second population, 50% of the plants produce 90 seeds each, while the remaining 50% produce 50 seeds each, the average population fitness, $\bar{W}_2$, equals $[(0.5 \times 90) + (0.5 \times 50)]/90 = 0.78$. Even though W_1 exceeds W_2, the average number of seeds produced by the first population is smaller than that produced by the second population (66.5 and 70, respectively).

Other measures of population fitness have been proposed. Among these are a population's degree of polymorphism, persistence in time, susceptibility to deleterious mutations, ability to increase in size, interspecific competitiveness, and extent to which developmentally anomalous individuals occur. Each of these measurements of adaptiveness has merit, but each poses practical and theoretical problems. The struggle to find an inclusive measure of population fitness continues, just as does the search for an inclusive measure of Darwinian fitness uniformly applicable to all species. In many respects, the two efforts are inextricably connected, because the various properties contributing to an individual's fitness are very often those required to maintain a population or an entire species.

The difficulty of measuring fitness becomes more complex if we consider altruistic or social behavior. William D. Hamilton (1936–2000) sought to analyze the evolution of social behavior and altruism by introducing a quantity called *inclusive fitness*. Inclusive fitness quantifies the fitness of an individual (sometimes referred to as the actor) by first removing the amounts of benefit or harm done by others (sometimes referred to as recipients) to the individual and then adding the amounts of benefit or harm the individual does to other individuals, which are weighted by the degree of the relatedness to the recipient (Hamilton 1964). Put differently, inclusive fitness is calculated by adding up all the fitness effects causally attributable to an actor and weighting each component by the relatedness between the actor and the recipient to yield the net effect of a social behavior on the actor's overall genetic representation in the next generation. Hamilton showed that under certain simplifying assumptions, selection can result in altruistic behaviors that yield the largest inclusive fitness. Hamilton's treatment leads to the prediction that a gene coding for a social behavior will be favored by selection if and only if the product of the coefficient of relatedness r between an actor and a recipient and the benefit b that the behavior incurs exceeds the costs c that it imposes on the actor (altruistic behavior is predicted to evolve if and only if $rb > c$). This has been called Hamilton's rule.

However, the concept of inclusive fitness implicitly assumes that an actor's fitness can be partitioned into separate quantities due to itself and to each of its recipients. Another difficulty is that to know the degree of relatedness there needs to be a one-to-one mapping of a

phenotype's social behavior onto its genotype. This presents a problem if the actor and recipient are not genetically related, as for example a flowering plant and its pollinator(s). On the positive side, the
concept of inclusive fitness can be used in game theoretic treatments
of evolution to ask questions, such as "What should I do to maximize
my inclusive fitness?," to obtain predictions about social behavior and
ecological associations.

Additive Genetic Variance and Heritability

Many factors contribute to the phenotypic variance observed for a
trait in a population. In general terms, however, the total phenotypic
variance of a trait can be subdivided into two very broad variance
components: the total genetic variance and the environmental variance. The total genetic variance results from the presence of different
genotypes in the population; the environmental variance results from
differences in the physical or biological environment. The extent to
which phenotypic variation is heritable is largely governed by the total
genetic variance of the trait, although heritability can be substantially
reduced when the environmental variance is exceptionally high. Here,
we will examine the components of total phenotypic variance. Special
attention will be given to the total genetic variance, because this variance can be further subdivided into components in ways that predict
the effect of selection on the trait. Specifically, the effect of selection
on a trait will be found to be proportional to the genetic variance associated with the average effects of substituting one allelic form of a
gene for another allelic form of the same locus. This component of
total genetic variance is called additive genetic variance. When the
additive genetic variance for a trait is high compared with the total genetic variance and environmental variance, the trait is highly heritable
and thus the effect of selection on the trait is high.

Consider the following data for two alleles, *A* and *a*, whose three
allelic combinations result in phenotypes differing in a quantitative
trait, say plant height:

Genotype	AA	Aa	aa
Frequency (f_i)	0.49	0.42	0.09
Phenotype (x_i)	20 cm	14 cm	10 cm

These data show that genetic variance for plant height exists—genotypes with the *A* allele are taller than those lacking this allele. They also indicate that the *A* allele is not completely dominant—the mean height of heterozygotic individuals is less than that of those with the *AA* genotype and more than that of those with the *aa* genotype. Thus the total genetic variance (s_T^2) for plant height consists of two components: (1) the variance resulting from substituting one allele for another, which is the additive genetic variance (s_A^2), and (2) the variance resulting from the partial dominance of the *A* allele over the *a* allele, which is called the dominance variance (s_D^2). Clearly, traits like plant height are subject to environmental and developmental variance. But for illustration we will assume that these components of variance are negligible such that $s_T^2 = s_A^2 + s_D^2$.

To calculate the total genetic variance, we must determine the average effect of each of the two alleles on the phenotype. For each allele, this average effect equals the sum of the number of the allele in each phenotypic class bearing it times the mean height of the individuals with the allele divided by the total number of the allele in the entire population. Because the population consists of diploid organisms, the average effect of *A* is given by $\bar{A}$ = [2(0.49)20 cm + 1(0.42)14 cm] / [2(0.49) + 1(0.42)] = 18.2 cm, and the average effect of *a* is given by $\bar{a}$ = [2(0.09)10 cm + 1(0.42) 4 cm] / [2(0.09) + 1(0.42)] = 12.8 cm. The difference between the average effects of the two alleles on plant height is 5.4 cm. To calculate s_T^2, we also must know the mean height for all the plants in the population. This equals the sum of the average heights of the three phenotypic classes weighted by their frequencies of occurrence. In other words, mean phenotype height = $\bar{x}$ = $\Sigma_i x_i f_i$ = 20 cm(0.49) + 14 cm(0.42) + 10 cm(0.09) = 16.58 cm. The total genetic variance of the population equals the sum of the frequency of occurrence of each phenotypic class times the square of the difference between the mean height of each phenotypic class and the mean height of the entire population. Thus, $s_T^2 = \Sigma f_i(x_i - \bar{x})^2 = 0.49(20 \text{ cm} - 16.58 \text{ cm})^2 + 0.42(14 \text{ cm} - 16.58 \text{ cm})^2 + 0.09(10 \text{ cm} - 16.58 \text{ cm})^2 = 5.73 \text{ cm} + 2.79 \text{ cm} + 3.89 \approx 12.42 \text{ cm}$. To calculate the additive genetic variance, we must also compute the frequencies of A and a in the population. Inspection of the data shows that the frequency of $A = f_A = [2(AA) + 1(Aa)]/2 = [2(0.49) + 1(0.42)]/2 = 0.70$. Likewise, the frequency of $a = f_a = [2(aa) + 1(Aa)]/2 = [2(0.09) + 1(0.42)]/2 = 0.30$. The additive genetic variance equals twice the sum of each of these two frequencies times the square of the average effect of each allele minus the mean height of all phenotypes. Thus, $s_A^2 = 2[f_A(\bar{A} - \bar{x})^2 + f_{aa}(\bar{a} - \bar{x})^2] = 2[(0.7)(18.2 \text{ cm} - 16.58 \text{ cm})^2 + (0.3)(12.8 \text{ cm} - 16.58 \text{ cm})^2] \approx 12.25 \text{ cm}$. Because $s_T^2 = s_A^2 + s_D^2$, it follows that $s_D^2 = 12.42 \text{ cm} - 12.25 \text{ cm} = 0.17 \text{ cm}$. In this example, the dominance variation, which also includes the variance resulting from the interaction between genes at other gene loci and the gene influencing plant height (that is, the variance due to epistasis), is small compared with the additive genetic variance.

We are now in a position to appreciate why compartmentalizing the total genetic variance into its various components is important. In the absence of environmental variance or developmental "noise," the heritability of a trait, denoted by h^2, equals the additive genetic variance divided by the total genetic variance observed for the trait. That is, $h^2 = s_A^2/s_T^2 = s_A^2/(s_A^2 + s_D^2)$. Put into words, h^2 is the difference between a trait measured for selected parents and the mean for the trait measured for the unselected population (*the selection differential*) divided by the difference between the offspring of selected parents and the previous generation (*the selection response*). The greater the value of h^2, the greater the difference between the trait measured for selected parents and for the whole population that will be maintained by the offspring of the selected parents. In our example, $h^2 = s_A^2/s_T^2 = 12.25 \text{ cm}/12.42 \text{ cm} = 0.986$, which indicates that selection for Aa genotypes will maintain the difference between the height of the selected individuals and the mean height of the unselected population. Accordingly, a plant breeder could cross Aa individuals with Aa individuals and expect a high response to selection. The same cannot be said for the following population:

Genotype	AA	Aa	aa
Frequency	0.49	0.42	0.09
Phenotype	14 cm	20 cm	10 cm

Here, unlike the previous example, Aa individuals are on average taller than those with the AA or aa genotype, which appears to be an example of heterozygotic advantage. Nevertheless, calculations quickly show that $s_T^2 = 11.89$ cm and $s_A^2 = 0.73$ cm, and so $h^2 = s_A^2 / s_T^2 = 0.06$. For this population, the selection response for crossing Aa individuals will be poor, although, assuming very low environmental variance, hybrids made between previously intensely inbred lines of plants could be used to increase particular quantitative traits (a tactic often used by plant breeders).

Changes in the additive genetic variance for traits can be used to evaluate the strength of selection. For example, there is little disagreement among biologists that plant herbivores or pathogens can respond adaptively to changes in the defense mechanisms of their host plants, but there is still debate over whether selection imposed by herbivores or plant pathogens is strong enough to cause plant traits to respond adaptively. If coevolution occurs at all, then the additive genetic variance of plant traits believed to respond adaptively to attack should change in response to the intensity of attack. By way of one example, Ellen Simms and Mark Rausher (1989) showed that the additive genetic variation for Darwinian fitness (measured as lifetime seed production) was eliminated when herbivores were excluded from natural populations of the morning glory *Ipomoea purpurea*. Studies like this one indicate that herbivory can impose selection strong enough to cause plant populations to adaptively diverge.

Changes in the additive genetic variance of traits can be used to evaluate a variety of adaptive hypotheses. But it is important to bear in mind that a trait for which high heritability has been reported in one population under one set of experimental conditions can have substantial differences in heritability among natural populations experiencing different environmental conditions. This follows from the fact that the total phenotypic variance seen in a particular population is the sum of the total genetic variance, the environmental variance, or s_E^2, and the variance resulting from developmental noise, s_N^2. Because $h^2 = s_A^2 / (s_A^2 + s_D^2 + s_E^2 + s_N^2)$, substantial increases in environmental variance will tend to diminish the heritability of a trait. Environmental variance can be much reduced in laboratory-grown populations, but it is often substantially large for populations growing under natural conditions.

Because there is no guarantee that populations with the same genetic composition will respond in the same way to different environments, there is no reason to believe that evolution is an ineluctably progressive process. This erroneous view that evolution is progressive emerges from a misunderstanding of R. A. Fisher's (1930) "fundamental theorem of natural selection," which states that the rate of increase in the mean fitness of a population is proportional to the genetic variance in fitness. Because a variance can never take a negative value, Fisher's theorem has been incorrectly interpreted to suggest that fitness will invariably increase so that evolution must be "progressive." However, Fisher's theorem holds true only if the relative fitness of genotypes remains constant, which hardly ever happens in real populations. Thus, there is no reason to expect selection to lead to progressively more fit or more complex organisms, although this may be the case for some lineages during certain episodes in their evolutionary history.

The Phenotype

The Hardy-Weinberg equilibrium equation shows, in concrete quantitative terms, how reproductive (Darwinian) fitness is defined and how selection can "adapt" a population to a particular environment by removing individuals carrying reproductively unfavorable genotypes.

Unfortunately, this equation can give the erroneous impression that selection acts directly on the genotypes within a population when in fact it operates on the corporeal manifestation of each genotype, the phenotype, which is the sum of the physical, metabolic, physiological, and behavioral traits that define the individual. For example, in our hypothetical population of white- and red-flowered individuals, selection acted against white-flowered individuals, not against the *a* allele per se, a point driven home by the fact that the reproductive fitness of all red-flowered individuals was equal, regardless of whether red flowers resulted from the *Aa* genotype or the *AA* genotype. Selection changes the gene pool of the population because it identifies the phenotypic traits that influence reproductive fitness. Selection against the phenotypes carrying these traits indirectly changes the gene pool only because a correlation exists between the phenotype and its genotype. The true measure of our understanding of how the genetic structure of a population changes owing to selection is our ability to relate particular genotypes to particular phenotypes and to uncover how the environment discriminates among phenotypes (box 3.2).

Even though selection pressures can discriminate among phenotypes with different relative fitness, three features limit our ability to map phenotypes onto their corresponding genotypes. First, many important phenotypic traits are quantitative rather than qualitative. A quantitative trait may be defined as any trait for which the average phenotypic differences among genotypes are much smaller than the variation among individuals within a genotypic class. Traditionally, it is assumed that quantitative traits are determined by a large number of genes, each having a small and independent effect on the phenotype. This multiple-gene-factor hypothesis indicates that a quantitative trait is governed by a polygene—a confederation of "small but equal effect" genes. Quantitative traits cannot be studied by simple Mendelian genetics and so pose practical limitations on mapping phenotypes on to their genotypes.

The second difficulty is that a single gene may influence more than one phenotypic trait. Such genes are called pleiotropic genes. The in-

Box 3.2. The Genotype-to-Phenotype Map

Similar genotypes can give rise to dissimilar phenotypes, and similar phenotypes can be produced by very dissimilar genotypes. Consequently, it can be very difficult to map genotypic-to-phenotypic correspondences and to predict the effects of selection on organisms competing for the same resources. Computer simulations have been used for this purpose (in lieu of direct empirical tests) employing algorithms whose predictions have been verified by means of large data sets.

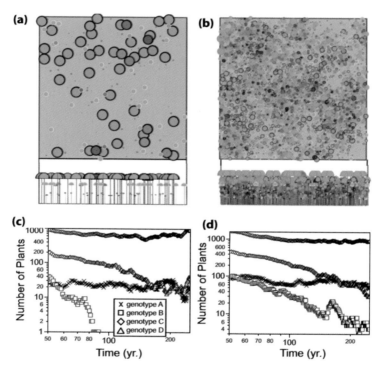

Figure B.3.2.a. Computer simulations of SERA showing competition for space and light among four different genotypes (see insert in c). In these simulations, a "genotype" is defined by the mathematical rules for the allocation of annual growth in biomass to the construction of leaves, stems, and reproductive propagules. A "phenotype" is defined based on the height, diameter, canopy spread, and propagule biomass a tree with a specific genotype produces. a–b. Polar and vertical views (upper and lower panels) of competition in an undisturbed and disturbed environment (a and b, respectively). c–d. Graphs showing the numbers of plants for each of the four genotypes as a function of years in competition. Notice that genotype B goes to extinction in less than 90 yrs. in an undisturbed environment (c), but persists in a disturbed environment (d). Although not visually evident, genotypes A and D are similar, but yield different phenotypes that nevertheless coexist.

One such algorithm, called SERA (spatially explicit reiterative algorithm), has been successfully used to predict the behavior of tree species differing in functional traits pertaining to the allocation of annually accrued biomass ("annual growth") to the production of new leaves, stems, and propagules (seeds and fruits) (Hammond and Niklas 2009). Conceptually, the traits that specify different allocation patterns may be thought of as "different genotypes," whereas the resulting morphology and life-history characteristics of the plant translate into "different phenotypes." SERA permits a researcher to stipulate these genotype-to-phenotype mappings and has the additional advantage of allowing a researcher to place these hypothetical trees in direct competition with their own kind or with different species under different environmental conditions (light and nutrient conditions, differing degrees of disturbance, different probabilities of seedling mortality and extinction, etc.).

Among the lessons drawn from SERA simulations, two are particularly germane to the concept of the genotype-to-phenotype map: (1) very small genotypic differences can result in very different phenotypes (as defined by tree height, trunk diameter, canopy spread, and biomass allocation to reproductive biomass and propagule size), and (2) very dissimilar phenotypes can be competitive equals, particularly when an environment experiences moderate to severe physical disturbance (fig. B.3.2.a). Another lesion pertains to selection: mortality is age-dependent and most severe during early growth (fig. B.3.2.b).

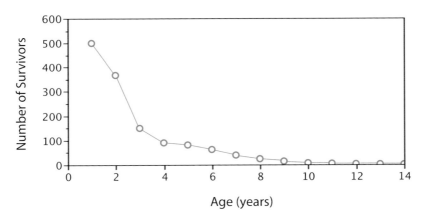

Figure B.3.2.b. A bivariate graph showing the number of surviving progeny produced by a single tree simulated by SERA. In 10 years, the number of survivors drops from 500 to 2. The mortality rate is indistinguishable from the mortality rates of real plants (see, for example, data reported by Van Valen 1975).

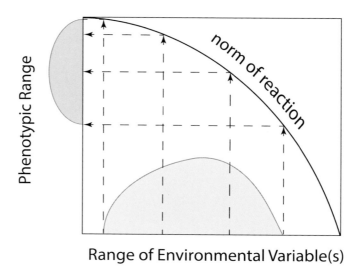

Figure 3.9. Schematic of the concept called the "norm of reaction" or the "reaction norm," which describes the range of phenotypes a single genotype can manifest in response to one or more environmental variables.

fluence of a pleiotropic gene on Darwinian fitness can be complex. In general terms, it will depend on whether the positive effects of the pleiotropic gene outweigh its negative effects. Suppose the dominant allele *A* coding for red flower color also codes for the production of little or no nectar. In these circumstances, the overall reproductive fitness of red-flowered plants would depend on whether insect pollinators are attracted more by flower color than by the rewards they receive for inadvertently carrying pollen from one flower to another.

The third difficulty in assigning a one-to-one correspondence between each phenotype and its genotype is that the appearance and structure of an individual organism can vary with local environmental conditions. The color of *Hydrangea* flowers depends on the acidity of the soil in which a particular plant grows and will change if the plant is moved from one soil type to another. The size of the leaves and stems of many plants depends on whether shoots are exposed to sunlight or wind. These and many other examples of "phenotypic plasticity" show that the phenotype is not always rigidly defined but may have a natural range of variation, called the norm of reaction. Because the norm of reaction can vary as a consequence of the environment, it is often difficult to determine the precise relation between a genotype and its range of phenotypic expression (fig. 3.9).

The norms of reaction for vegetative and reproductive traits may differ in ways that make perfect intuitive "adaptive sense." The norm of reaction tends to be comparatively broad for vegetative organs (leaves, stems, and roots), and remarkably different-looking pheno-types can be produced by the same genotype depending on local con-ditions. This makes adaptive sense because most plants are sedentary organisms and many have a comparatively long life expectancy. Be-cause most plants cannot move from the location where they began their life and must continue to vegetatively survive in an environment that can change from year to year or decade to decade, it is logical that vegetative organs have broad norms of reactions. Also, the same kind of vegetative organ may continue to function adequately despite what often appear to be significant phenotypic differences. Whether the blade of a leaf has two, three, or four lobes may not significantly alter its ability to intercept sunlight. In contrast, the norm of reaction for reproductive traits tends to be narrow, so that a given genotype results in a very specific phenotype. This feature explains why reproductive traits are favored by systematists when constructing phylogenetic hypotheses. It may be that reproductive "conservatism" is a hedge against subtle changes in reproductive organs that can have profound negative effects on reproductive success. A change in the hue, shape, or size of a flower may subtract from its ability to attract a pollinator and so reduce the number of seeds it produces. More dramatic depar-tures in the appearance of reproductive organs may make them totally useless. By the same token, a random mutation may dramatically in-crease reproductive fitness and so inspire bursts in plant evolution.

Although the norm of reaction of vegetative organs may be broad in contrast to that of reproductive organs, the ability of a phenotype to successfully reproduce depends on its ability to vegetatively survive and grow and, in turn, on the plant's ability to invest metabolites in

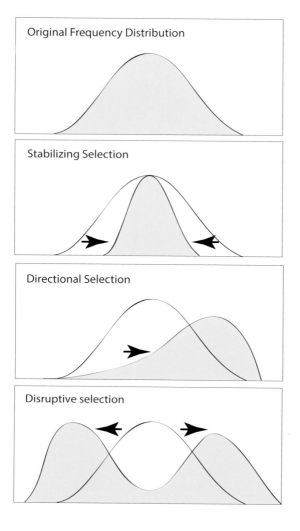

Figure 3.10. Schematics for three kinds of selection regimes that can change the frequency distributions of genotypes in a population. A normal frequency distribution is used to illustrate the original frequency distribution under no or little selection (top panel). Stabilizing selection results when selection eliminates outliers (second panel). Directional selection occurs when one of the extremes in the distribution is eliminated (third panel). Disruptional selection results when intermediates in the population are selected for elimination (bottom panel). Note that selection regimes can change over time. If they occur frequently, they can have the same effect as disruptional selection.

the formation of reproductive organs and the progeny therein. Clearly, just as the genotype is the sum of all its genes, the phenotype is also the sum of its many vegetative and reproductive traits, each of which can contribute to the reproductive fitness in different, albeit interrelated, ways. Although it may be easier (more convenient) to focus on one or two "obvious" reproductive traits to evaluate how much reproductive fitness varies as a function of variations in these traits, the entire phenotype is the vehicle of evolution.

Modes of Selection

Selection can change the prevailing phenotype of a population in one of three broad modes: stabilizing selection, directional selection, or diversifying (disruptive) selection (which on a large scale is called adaptive radiation). Each mode depends on how the environment changes, as may be shown by plotting the number of individuals sharing the same phenotype against the different phenotypic categories within a population (fig. 3.10). Stabilizing selection results when the interactions between the plant and its environment remain constant and natural selection removes the extreme phenotypes, leaving behind intermediate phenotypes. Stabilizing selection therefore reduces the phenotypic variation in the population. For example, plants with intermediate seed sizes may be selected for when the nutrients stored in very small seeds are insufficient to establish seedlings and when very large seeds become too obvious a source of nutrition to escape herbivory. The extreme phenotypes within populations have traditionally been viewed as homozygous because the reaction of homozygotic individuals to the environment tends, on the average, to be more extreme than that of heterozygotic ones. By causing homozygotic individuals to take on extreme phenotypes, environmental variation tends to expose them to intensive stabilizing selection.

Directional selection results when individuals expressing a particular trait (to either the greatest or the least extent) are preferentially subjected to greater selection. This mode of selection, which tends to shift the frequency distribution of phenotypes to one or the other end of the spectrum (and so favors originally rare phenotypes), can continue until the genetic diversity within a population becomes exhausted. Directional selection and the consequences of increasing genetic uniformity are nicely illustrated by considering the yield of crop plants, like oats, subjected to continuous inbreeding by plant breeders. During the early 1900s, the yield of oats (bushels/acre) in the United States was raised by identifying superior varieties from around the world and importing seed. During the 1920s and 1930s, however, the yield plateaued dramatically when plant breeders sought to improve pure inbred (homozygous) lines by artificial selection. From the mid-1930s through the mid-1960s, the yield of oats was increased once again by programs based on hybridization and the selection of superior individuals among the progeny of successive generations. During these three periods, the rates of improvement in bushels/ acre/year were 1.8, 0.1, and 1.5 (Sprague 1967). These data show that the effect of directional selection decreases as genetic materials become uniform.

The final mode of phenotypic selection, called diversifying or disruptive selection, occurs when intermediate phenotypes have a selective disadvantage relative to those of both extremes. Biologists repeatedly emphasize environmental heterogeneity or "graininess" and how well organisms can adapt to different "microenvironments" that subject individuals within the same population to potentially discordant selection pressures.

The three modes of selection are not mutually exclusive. Successive generations may undergo one mode of selection and then another in response to shifting patterns of environmental change. Phenotypic variation within a population subjected to directional selection will decrease and may eventually become stabilized. Likewise, diversifying selection favoring phenotypic extremes due to environmental heterogeneity may come to closure with stabilized phenotypes, or an environment may gradually change in one direction and then gradually change in the reverse direction, as from dry to wet and then wet to dry. If so, then directional selection will shift the mean phenotype, initially favoring one extreme and later the other. Thus, the prevailing phenotype that originally conferred the highest fitness may languish and die as its environment changes, while previously less fit phenotypes prosper and increase in number.

In this sense, evolution can involve a peculiar kind of hide-and-seek where a randomly changing environment can shift, obscure, or even eliminate previously prevailing conditions, while to survive and reproduce the organism must simultaneously pursue the niche to which it was previously adapted. Herein lies adaptive evolution's peculiar danger. By the time an adaptation to one set of environmental conditions takes place, the environment may have already changed, making the adaptation irrelevant or, worse, maladaptive. Indeed, this explains why few if any species are "optimally" adapted. Suppose that under directional selection a population evolves tubular flowers whose length is best suited to a particular insect pollinator with a commensurately long tongue. The same population would become poorly adapted if circumstances favored insects with short tongues and would be at risk should the long-tongued insect became extinct, perhaps owing to a sudden dramatic change in the environment. The failure of a population to adaptively evolve at the rate and in the direction an environment changes poses a danger because it could lead to the eventual extirpation of the population. That some environments change rapidly or unpredictably and that the adaptive state of an organism may lag significantly behind a mercurial environment are possible explanations for the extinction of entire species or higher taxa.

One solution to rapidly changing environments is to have a very short reproductive cycle (part of the r-selection stratagem), in which the genetic composition of a population can be altered at the same pace as the environment changes. Small organisms tend to have shorter reproductive cycles and higher mutation rates than larger organisms. Natural selection could also act directly on the development of an organism to favor individuals with a growth strategy capable of dealing with rapid environmental change. Many kinds of plants have developmental patterns evoking different phenotypes, each of which may be well suited to a different set of local environmental conditions. This "phenotypic plasticity" has already been mentioned in relation to the concept of the norm of reaction (and alluded to in the Introduction). The vegetative organs of plants were said to typically show a broad norm of reaction, so that very different phenotypes result depending on the local environmental conditions attending the growth and development of leaves, stems, and roots. This is particularly advantageous to a sedentary organism like a terrestrial plant that is incapable of leaving an inhospitable environment for a more advantageous one.

Another possible solution to highly variable environmental conditions is to be a "phenotypic generalist," capable of coping almost equally well with a fairly broad range of environmental conditions. This strategy entails the whole phenotype rather than the nature of the constituent organs of the individual. Such "whole organism" adaptations may involve the induction of enzymes, switching from one physiological pathway to another, alterations in phenology, and so forth. Another solution for dealing with a variable environment is dormancy. Many species of unicellular algae suspend their metabolic or sexual reproductive activities when the environment becomes too stringent and form resting cysts or encapsulated dormant spores or zygotes. Changes in patterns of reproduction in response to changes in the environment are not uncommon. Likewise, the seeds of many terrestrial plants may remain dormant for decades, forming "seed banks," and can reestablish a population when environmental conditions conducive to growth return.

Yet another way of coping with a changing environment is to physically "track" the conditions that best suit the organism. On a small geographic scale, many kinds of plants explore the habitat for sunlight, water, and minerals by means of their growth habit. A broad spectrum of plants have a rhizomatous growth habit; that is, they form horizontally oriented stems that can grow either below or above the ground surface. New leaves and stems are produced at the growing tip of the rhizome, while older parts of the rhizome may rot away. By branching ahead and fragmenting behind, a rhizome can grow forward and spread outward, thereby colonizing new areas while leaving parts of itself behind in previously occupied locations. Much the same thing is accomplished by plants like the banyan tree (a species of *Ficus*), whose aerial branches produce prop roots that expand in girth and serve as vertical supports functionally equivalent to vertical stems (fig. 3.11). When the main trunk of a banyan tree dies and rots away, the aerial spreading branches of the original tree may separate and "walk away"

Figure 3.11. The aerial roots of a *Ficus* tree look very much like stems. They emerge from the lower surfaces of branches as slender flexible roots that grow downward. When they enter the soil, they contract and mechanically function like guy-wires. They subsequently develop wood and gradually mechanically function as columnar support members. It is not uncommon for the trunks of some trees to decompose entirely. The living canopy of these trees is elevated aboveground by an extended aerial root system. Fragmentation of the aerial branching system can produce genetically identical plants that "walk" away from one another in different directions to explore new habitats. The largest *Ficus benhalensis* is located in the Acharya Jagadish Chandra Bose Botanic Garden, near Kolkata, India. It lacks its original trunk and occupies 14,500 square meters (about 1.5 hectares or 4 acres). It has over 3,300 aerial roots reaching the ground surface.

in different directions on their massive, stilt-like prop roots. Another way plants migrate is by means of spores, seeds, and fruits that often can be transported remarkable distances by wind, water, or animals. Plant populations languishing in one location because of deteriorating environmental conditions can colonize more amenable locations by long-distance dispersal, while simultaneous asexual and sexual reproduction permits the simultaneous exploration of new habitats and the retention of favorable old ones.

Some of these strategies for dealing with changing environment may have deleterious long-term consequences. It is possible that the phenotypic generalist is a "jack of all trades but master of none"—that it may survive under different conditions but be less successful when pitted against specialist species. In theory, organisms with broad norms of reaction (or those that go dormant when conditions become unfavorable) can be at risk because they are shielded from the potentially beneficial long-term effects of intense selection. Unconditional deleterious mutations may accumulate in a population experiencing little or no selection, only to surface with catastrophic results when the population is later subjected to selection. However, the evidence that phenotypically plastic and thus generalist species are less able to compete with specialized species is weak at best, suggesting that phenotypically plastic organisms may hold their ground when placed in direct competition even with highly specialized species. Noting that in the absence of sufficient genetic variation even a gradually changing environment may forecast a decline in the overall fitness of a specialist species, leading to its ultimate demise, we see that the ability of a single genotype to produce functionally appropriate phenotypes in response to different environmental conditions may confer a long-term evolutionary benefit.

If, as some believe, phenotypic plasticity is an adaptation sensu stricto, then it must be a heritable property subject to selection. The advantages of phenotypic plasticity are obvious—it provides an opportunity for the same or closely related genotypes to survive and reproduce in very different environments. In contrast, the modus vivendi by which phenotypic plasticity evolves is still a matter of conjecture, because a curious tension exists between evolutionary forces that act to integrate an organism's genome and other forces that act to partially dismantle and adaptively reshape the genome in the face of environmental challenges.

The traditional view is that balanced systems of harmoniously interacting (coadapted) genes are required for the survival of the individual. Each individual is genetically tethered to its ancestors and thus is afforded a well-tested lifeline it would be foolish to abandon. Nevertheless, even a coadaptive gene network must change to some degree if populations are to evolve and cope with new environmental conditions. Thus a remarkable dynamic exists between selection pressures that act to build harmonious balanced systems of interacting genes and those that act to modify these systems to cope with new conditions (Waddington 1942; Schmalhausen 1949; Mayr 1963; Bonner 1988). This dynamic is underwritten by *epistasis*—a condition where the phenotypic expression of alleles at one locus depends on the alleles present at one or more other gene loci. Epistasis is profoundly important because it shows us that the expression of genetic and thus phenotypic variation is under a high level of control. Epistatically interacting genes can engender phenotypic uniformity despite underlying genetic diversity, but they can also give rise to phenotypic variation whenever large networks of formerly interacting genes are broken up or "modularized" into smaller networks, as when a species is subjected to stabilizing selection for some of its networks of genes and directional selection for other sets of epistatic genes. The tension between evolutionary forces that hold some coadapted gene networks together and dissociate other networks can result in what may be crudely thought of as "genomic mosaic" evolution. That the genomic architecture of most organisms evinces some degree of modularity is shown by the ability of most organisms to adapt to different environments, because an epistatically "megalithic" genome would resist evolutionary change—it would be virtually impossible to move by virtue of its sheer cohesive inertia. Because most organisms are capable of adapting to new environmental conditions, some genomic modularity must exist in most organisms. The way phenotypic plasticity evolves likely has its roots in how epistatic gene networks can become modularized when different selection forces act on different phenotypic traits. Just as some phenotypic traits that were once variable can become fixed, while others may become tightly linked with other characters, still other phenotypic characters can gain variability by compartmentalizing large gene networks into smaller, semiautonomous ones.

The modularity of coadapted gene regulatory networks could and probably does permit considerable variation within individuals, especially for plants. Indeed, most plants exhibit a fascinating phenotypic modularity that may well correspond to their genetic modularity. This possibility has received comparatively little attention because most models for the evolution of adaptive phenotypic plasticity as well as empirical studies of this effect tend to dwell on the role of coarse-grained environmental variations as the underlying selective agent (Via and Lande 1985; Van Tienderen 1991; Weis and Gorman 1990; Andersson and Shaw 1994). Here coarse-grained environmental variation means that each individual grows and develops essentially under one set of environmental conditions. Much less consideration has been given to fine-grained environmental heterogeneity, in which an individual experiences many different environmental conditions during its lifetime. Although the mature form of many kinds of animals is achieved during a comparatively brief period of development early in the life cycle (coarse-grained environmental heterogeneity), most multicellular plants continue to grow and develop for very long periods and thus typically experience many different physical and biological conditions (fine-grained environmental heterogeneity). During its lifetime a single plant may produce millions of leaves and stem internodes, and the potential for morphological and physiological differences among these organs (within-individual phenotypic plasticity) can be highly adaptive. Perhaps for this reason, it is not surprising that some of the most dramatic examples of phenotypic plasticity are drawn from the plant world.

The Fitness Landscape

Perhaps the most elegant way of visualizing most of what has been said up to now was offered by Sewall Wright, who envisioned the

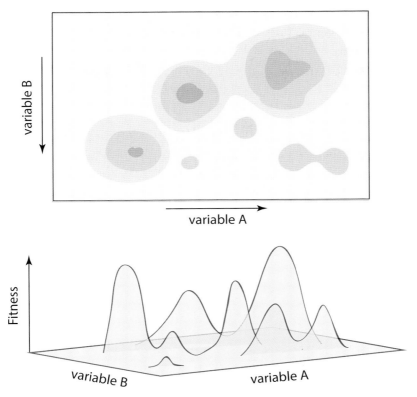

Figure 3.12. Schematic of Sewall Wright's concept of a fitness landscape as a metaphor for adaptive evolution. As originally proposed, the landscape depicts the fitness of all genotypic permutations in a population. In this schematic, the landscape is described by two variables (A and B), which could be genotypic or phenotypic. The elevation in the landscape is defined by the fitness of each permutation. "Peaks" denote regions of high fitness; "valleys" denote regions of very low fitness. Individual populations (not shown) can be depicted as "pancakes" draped over a peak or within a valley, since each population contains a range of genomic and phenotypic variation.

consequences of natural selection as an adaptive "walk" over a "fitness landscape" (fig. 3.12). Wright defined the dimensions of the landscape, its breadth and depth, in terms of all possible genotypic combinations within a population. Supposing that each gene has two alleles, the simplest landscape is described by a set of orthogonal axes (one for each allele of each gene). If we draw contour lines around different genotypes conferring the same or nearly the same relative fitness on their phenotypes, the landscape becomes a flat "fitness map." The fitness of all possible genotype combinations may also be plotted along a vertical axis so that the landscape becomes three-dimensional, with fitness "valleys" occupied by the least fit genotypes within the population and "peaks" occupied by the most fit. Because over many generations selection eliminates less fit genotypes in valleys and because the genotypes on the slopes of adaptive peaks are more fit, the genetic composition of the population "walks" toward and up fitness peaks. Because the vehicle of this walk is the phenotype rather than the genotype, the fitness landscape can be recast in terms of all the conceivable phenotypes produced by the population.

Each walk on the fitness landscape is a consequence of the combined effects of random forces (mutations) and nonrandom forces (selection). These two factors are not necessarily antagonistic. Random events are just as likely to push a walk up a peak as to direct it toward regions of lower fitness. Therefore an adaptive walk is the consequence of the simultaneous operation of random events and selection. Five additional points are equally important. First, walks over fitness landscapes are not a conscious process—a population does not "strive" toward adaptive peaks, nor does it "flee" lethal valleys. Walks are propelled by random effects and selection pressures, neither of which can divine the location of fitness valleys and peaks. Second, some adaptive walks may be impossible because selection can act only on the phenotypic variants within a population, and some conceivable phenotypes may never be developmentally realized by individuals. Third, although not all phenotypes may be possible, dissimilar phenotypes may confer similarly high fitness. Consequently a fitness landscape can have manifold adaptive peaks. Fourth, the metaphor of an adaptive landscape is a gross oversimplification. Each phenotype is the sum of many highly integrated functional traits, and each trait may contribute to the total fitness of an individual organism. However, each trait is also affected by all other traits to various degrees. So the phenotypic landscape is more realistically rendered as a multidimensional space. And fifth, in addition to its spatial complexity, the fitness landscape changes over time as the environment changes biologically and physically. As it evolves, the population changes, and so the fitness landscape changes because the prevailing phenotype changes. As the physical environment changes, the relative fitness of an individual changes.

Thus, the occupants of adaptive peaks are never "perfect" because the environment is always changing, because selection cannot create new phenotypes, and because the variation that exists in a population is the result of chance genetic mutation and recombination, the initial genetic composition of a colonizing population, and other factors. Likewise, the phenotypic variation within the inclusive population of a species reflects an evolutionary history of descent with modification through a long line of ancestors. This legacy cannot be abandoned willy-nilly. To be sure, how far adaptive walks are obstructed by genetic, developmental, or other barriers is relative rather than absolute because it undoubtedly varies among different kinds of organisms and changes over evolutionary time for each kind. The traditional view is that plants are developmentally more "plastic" than most animals (Scharloo 1991), perhaps because they tend to be sedentary organisms that cannot leave a difficult environment, or because plant metabolism is more closely attuned to the physical environment than animal metabolism, or because plant development relies more heavily on environmental cues than animal development. Another important feature is that plants may not be able to achieve the precise orchestration of development seen in animals. Plants are composed of fewer different cell types than most animals, and cells are not free to migrate within the plant body as happens in animal development. Whatever the reasons, the developmental plasticity of plants is so remarkable that one has the impression that their adaptive walks may be far less impeded by genetic or developmental barriers than those of animals. Nevertheless, it remains that the fitness landscape of plants as well as animals is very much like an obstacle course—some walks are more difficult than others, some may be temporarily blocked, and others may be forever closed.

Yet another limitation on adaptation is that each organism must perform many diverse tasks to grow, survive, and reproduce. Undoubtedly some of these tasks have different phenotypic requirements. For example, to grow and reproduce, plants require sunlight. As the surface area projected toward the sunlight increases and the volume bounded by this surface area decreases, the efficiency of light interception increases. Therefore the biological obligation of plants to intercept sunlight is best fulfilled by phenotypes possessing very high ratios of surface area to volume. These phenotypes are easily maintained in an aquatic habitat, where plants are continuously bathed in water and structurally supported by it. On land, water is lost through the external surfaces exposed to the drier atmosphere. Because the amount of water lost through evaporation decreases as the surface area exposed to the atmosphere is reduced, land plants typically have highly reduced leaf areas, particularly in arid habitats. And in general terms the biological obligation to conserve water is best fulfilled on land by plants with extremely low ratios of surface area to volume. An extreme reflection of this response is seen in some species of cacti whose leaves are reduced to nonphotosynthetic spines and whose succulent stems have a low ratio of surface area to volume.

Another example of conflicting phenotypic requirements is seen when light interception and mechanical stability are jointly considered. All plants depend on sunlight for growth, and all terrestrial plants must resist the force of gravity to grow upright. In general terms, light interception is maximized when photosynthetic organs are oriented nearly horizontal to the sun. But this orientation also maximizes the bending forces in stems and leaves. When these forces exceed the strength of leaf and stem tissues, leaves and stems break. Therefore the orientation that maximizes light interception tends to be exactly the opposite of the orientation that minimizes the risk of mechanical (bending) failure. Over the course of their evolution, plants have managed to reconcile these and other conflicting phenotypic requirements by evolving "middle ground designs," that is, phenotypes that work well enough, though not as well as they conceivably could if only one task determined survival and reproductive success.

Although organisms are not machines, the relation between the number of functional tasks an organism must perform to survive and reproduce and the number of phenotypes that confer the same overall fitness is somewhat similar to the relation between the number of equally efficient designs for an engineered artifact and the number of tasks the artifact must perform. Engineering theory and practice show that as the number of tasks an artifact must perform increases, the number of equally efficient designs also increases (this will be discussed in detail in chapter 8). At the same time, the efficiency with which any task is performed decreases as the number of tasks increases. If these generalizations hold true for organisms, even in part, then as the number of biological tasks an organism must perform increases, the number of adaptive peaks on a fitness landscape will increase, while the heights (relative fitness) of these peaks will decrease. These expectations are intuitive, at least from an engineering perspective, because, on the one hand, each function likely has a specific design specification that maximizes efficiency in performing the task and, on the other hand, reconciliations among conflicting, highly specific design requirements cause each task to be performed less efficiently. Although there is no a priori reason organisms should subscribe to engineering theory and practice, it does seem reasonable to suppose that phenotypic rapprochements exist in the sense that the phenotype has come at some cost in terms of lower efficiency in performing its many biological tasks. If this supposition is true, and if it is also true that every type of organism is a "multifunctional device," then most suffer the same disadvantage to some degree, and so the evolutionary playing field is level for all.

Curiously, there may be an evolutionary advantage to being more biologically complex. Provided the lessons of engineering carry over into biology, the number of adaptive peaks in a fitness landscape will increase while the topography of the landscape becomes less severe as the number of biological functions an organism must reconcile increases. If so, then adaptive walks may become easier as complexity increases, in the sense that differences in the fitness of phenotypes become less pronounced. Regardless of how we try to visualize adaption, there are two necessary conditions for adaptation to occur: (1) selection must be able to distinguish among the individuals within a population, and (2) selection must not act within individuals. Indeed, these two stipulations go a long way to define what a biological "individual" really is.

Testing Adaptive Hypotheses

The most frequently used method of constructing a phylogenetic hypothesis is cladistic analysis. The principal objective of cladistics is to group species or higher taxa sharing a common ancestor. The hypothesis based on cladistic analysis is depicted as a bifurcating "tree" called a cladogram (Hennig 1966; Wiley 1981). The branch points on a cladogram are called nodes, and taxa linked to the same node are hypothesized to have a common ancestor. The branching topology of a cladogram is determined by the distribution of shared, derived characters among the taxa included in the analysis. Although systematists have an intuitive and practical grasp of what is meant by "character," numerous attempts have been made to formally define this term. At the risk of oversimplification, a character may be thought of as any heritable trait that serves to identify an organism. Each character must have two or more heritable states. For example, "floral symmetry" has at least two character states, "radial symmetry" and "bilateral symmetry." Which among alternative character states is "ancestral" and which is "derived" is the key issue cladistic analyses attempt to resolve. Ancestral and derived character states may be determined from developmental and historical information, or they may be inferred by designating the character states of each node in the cladogram that minimize the number of character transformations (character reversals) required to explain the observed patterning of character states

among descendant taxa. Either method serves to identify the character states of the ancestor shared by closely related taxa.

A fundamental principle governing the assignment of character
states to the nodes of a cladogram is *parsimony*, the assumption that
the smallest number of character state transformations needed to
construct a phyletic hypothesis most likely results in the most historically likely (simplest) phylogenetic reconstruction. In a sense Occam's
razor is used to uncover the evolutionary scheme of relations among
taxa that is most likely to be correct. However, character reversals can
occur and this is one of the important reasons why many features are
used to construct cladograms and that cladistic hypotheses are more
robust when they are based on large data sets. Nonetheless, the principle of parsimony presumes that the rate of overall character change
within lineages is slow relative to the rate of speciation. Only in this
way can a species retain a phyletic "memory" of its ancestral character
states. If the rate of character change is high relative to the number of
speciation events, the phyletic memory of a lineage will progressively
"decay" with time, making the retrieval of phyletic relations more difficult.

By way of illustration, suppose that the phyletic relations among
five species, A–E, are reconstructed based on the number of floral
parts, the extent to which they are fused, and the type of floral symmetry (fig. 3.13). Suppose A–E all have floral parts numbering five,
but that two species, say A and B, have radially symmetrical flowers
with unfused petals, while species C–E have bilaterally symmetrical
flowers with varying degrees of laterally fused petals. In the absence
of other characters, a parsimonious phyletic hypothesis is that A–E
shared a common ancestor that had flowers with unfused floral parts
arranged in fives (a pentamerous flower) and radial symmetry; A and
B shared a common ancestor with the same characteristics; that C, D,
and E shared a common ancestor that had pentamerous flowers with
unfused floral parts and bilateral symmetry; and that D and E shared
a common ancestor that had pentamerous bilateral flowers with fused

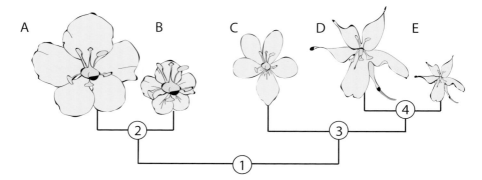

Ancestor 1: parts in fives + unfused parts + radial symmetry
Ancetsor 2: parts in fives + unfused parts + radial symmetry
Ancestor 3: parts in fives + unfused parts + bilateral symmetry
Ancestor 4: parts in fives + fused parts + bilateral symmetry

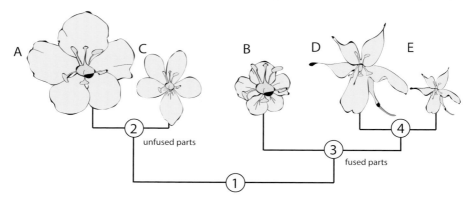

Ancestor 1: parts in fives + unfused parts + radial symmetry
Ancetsor 2: parts in fives + unfused parts
Ancestor 3: parts in fives + unfused parts + radial symmetry
Ancestor 4: parts in fives + fused parts + bilateral symmetry

Figure 3.13. Cladistic (phylogenetic) relationships among five hypothetical species (A-E) based on the characteristics of their flowers (upper cladogram) versus those based on mapping floral characteristics on a molecular-based cladogram (lower panel). The characteristics of each of the four last common ancestors (at nodes 1-4) as inferred by the two analyses are listed beneath each diagram. Note that inferences regarding the floral character states (for example, radial versus bilateral symmetry) change significantly depending on which cladogram one examines.

floral parts. But this cladogram is a "closed system" in the sense that there is no a priori way to infer that this hypothesis is correct. That is, there is no compelling reason to assume that species with radially symmetrical and unfused flowers are more ancient than species with bilaterally symmetrical flowers with fused parts. The inclusion of additional taxa could "open this system" to further analysis shedding light on whether radial symmetry and fused floral parts are the ancestral or derived character states. Nonetheless, because A and B are linked to the same node of the cladogram and because both have radially symmetrical flowers, the common ancestor to A and B is inferred to have the same character states. Likewise, the common ancestor of C–E is inferred to have had bilaterally symmetrical flowers with fused petals.

This phyletic hypothesis is intentionally crude. It is based on only a few characters (floral symmetry, fusion, and number of parts) and could dramatically change in light of how other characters (and their character states) are distributed among species A–E. Suppose that molecular data were available such that morphological features could be "mapped" onto the molecular-based phylogeny and suppose we found that species A and C shared a last common ancestor and that species B, D, and E shared a last common ancestor. In this hypothetical case (see fig. 3.13), we would have to change our hypotheses about the character states of the ancestors of the nested species in this cladogram.

The foregoing hypothetical case is intended to emphasize that cladistic (phylogenetic) hypotheses depend on the characters, the character states, and the species used in an analysis. The incorporation of new species and new data about them can result in significant changes in the inferences we draw about the ancestral states of organisms. Another concern is that the characters selected to construct a hypothesis may not be independent of those used to construct the adaptive hypothesis. The same danger applies to the characters used to construct the cladogram. We have already seen that the phenotype is the physical manifestation of coadapted gene complexes for which small genetic alterations can have potentially manifold or cascading effects. Gene hitchhiking is always a concern since some character transformations may be autocorrelated as a consequence of this phenomenon. Thus, even when constructing a cladogram for the sole purpose of inferring phylogenetic relations, the independence of characters and character states must be approached with caution. In terms of testing adaptive hypotheses, we must be aware of how far adaptive characters are associated with one another in addition to the rate of the branching episodes of speciation in a particular lineage (cladogenesis). Reconstructed phylogenies will shed light on adaptive hypotheses only when the selective associations among characters are weak and the rate of cladogenesis is very high. Notice that the test for an adaptive hypothesis is that the predicted historical sequence of adaptive characters complies with the history of speciation inferred from a cladogram. If the association among adaptive characters is very strong (presumably because of strong selection), adaptive characters will tend to appear concurrently at the same nodes in a cladogram. This may obscure the "sequence of appearance" of adaptive characters, with the curious result that adaptive hypotheses dealing with characters strongly associated owing to selection will tend to be rejected more frequently than hypotheses dealing with characters that are weakly associated.

Yet another caveat is that a phyletic pattern of speciation and extinction may exhibit what appears to be an "adaptive" evolutionary trend even when many species within a lineage counter the trend. Evolutionary trends of this sort may result when selection acts on emergent properties of species rather than on the properties of individuals (this is called *species selection*; see chapter 6). The species within a lineage that survive the longest may have a higher probability of producing new species than do species with shorter durations, and therefore the longest-lived species within a lineage may determine the direction of major evolutionary trends (fig. 3.14). (The differential speciation envisaged by this hypothesis is analogous to differential reproductive success in microevolution.) Regardless of whether species selection occurs, attempts to extract a single evolutionary procession of species from an incomplete fossil lineage is very much like trying to describe

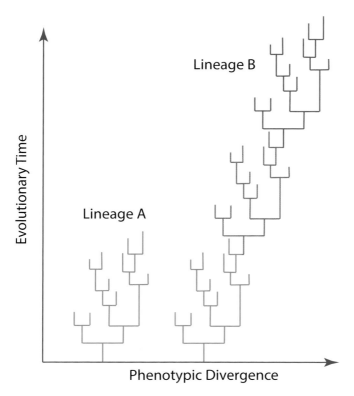

Figure 3.14. Schematic of a pattern resulting in differential selection and extinction that appears to manifest directional phenotypic evolution within a lineage of clade (species selection). Each vertical line in the diagram represents the history of a single species (termini denote species extinctions); each horizontal line depicts a speciation event. As a result of differential speciation and extinction, an evolutionary trend may appear and may be incorrectly construed as evidence for an adaptive evolutionary trend. In the example depicted here, lineage A appears not to manifest a trend in phenotypic divergence as the lineage evolves through time. In contrast, lineage B appears to evolve in the direction of greater and greater phenotypic divergence. Yet, as shown by the different colored lines, lineage B is reiterating the same pattern as lineage A.

the shape of a tree by tracing the branching pathways from a few twigs to the base of the trunk (the perception of a tree's shape is governed by both the number and the position of the terminal twigs selected for study). In like manner, in cladistic analyses incorporating fossil species, including or excluding certain species can produce different phylogenetic hypotheses.

Suggested Readings

Andersson, S., and R. G. Shaw. 1994. Phenotypic plasticity in *Crepis tectorum* (Aster-
    aceae): Genetic correlations across light regimes. *Heredity* 72: 113–25.
Bonner, J. T. 1988. *The evolution of complexity.* Princeton, NJ: Princeton University
    Press.
Darwin, C. 1859. *On the origin of species by means of natural selection.* London: John
    Murray.
Hamilton, W. D. 1964. The genetical evolution of social behavior, I & II. *J. Theor. Biol.*
    7: 1–52.
———. 1970. Selfish and spiteful behavior in an evolutionary model. *Nature* 228:
    1218–20.
Hammond, S. T., and K. J. Niklas. 2009. Emergent properties of plants competing
    in silico for space and light: Seeing the tree from the forest. *Amer. J. Bot.* 96:
    1430–44.
Hartl, D. L., and A. G. Clark. 1997. *Principles of population genetics.* Sunderland, MA:
    Sinauer.
Hennig, W. 1966. *Phylogenetic systematics.* Urbana: University of Illinois Press.
Knoll, A. H. 2003. *Life on a young planet: The first three billion years of evolution on
    earth.* Princeton, NJ: Princeton University Press.
Mable, B. K. 2004. "Why polyploidy is rarer in animals than in plants": myths and
    mechanisms. *Biol. J. Linnean Soc.* 82: 453–66.
Mayr, E. 1963. *Animal species and evolution.* Cambridge, MA: Harvard University
    Press.
Scharloo, W. 1991. Canalization: Genetic and developmental aspects. *Ann. Rev. Ecol.
    Sys.* 22:265–94.
Schmalhausen, I. I. 1949. *Factors of evolution: The theory of stabilizing selection.*
    Chicago: University of Chicago Press.
Schopf, J. W. 1999. *Cradle of life.* Princeton, NJ: Princeton University Press.
Simms, E. I., and M. D. Rausher. 1989. The evolution of resistance to herbivory
    in *Ipomoea purpurea.* II. Natural selection by insects and costs of resistance.
    Evolution 43:573–85.
Sprague, G. F. 1967. Plant breeding. *Ann. Rev. Genet.* 1: 269–94.
Sun, G., Q. Ji, D. L. Dilcher, Q. Zheng, K. C. Nixon, and X. Wang. 2002. Archaefructa-
    ceae, a new basal angiosperm family. *Science* 296: 899–904.
Van Tienderen, P. H. 1991. Evolution of generalists and specialists in spatially
    heterogeneous environments. *Evolution* 45: 1317–31.
Van Valen, L. 1975. Life, death, and energy of a tree. *Biotropica* 7: 259–69.
Via, S., and R. Lande. 1985. Genotype-environment interaction and the evolution of
    phenotypic plasticity. *Evolution* 39: 505–22.
Waddington, C. H. 1942. Canalization and development and the inheritance of
    acquired characters. *Nature* 150: 563–65.
Wallace, A. R. 1889. *Darwinism: An exposition of the theory of natural selection, with
    some of its applications.* New York: Macmillan.
Wallace, B. 1981. *Basic population genetics.* New York: Columbia University Press.

Weis, A. E., and W. L. Gorman. 1990. Measuring selection on reaction norms: An explanation of the *Eurosta-Solidago* system. *Evolution* 44: 820–31.

Wiley, E. O. 1981. *Phylogenetics: The theory and practice of phylogenetic systematics*. New York: John Wiley.

Woese, C. R. 1987. Bacterial evolution. *Microbiol. Rev.* 51: 221–71.

4

Development and Evolution

"It seems, as one becomes older,
    That the past has another pattern, and ceases to be
    a mere sequence—
    Or even development: the latter a partial fallacy
    Encouraged by superficial notions of evolution,
    Which becomes, in the popular mind, a means of
    disowning the past."
    −T. S. ELIOT, *The Dry Salvages* (1941)

The goal of this chapter is to explore plant evolution from an evolutionary-developmental perspective. As we will see, this perspective employs a hierarchical approach that emphasizes the need to integrate the interactions of many different levels of biological organization. Historically, much of biology sought to decompose complexity into its constituent parts—organisms were studied as a collection of organs, organs as a collection of tissue types, etc. But it soon became apparent that these constituent parts needed to be concomitantly reassembled and integrated to answer the major questions and problems in biology. Both approaches—the reductionist and the constructionist approach—are necessary when dealing with hierarchical systems because we can reassemble and understand the whole only if we understand the parts and because the functions of the parts depend on how the whole is constructed. As we will see in this chapter, this bio-egalitarian view of life is nowhere better seen than when we examine how genes, gene regulatory networks, gene products, and intrinsic and extrinsic signaling systems operate to assure the normal course of growth, development, and reproduction.

Nevertheless, uncovering the many synergistic layers of development and learning how they are integrated in time and space remain daunting challenges even in the thrilling age of molecular biology. It is almost tautological to say that the metabolic and genomic "machinery" of development is exquisitely complex, especially in light of the concurrent molecular signaling required to keep this machinery operating in a coordinated manner across the many levels of biological organization. For example, we have long known that hormone signaling elicits rapid cellular responses predicated on preexisting metabolic and physical conditions and that it can also evoke delayed differential gene transcription, which can dramatically alter preexisting conditions. However, developmental signaling has been shown to involve the translocation of messenger RNAs within the living substance of cells (symplastic translocation) to targeted sites of action as well as numerous interactions among a plethora of very small and thus difficult to detect molecular species, as for example polyamines and protein-folding chaperones (FK506-binding proteins). Indeed, as biologists continue to explore the multifarious lines of plant molecular communication, the emerging picture is becoming more and more complicated.

Various attempts have been made to explain this complexity in ways that are both conceptually robust and easily communicated without loss of information. In this respect, many biologists are converging on the pictorial and analytical conventions used by engineers and computer scientists who regularly deal with large networked systems. Specifically, signal pathways are being rendered in ways that are strikingly similar to logic circuits "hot-wiring" signal-activated metabolic or genomic subsystems. The circuit/subsystem paradigm has merit. Seemingly insuperably complex phenomena can be diagrammed or predicted. As yet undetected components in these diagrams can be identified, often by visual inspection. Missing elements can be sought either experimentally or analytically. Complete diagrams can be modeled mathematically and manipulated to predict the consequences of changing one or more circuit or subsystem components, and the predicted responses can be evaluated by experimental manipulation of real biological systems. Finally, once verified, models for subcellular and cellular developmental phenomena can be grouped by progressive combination into larger models treating the behavior of higher levels of organization.

This paradigm is also attractive when viewed from the perspective of evolutionary biology. In theory, ancestral patterns of development can be "switched off" by suppressing one or more signals or eliminating one or more systems just as they can be evolutionarily transformed by establishing new signals or systems, or redirecting old ones. Genomic changes are required in either case, but large portions of the machinery of development can remain intact and lie dormant for generations only to be resurrected from the "dead" unless corroded by extensive random mutations. Indeed, when development is viewed in terms of its "logic" and "machinery," Dollo's law, which states that complex morphological features are unlikely to be evolutionarily regained in a clade once they are lost, appears terribly misleading, because it ignores the propinquity and likelihood of repairing ancient developmental signaling pathways. Nonetheless, the use of the circuit/subsystem paradigm in biology is not well established nor is it without potential pitfalls. Therefore, as auditors of morphogenesis, we must understand how logic circuits and signal-activated subsystems work. It is also necessary to evaluate their utility when confronted with real developmental phenomena.

This chapter is outlined around two goals. The first is to provide a conceptual review of logic circuits and signal-activated subsystems in a manner that can be wedded to our current understanding of molecular developmental biology. The second goal is to explore a few selected developmental phenomena to look for major themes in developmental biology as they pertain to plants and other eukaryotic organisms.

Development as a Logic Circuit

Developmental systems have many features in common with one another and thus appear to follow some general rules submerged in a sea of details that can obscure their generalities. Regardless of their evolutionary history, they receive and process information by responding to external and internal signals, and they typically compensate or at least respond adaptively when these signals deviate from the norm. In this respect, developmental systems appear to be "wired" very much like electrical circuits are wired to shuttle and coordinate signals. This metaphor conveys the notion that developmental systems use signals that must be processed in a predictable, *bio*-logical manner in much the same way logic gates or logic circuits process input signals to obtain output signals. This metaphor is instructive in many ways. For example, very simple logic circuits can result in astronomically large combinatorial numbers of output signals (which raises questions about simple versus complex systems), and very different logic circuits can process the same input signals to obtain the same results (which raises questions about whether homology resides at the molecular or morphological level of organization).

Although a synoptic treatment of logic circuits and how they work is well beyond the scope of this chapter, the concept and its implications are more important. However, at a very basic level, there are only two kinds of logic circuits: combinatorial circuits, in which the output signal depends exclusively on the instantaneous value of the input signal, and sequential circuits, in which the output signal also depends on the history of prior inputs. Both types conceptualize a signaling pathway as an electrical circuit containing one or more switches. The "logic" of a circuit is the formal algorithm that describes the conditions

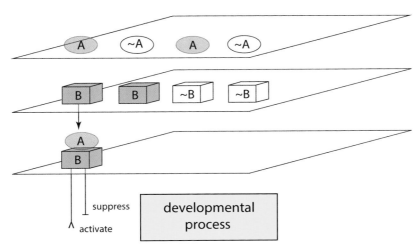

Figure 4.1. Schematic of a hormone-mediated signal transduction pathway that requires a single hormone A and a single receptor molecule B to become operational to either activate or suppress a developmental process. In the parlance of a logic circuit, this pathway has two switches (A and B) that must be turned on (denoted as A · B). If either switch is turned off (A · ~ B or ~A · B), or if both are turned off (~A · ~B), the pathway is turned off. Note that the signal transduction pathway can either suppress or activate the developmental system in this diagram. For a biological example, see fig. 4.5.

(logical propositions) that dictate whether a signal passes through a circuit. These conditions are depicted as "switches" (denoted by letters A, B, . . . , Z) each of which is either "on" if its condition holds true or "off" if its condition is not true (for example, A and ~A, respectively).

Consider a hormone-mediated signal transduction pathway (fig. 4.1). The hormone A and its receptor molecule B are required to switch on a signal transduction pathway (that is, a series of operations that evokes the response the signal is intended to produce). In this simple example, there are two initial switches (biological conditions) that govern the passage of the pathway of the signal, and both must be turned on for the pathway to operate. Symbolically, these conditions are A · B, where · denotes "and." If either switch is turned off (A · ~ B or ~A · B), or if both are turned off (~A · ~B), the signal transduction pathway is closed.

Parallel and serial circuits exist (fig. 4.2). The latter are not uncom-

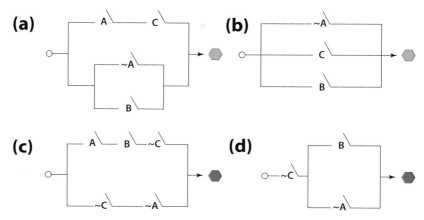

Figure 4.2. Four representative parallel logic circuits that illustrate how an input signal can evoke an output signal. Each circuit is a metaphor for how a developmental input signal (a hormone, compression stimulus, transcription factor, etc.) can evoke a series of actions/reactions (shown as switches) that ultimately gives rise to an output signal (cell wall loosening, cell division, gene activation or repression, etc.). In each circuit, the input and output signals are depicted as open circles and colored hexagons, respectively. Switches are individually identified by capital letters. Open switches are depicted as _; closed switches are identified by ~ placed before a capital letter. Although these circuits may appear to be different from one another, circuit (a) is logically equivalent to circuit (b), and circuit (c) is logically equivalent to circuit (d), in terms of the ability of a circuit to produce a given signal. That is, circuits (a) and (b) have the same signal output repertoire, and circuits (c) and (d) have the same signal output repertoire. Thus, in theory, very different switching (developmental) systems can rely on the same input signals to produce the same or different output signals.

mon, particularly in some biosynthetic pathways such as the flavonoids. However, they are rather simple. In contrast, parallel circuits are far more interesting because they provide manifold responses to the same signal depending on current conditions, since parallel circuits allow an initial input signal to flow through two or more circuit pathways, permitting two or more output signals at each terminus and thus, theoretically, four or more combinatorial responses from the machinery the circuits operate. Consider a parallel circuit constructed with three switches, A, B, and C. This trifurcate signal pathway (symbolized as A v B v C, where v denotes "or") has eight possible combinatorial responses—that is, $2^{N=3} = 8$, where N is the number of switches. Responses coordinated by parallel logic circuits can produce continuous variation in response to the passage of a single input signal if (1) they contain even a modest number of switches, (2) some switches activate or suppress other switches, (3) a logic circuit has two or more input signals, or (4) the output signals interact combinatorially. For example, there are 1,024 possible responses when $N = 10$, and there are 1,048,576 possible responses when $N = 20$. Finally, if on-off switches respond to more than one signal, the number of possible output signals S is given by the formula $S = 2^{2^N}$, where N is the number of input signals to which switches can respond. For example, if there are eight input signals to which switches respond, the number of output signals exceeds 1.0×10^{77}, which is more than the number of organisms that have ever existed on Earth.

Although a complex logic circuit can be simplified mathematically, four caveats are evident when biological systems are approached in this way: (1) there is no a priori method to determine which among logically equivalent circuits is biologically real, (2) incomplete signaling pathways may appear to "work" when diagrammed (missing components are not invariably obvious), (3) parallel logic circuits may obtain invariant output signals that give the appearance that input signals pass through serial switches, and (4) nothing in a logic circuit per se indicates how long a switch is turned on or off, how long a genomic or metabolic product lasts (the temporal components of signaling are lost), or whether a switch "makes a good connection." This last point is important because most developmental phenomena are perceived as a series of continuous transformational events, whereas the logic circuit paradigm operates using "either/or" propositions. This apparent paradox results from the different time-scales over which development and its molecular signaling systems operate. Provided that a complex binary system operates combinatorially over short timescales, it can appear to be seemingly seamlessly transformative, much as the plucking of harpsichord strings creates a melody (and, whereas the number of strings is finite, the number of possible melodies is effectively boundless).

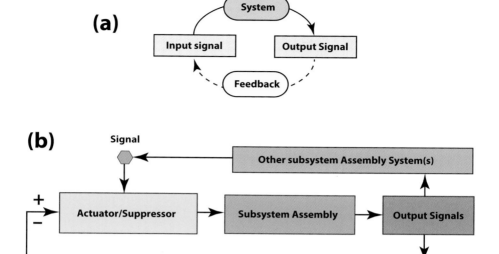

Figure 4.3. Schematics of a simple signal (error)-activated system in which the output signal is used to modulate an input signal. a. At the simplest level, input-output signals comprise a loop with a processing system between the two signals that relays the signal and a feedback mechanism that conveys the output signal. b. When conceptually explored more deeply, the signal-activated system has four essential components: (1) an actuator/suppressor to convert an external signal into an internal signal, (2) the assemblage system that is controlled (the subsystem assembly), (3) one or more feedback elements to direct the immediate output signal back to (4) a comparator to measure the difference (error) between the actual and the desired output of the subsystem.

Signal-Activated Subsystems

Even the most casual review of logic circuits shows that seemingly very simple logic circuits are more than sufficient to produce a large number of different combinations of signals. However, taken in isolation, a "signal" has no meaning unless it is interpreted or used in some fashion. Therefore, to be useful, logic circuits must be wedded to some sort of system or group of subsystems they supervise. The simplest system is a signal-activated system that is error activated. That is, the output signal is used to modulate the input signal (fig. 4.3a). This feedback loop configuration has four essential components: (1) a comparator to measure the difference (error) between the actual and the desired output of the subsystem, (2) an actuator/ suppressor to convert the error-signal into an internal signal, (3) the actual machinery or assemblage that is controlled (the subsystem assembly), and (4) a feedback element to direct the immediate output signal of the assemblage back to the comparator (fig. 4.3b). In some respects, the feedback element is the most important of these four components, because its influence (and thus presence) can be rapidly diagnosed, thereby permitting an experimentalist to determine whether a particular biological phenomenon is appropriately treated as an error-activated subsystem. Technically, feedback is defined as that property of a closed-loop system that permits the comparison of the output signal (or some other variable controlled by the subsystem) to the input of the subsystem (or an input to some other internal component) so that the control action is some function of the input-to-output ratio. Importantly, the linkage of two or more subsystems confers two emergent properties: (1) the ability to achieve global stability, and (2) recursive combinatorial regulation, which confers homeostasis. Much like an open chemical system, any network composed of two or more error-actuated subsystems rapidly achieves a steady state regardless of its initial conditions (such as reactant-product concentrations). Likewise, recursive combinatorial regulation is important. This property refers to the synergistic feedback signaling of numerous components that can permit a network to repeatedly cycle through a programmed series of transformations.

These features are illustrated by returning to the logic circuit paradigm and a simple circuit consisting of three switches (1, 2, and 3), one of which is governed by an "and" function (switch 1) and two of which are governed by "or" functions (fig. 4.4). In this example, each switch receives signals from the other two, examines the combined input signals, checks its logic "rules," and assumes its designated state in synchrony with the other two switches to achieve an instantaneous configuration. This process is repeated during which each configuration updates all the switches such that recursive cycles reconfigure

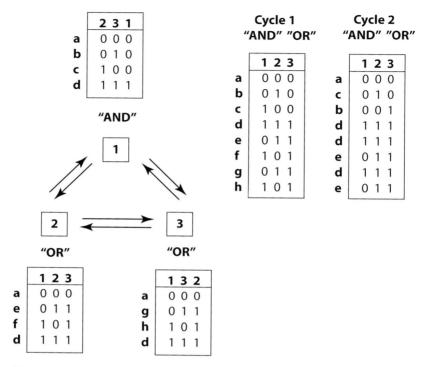

Figure 4.4. Emergent properties of networked systems illustrated by a logic circuit consisting of three switches (1, 2, and 3), one governed by an "and" function (switch 1) and two governed by "or" functions (switches 2 and 3). Left: Each switch assesses the condition of the other two, and alters its binary state as indicated in the rectangular tabulations where the response of a particular switch is shown in the left-hand column and the activity of the other two is indicated in the other columns (read each line from left to right). Right: Each of the eight different combinations are denoted by a–h. The switch-permutations in the first cycle of operation (Cycle 1) evoke a second cycle of combinatorial reassessment (Cycle 2).

the responses of the system driven by the circuit. In this way, the circuit-driven system achieves a predictable series of transformations (fig. 4.4). This system is deterministic in the sense that the series of transformations it achieves is finite and repetitive (at some point the input signals return to their initial conditions and the entire series of transformations is repeated). An obvious analogy is the cell cycle.

Notice that in this three-switch example, there is no single "master" switch once the system is set into operation. The network governed by the three-switch logic circuit as a whole requires "initiation"—that is, one switch must receive a signal and temporarily function as an epistatic actuator. However, when switched on or off, this switch sparks the operation of the entire network, suppressing or activating one or more of the other two switches. Indeed, an "incompleteness theorem" governs the working of networked systems. The operation of a subsystem cannot be fully diagnosed in isolation of the operations of the other subsystems to which it is networked. This theorem can be proved mathematically, but it is illustrated by the example presented in fig. 4.4.

The incompleteness theorem is also illustrated by considering the biology of indole-3-acetic acid (IAA) as a signaling molecule. Although not the authors of its chemical designation, Charles and Francis Darwin (1848–1925) are widely credited to be the first to hypothesize that the growing tips of plants responded to the direction of light by means of some sort of chemical messenger. In 1913, Peter Boysen-Jensen (1883–1959) subsequently showed that this chemical was mobile and not transfixed. In 1926, Frits Went (1903–1990) named the chemical auxin and pronounced it to be a growth-promoting plant hormone. Thousands of papers have been published since then, and yet we still know comparatively little about how IAA signaling evokes the many genomic and developmental responses that it does in the flower plants and many other plant lineages (table 4.1). Indeed, it appears to be involved in almost every facet of plant growth, which evokes the quaint aphorism "If it does everything, it does nothing." For this and other reasons, IAA provides an excellent entry point with which to explore plant development.

In the following sections, we will explore (1) how this signal is transported, (2) the IAA transduction pathway, and (3) the developmental subsystems it actuates. We will also consider how these subsystems are networked by examining IAA-mediated cell wall loosening, which is essential to plant growth and development.

Table 4.1. Examples of developmental processes affected by IAA-mediated signaling

Positive phototropism	The lateral transport of IAA stimulates differential cell growth and leads to curvature toward the higher light intensity.
Negative gravitropism	Experiments indicate that IAA is produced by the root cap and a root growth inhibitor involved in the downward growth of root apical meristem.
Apical dominance	The transport of IAA downward from the shoot apical meristem inhibits the growth of axillary buds (a phenomenon called "apical dominance").
Delay of leaf abscission	IAA transport from the leaf lamina inhibits the formation of the abscission layer. Leaf senescence is associated with a drop in IAA levels that stimulated the formation of this layer of cells.
Floral bud development	The polar transport of IAA in the inflorescence is required for normal floral bud development in *Arabidopsis*.
Vascular differentiation	High IAA concentrations induce the formation of xylem and phloem.
Phyllotaxic patterns	The patterns of IAA efflux and influx within the shoot apical meristem correlate with the patterns of leaf primordial formation and thus leaf patterns (phyllotaxy).

Note. IAA, indole-3-acetic acid.

IAA Transport

Indole-3-acetic acid (IAA) provides an excellent example of a signaling molecule for four reasons: (1) it is utilized as a signaling molecule by diverse organisms including bacteria and the red and brown algae as well as the Viridiplantae, (2) endogenous IAA is involved in a vast range of developmental processes (cell wall loosening and expansion, phototropism and gravitropism, shoot apical meristem dominance, lateral and adventitious root formation, leaf abscission, floral bud and fruit development, and the differentiation of the vascular tissues), (3) IAA biosynthesis, directional transport, signal perception, and transduction, and cell- or tissue-specific responses triggered by

IAA have been extensively studied at the molecular, cellular, and organismic levels, and (4) the interrelations among these hierarchical levels involve numerous protein complexes operating at different developmental levels that serve as a paradigm for plant development and evolution in general.

Auxin influx occurs either by the passive diffusion of the protonated form (IAAH) across the cell membrane, or by the active transport of the dissociated, anionic form (IAA–) by a permease $2H^+$-IAA co-transporter, as for example AUX1. Once within a cell, IAA is transported in a polar or in a non-polar manner. Auxin polar transport involves an IAA-influx carrier protein encoded by the *AUX1* gene, whereas IAA efflux involves the activity of at least two membrane-bound and membrane-associated proteins. One of these is a trans-membrane transport protein encoded by members of the *PIN* gene family that encode for PIN proteins with ten trans-membrane segments and a large hydrophilic loop. The second category of proteins called IAA-inhibitor-binding proteins perform a regulatory function in response to endogenous, naturally occurring substances, such as flavonoids. These proteins, which were originally described as NPA-binding proteins, are ATP-dependent transporters belonging to the phosphoglycoprotein B subclass of the large superfamily of ATP binding cassette (ABC) integral trans-membrane transporter proteins. The ABCB proteins function as ATP-dependent amphipathic anion carriers involved in auxin efflux. One class of permeability glycoproteins (PGPs) is represented by the AtPGP1 catalyze auxin export, while another class with at least one member, AtPGP4, appears to function in auxin import. Evidence that dephosphorylation and phosphorylation regulate the activity of IAA efflux carrier proteins comes from a variety of sources and indicates that these mechanisms can be used to regulate how and where IAA can be transported to various plant cell- or tissue-types. For example, the *ArabidopsisrootscurlinNPA1* gene (*RCN1*) encodes a regulatory subunit of proteinphosphatase 2A (PP2A) that is expressed in seedling root tips, lateral root primordia,

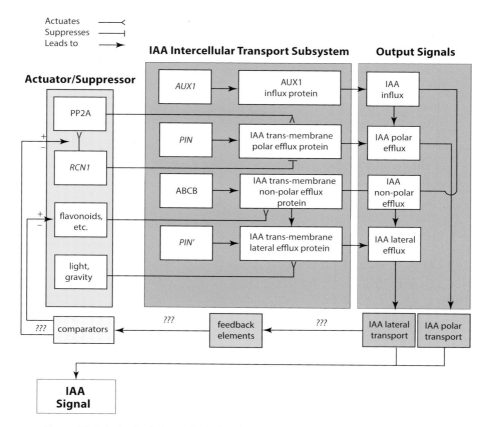

Figure 4.5. A logic circuit sign-activated rendering of indole-3-acetic acid (IAA) lateral and polar transport within multicellular land plants based on studies showing that IAA polar transport involves an IAA-influx protein carrier encoded by the *AUX1* gene, whereas IAA efflux involves the activity of at least two membrane-bound and membrane-associated proteins. One of these is a trans-membrane transport protein, which is encoded by members of the *PIN* gene family. The second category of proteins is the ABCB transporter proteins. In terms of the incompleteness theorem, notice that a number of internal and external signals participate as actuators and suppressors, as for example PP2A (protein phosphatase 2A) and light and gravity.

the pericycle, and the stele. The *rcn1* mutant exhibits reduced PP2A activity and defects in IAA-mediated responses involving anisotropic cell expansion.

These and other details are summarized in fig. 4.5, which offers a logic circuit as a schematic for IAA lateral and polar transport in the form of a signal-activated subroutine. This schematic is by no means complete in every detail, and there are many aspects of IAA transport that are still actively debated and thus require experimental exploration. This rendering is presented for two reasons only: (1) it provides a concrete example of how a developmental phenomenon can be diagrammed as a logic circuit signal-activated system, and (2) it illustrates the incompleteness theorem (notice that a number of external and internal signaling systems participate as actuators or suppressors, and that the feedback elements and comparators in this system remain unknown or are problematic).

IAA and Cell Wall Loosening

As noted, without some concrete examples, the application and implications of logic circuits and signal-activated subsystems to understand development appear to be theoretical. To correct this, we rendered the lateral and polar transport of indole-3-acetic acid (IAA) in multicellular land plants as a logic circuit and signal-activated subsystem (see fig. 4.6). However, this diagram says nothing about how IAA functions as a signal. Accordingly, we can now consider how the IAA signal participates in signal transduction pathways by way of illustrating its involvement in cell wall loosening. This process is particularly relevant to the topics treated in this chapter because it is the mechanism by which plant cells grow in size, which is a requisite for plant development and morphogenesis.

The mechanisms for IAA perception and response are poorly understood in part because more than one response pathway may be activated or repressed by a single signal-receptor mechanism. However, IAA-induced rapid responses, such as cell membrane hyperpolarization, help to identify the mechanism of IAA perception, whereas delayed IAA-mediated responses, such as gene transcription and cell expansion, help to evaluate long-term responses. Studies of rapid IAA-mediated responses have identified a IAA-binding protein (ABP1), which is identified to contain an endoplasmic reticulum lumen-retention sequence, which provides some evidence that IAA perception and signal transduction do not reside exclusively on the cell membrane, but also within the cell.

Technologically divergent studies provide convincing evidence that ABP1 mediates cell perception of IAA and subsequent signal transduction. Cell membrane hyperpolarization assays in tandem with antibodies directed against ABP1 indicate that the native protein rapidly transduces the IAA-mediated signal to effect membrane electrical potential. This and other lines of evidence indicate that IAA binds to and changes the membrane-bound ABP1 configuration, which then interacts with docking proteins or ion channels to transmit a signal rapidly, as for example the rapid upgrade of electrogenic plasma membrane-bound ATPase activity, the acidification of the cell wall, and the subsequent activation of one or more cell wall polysaccharidases to cleave load-bearing cell wall bonds allowing for plastic and thus permanent cell wall expansion (fig. 4.6 a). Delayed IAA-mediated signal transduction events are more complex and thus less well understood (fig. 4.6 b). Long-term increases in ribosomal RNA and ribosome accumulation correlate with a three- to eight-fold increase in translatable ribosomal mRNA, and as many as 40 mRNAs may be up- or down-shifted within 5 h after treatment with IAA. Some of these gene products may be involved in the synthesis of cell wall components, whereas others may mediate the vesicular transport and delivery of these components to the cell wall.

Once IAA enters a cell it is capable of modulating the transcription of a large number of genes. The most intensively characterized IAA response cascade involves proteins transcribed by three genes: *auxin/idole-3-acetic acid (AUX–IAA)*, *auxin response factor (ARF)*, and *transport inhibitor response1/auxin signaling f-box protein1–3 (TIR1-AFB1–3)* (fig. 4.6 b). At low IAA concentrations, AUX–IAA form dimers with ARF transcription factors that bond to consensus TGTCTC auxin-repressive promoter elements. These AUX–IAA/ARF protein complexes block the transcription of early IAA response genes. At higher concentrations, IAA interacts with *TIR1-AFB1–3* receptors, which are an integral part of the *SKP1-CULLIN-F-BOX PROTEIN (SCF)* complex,

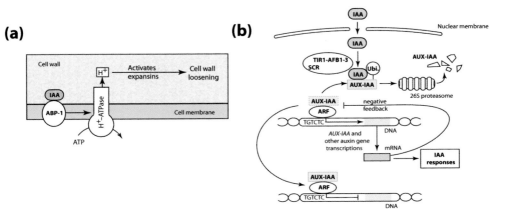

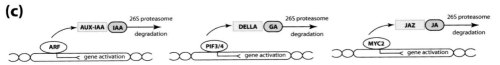

Figure 4.6. Schematics of indole-3-acetic acid (IAA) rapid responses, such as cell membrane hyperpolarization, such as that associated with cell wall loosening (a), delayed IAA-mediated responses, such as gene transcription (b), and the similar logic in activating gene expression by degrading a transcriptional repressor (c). a. Schematic for ABP-1/IAA binding activation of V-type H⁺-ATPases that promote conformational changes in cell wall proteins like extensins that lead to wall loosening. b. A model for IAA-mediated gene transcription that involves proteins transcribed by three genes (*AUX-IAA*, *ARF*, and *TIR1-AFB1-3*). At high concentrations, IAA interacts with TIE1-AFB1-3 receptors (which are part of the SCF complex) and targets bound and ubiquinated AUX-IAA for degradation by the 26S proteasome. At low concentrations, AUX-IAA proteins form dimers with ARF transcription factors bound to auxin-repressive promoter elements (TGTCTC). These AUX-IAA/ARF protein complexes block the transcription of early auxin response genes. In conjunction with a negative feedback loop due to the expression of *AUX-IAA*, early auxin gene expression results in a variety of cell- or tissue-specific IAA responses (such as cell wall loosening). See text for additional details. c. Schematic comparison of three plant hormone mediated signaling systems (indole-3-acetic acid, IAA; gibberellic acid, GA; and jasmonic acid, JA), the transcription repressors they affect (AUX-IAA, DELLA, and JAZ), and the transcription factors that are activated (ARF, PIF3/4, and MYC2) once the repressor is ubiquitinated and degraded by the 26S proteasome.

and targets bound and ubiquitinated *AUX–IAA* for degradation by the 26S proteasome, thereby activating *ARF*, which activates early IAA gene expression. Early IAA gene expression obtains a variety of cell- or tissue-specific IAA responses, along with a negative feedback loop due to the expression of *AUX–IAA*. It is worth noting that the logic circuit of the IAA signal cascade is reiterated in other plant hormone signaling cascades, each of which involves the degradation of a transcription factor repressor by means of ubiquitination and subsequent 26S proteasomal degradation (fig. 4.6 c).

The signal transduction pathway for cell wall loosening (and IAA transport) is rendered in fig. 4.7 as a logic circuit signal-activated subsystem, which is wedded to the logic circuit signal-activated subsystem for IAA lateral and polar transport. This circuit is probably very ancient, extending back to the earliest green algae in the case of the land plants. The membrane-bound ABP1-IAA conjugate is diagrammed as the actuator/suppressor switch for ATPases. Once activated, the cell wall is acidified, cell wall bonds are relaxed or broken (a family of proteins, called expansins, are involved in this process), and an increase in turgor pressure resulting from an influx of water drives cell expansion (not shown). The ABP1-IAA switch is also diagrammed to trigger delayed cytoplasmic and genomic responses involving the synthesis and delivery of cell wall components. This logic circuit diagram shows that sustained osmoregulation and cell wall loosening are required for continued cell expansion. The diagram also shows that the feedback loop and comparator for the output signal of the cell expansion machinery is unknown and must be sought experimentally. The IAA degradation, the down-regulation of solute concentrations, the synthesis of new cell-wall-binding polymers, the reorientation of cellulose microfibrils, the deposition of secondary wall layers, and the degradation of wall-loosening enzymes are among the many viable candidates for these missing network components.

The Subsystem Incompleteness Theorem

Thus far, we have considered how developmental processes can be conceptualized as logic circuits "wiring" signal-activated subsystems as illustrated by IAA-mediated induction pathways and cell wall loosening. The primary objective of this exercise was to illustrate the in-

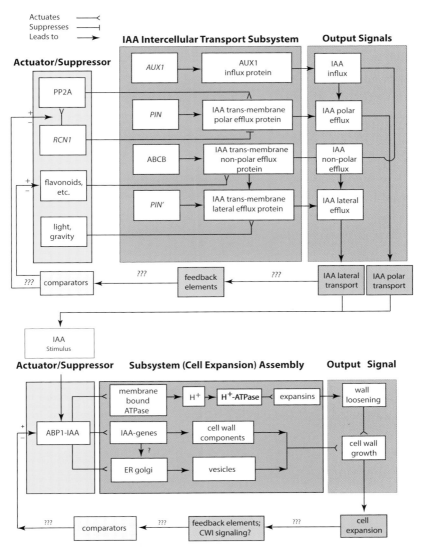

Figure 4.7. Logic circuit and signal-activated subsystem for lateral and polar IAA transport (see fig. 4.6) wedded to a logic circuit and signal activated subsystem for cell wall loosening and expansion. Comparable components in the two signal-activated subsystems are color coded. See text for details.

terconnectivity of developmental processes that manifest a collective phenomenology. As we have seen, although we can treat each component separately as a subsystem, we cannot understand the collective by this reductionist approach. If this lesson has escaped attention, consider that cell wall loosening and growth in size, which is arguably one of the most fundamental aspects of plant growth and development, involves numerous other subsystems yet to be treated (or understood). For example, the cells of eudicots are glued together by pectins, and studies of flowering plants show that methylesterases and Ca^{2+} affect the mechanical properties of pectins, alter the degree of cell wall hydration, and thus control to some degree the ability of cells to expand.

Pectins form a functionally and structurally diverse class of galacturonic acid-rich polysaccharides that can undergo significant modifications in their physicochemical properties. Recent attention has focused on homogalacturonan demethylesterification catalyzed by the ubiquitous enzyme pectin methylesterase (PME) as an important component in the control of cell wall hydration, expansion, and growth. Reconstructions of systems composed of cellulose and pectin, using cellulose-producing bacteria, document that pectin enhances the extensibility of the composite, even after the removal of these "soft" wall polymers. Several studies have shown that cross-linking of pectin subdomains by boron is necessary for wall biogenesis and organ growth. Perhaps more important is the fact that dynamic modifications of homogalacturonan (HG), a component of the pectins within the matrix of the wall, play important roles during cell growth (fig. 4.8).

Figure 4.8. (*opposite page*) Schematic for the relationships among three logic circuits and the signal-activated subroutines they wire related to the transport of an IAA signal to mediate cell wall loosening. Note that feedback loops for some of the components are not diagrammed because they remain unknown. Nevertheless, these loops must exist since components in these developmental processes must have a "stop" as well as a "go" signal. Comparable components in each of the three subsystems are color coded. See text for details.

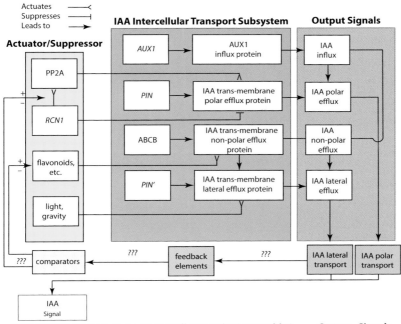
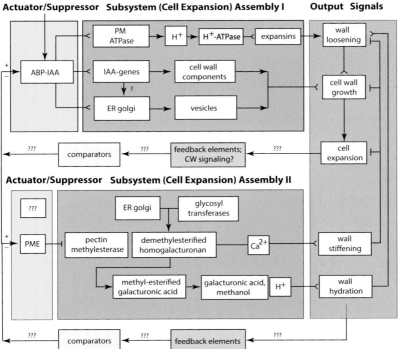

Specifically, HG is polymerized in the cytoplasmic Golgi apparatus by glycosyl transferases. After substitution with methyl groups at
the C6 position, the material is deposited into the walls in a methyl-
esterified state. Once outside the cell, PME can remove methyl units,
which releases free carboxylic acid groups, methanol (which can be
consumed by epiphytic bacteria), and protons. Depending on pH, ion
availability, and other physiological factors, PME activity also results
in a variety of different methyl esterification configurations. For example, in hypocotyls and pollen tubes, a low percentage of pectin
methyl esterification is correlated with a decline in cell wall extensibility and the cessation of elongation growth. In addition, Ca^{2+}-mediated
cross-linking of dimethyl-esterified pectin may contribute to cell wall
stiffening by reducing the ability of cellulose microfibrils to shear in
the cell wall matrix.

Alternatively, pectin dimethyl esterification can also provide
a mechanism for wall loosening by facilitating the degradation of
homogalacturonan via polygalacturonase. What can be said with some
certainty is that dimethyl esterification expedites cell wall hydration
and decreases cell wall pH, both of which can significantly alter the
mechanical properties of cell walls. The physiological versatility of
PME-mediated dimethyl esterification appears to be ancient among
the chlorobionta, since HG methyl esterification epitopes are reported
to exist in unicellular charophycean algae (such as *Penium margarita-
ceum*).

All of the preceding is summarized as a logic circuit signal-activated
subsystem that is wedded to a comparable diagram for IAA-mediated
cell wall loosening and IAA lateral and polar transport in fig. 4.8. For
example, sustained osmoregulation is required to produce turgor pressure, which provides the motor force in cell wall expansion, and all of
these processes must be regulated in ways that are compatible with
the cell cycle. The purpose of this diagram is merely to emphasize,
once again, that each developmental subsystem thus far discussed
is part of a greater whole about which we continue to discover new
interrelationships and new components. An additional feature that

may not be immediately apparent is that the linkages among these subsystems produce emergent phenomena, as for example the ability of cells to slide past one another as the pectins holding cells together undergo a solid-to-viscous phase transition.

The Gene Regulatory Network Paradigm

The role of transcription factors as "master regulatory switches" in gene regulatory networks is a pervasive paradigm in the literature discussing developmental biology. This paradigm draws attention to at least six molecular mechanisms for phenotypic evolution: (1) gene array duplication and subsequent sub-functionalization, (2) changes in the spatial expression patterns of preexisting gene arrays, (3) homeodomain protein sequence alterations, (4) modifications of DNA binding domains, (5) alterations in downstream regulated gene networks, and (6) changes in upstream regulatory genes. Even when the mode of action and the spatial domain of gene expression remain unchanged, this paradigm implies that modifications in the interactions between regulatory and downstream target genes participate in significant phenotypic evolutionary change. By the same token, the literature treating plant development has identified four developmental "rules" largely based on flowering plants as model systems (such as *Zea mays* and *Arabidopsis thaliana*): (1) dedicated gene expression determines cell, tissue, and organ identity, (2) cell fate is determined by position and not by clonal history, (3) developmental and signal transduction pathways are typically controlled by large gene regulatory networks, and (4) development is typically modulated by cell-to-cell signaling. Finally, in this context, it is noteworthy that a broad spectrum of molecular phylogenetic studies dealing with very different plant lineages and clades reveals that a limited number of genomic/developmental subsystems or "modules" are very ancient and have been rewired in ways that engendered, or at least participated in, major morphological transformations.

At issue is whether the gene regulatory network is deterministic in terms of cell, tissue, or organ fate determinancy. A fundamental assumption of contemporary developmental biology is that gene regulatory networks (GRNs, herein defined as circuits of interacting transcription factors and their cis-acting regulatory elements) are primary mechanisms controlling development. According to this assumption, at any time, the relative levels of transcription factors in an extended network determine the progress of development by regulating downstream genes. This conception of gene control in multicellular organisms, which was formulated in several related versions a half century ago, proposes that GRNs are deterministic dynamical systems exhibiting multiple stable states. The theoretical foundations of this framework can be traced to studies of the bi-stable gene regulatory switch between the lytic and lysogenic states of the lambda phage in *Escherichia coli*, and have been generalized and applied to the larger and more elaborate GRNs of eukaryotes in the form of models ranging from discrete Boolean networks to continuous systems of ordinary differential equations. A commonality among these models describing cell differentiation is the assumption that gene products (proteins, particularly transcription factors) have specific identities and connectivity relationships to one another in the GRNs in which they function. According to this view, variation in the outcome of the function of GRNs (alternative cell, organ, or tissue types) arises from nonlinearities and stochastic effects to which such complex, deterministic systems are subject.

This paradigm has been extended to other gene expression mechanisms that have been characterized since the GRN dynamics model was first proposed. Among these mechanisms is the alternative splicing (AS) of pre-mRNA exons and introns to assemble different proteins, a process that permits variation in the functionalities of subsets defining components of GRNs (as for example, transcription factors) at the level of RNA processing. Although the mixing and matching of basic system components has no direct counterpart in a truly deterministic GRN model, the factors controlling AS can be viewed as having well-defined functionalities, since GRN dynamics permit different cell fates in a combinatorial deterministically prescribed manner. Likewise, the modulatory effects of riboswitches, microRNAs, and the enzymes that mediate post-translational modifications (and that can silence genes) can be viewed as adding a complicating, yet still deterministic, set of regulatory mechanisms (box 4.1).

However, this deterministic perspective must contend with evidence that the majority of eukaryotic transcription factors contain intrinsically disordered protein (IDP) domains that comprise almost two-thirds of their sequences and with the fact that the conformations of these domains, and hence their functions, are contingent on the intra-and extracellular environments in which these proteins function. Consequently, the specificity of the binding of most regulatory transcription factors to cis-regulatory elements (non-coding DNA that regulates gene transcription), as well as their partnering with other factors mediating conditional responses to cellular physiological status, are context dependent and subject to change even in the absence of genetic or epigenetic alterations. Importantly, the functions of IDPs are modulated further by both alternative splicing (AS) and post-translational modifications (PTMs), especially phosphorylation. For example, AS, IDPs, and PTMs are known to act synergistically in modulating the activities of the tumor repressing transcription factor p53 and to underlie the functions of several other proteins crucial for the evolution of multicellular organisms.

The combined functional consequences of AS, IDPs, and PTMs make modeling GRN dynamics as strictly deterministic systems incomplete at best. If transcription factors do not have fixed cis-acting regulatory element targets, but rather can alter their specific identity and network-topological status within a given GRN depending on other proteins in the nucleus and external environmental factors, it follows that GRNs can no longer be viewed as deterministic systems in a strict physical or mathematical sense (fig. 4.9). If this conceptualization is correct, the incorporation of AS, IDP, and PTM and their

---

Box 4.1. Riboswitches, microRNAs, and Gene Regulation

Riboswitches and microRNAs provide two mechanisms for post-translational gene regulation, and thereby challenge the deterministic gene regulatory network paradigm. Empirical and theoretical discoveries now make it clear that RNA can control its own transcription and thus regulate gene expression patterns after translation. The molecules that have this regulatory role in the expression of downstream gene products are called riboswitches. A riboswitch is a regulatory segment of a messenger RNA that typically binds to a small molecule (a ligand) and in doing so regulates the production of the protein it encodes, either suppressing or activating the production of the protein. The discovery of riboswitches expanded our understanding of mRNA functionalities (the encoding of proteins, the catalysis of chemical reactions, and binding to other RNA or protein molecules) and strengthened the "RNA world hypothesis" (see chapter 1).

Conceptually, a riboswitch consists of two functional domains, an aptamer and an expression platform (the antiterminator stem-loop) (fig. B.4.1). The expression platform undergoes structural changes in response to the aptamer when it binds with a ligand. The expression platform typically turns gene expression off, although some riboswitches are known to turn gene expression on. Most riboswitches are known from eubacteria. However, one form of a functional riboswitch, called the thiamine pyrophosphate (TPP) riboswitch, occurs in plants and some fungi (for example, *Neurospora crassa*). The TPP riboswitch is an active form of vitamin B_1 (thiamine).

MicroRNA (miRNA) is a small noncoding RNA found in plants, animals, and some viruses. miRNAs post-transcriptionally regulate gene expression by base-pairing with their complementary mRNA sequences and thereby silencing them by one of three processes: (1) mRNA cleavage, (2) destabilizing the mRNA, or (3) reducing translational efficiency. Plant miRNAs manifest high binding specificity to their complementary mRNAs and typically induce mRNA cleavage. This contrasts with animal miRNAs, which have a reduced capacity for pairing with their mRNA targets, although the human genome encodes

---

well-documented synergistic interactions into an expanded (and thus more computationally sophisticated) approach is required to provide deeper insight into developmental biology and evolution.

The following sections present evidence that AS, IDP, and PTM promote alternative, context-dependent GRN states, and thus serve a critical role in a broad range of cellular responses, including cell fate specification. Evidence is also presented that these three components are ancient in eukaryotic GRNs, a speculation driven by the observation that early divergent unicellular eukaryotes achieve

≈1,500 miRNAs, each with the potential to bind with hundreds of different mRNAs. This contrast and others foster the speculation that miRNAs evolved along different pathways after the divergence of plants and animals from a last universal common ancestor. Evidence is mounting that some evolutionarily rapid morphological transformations are associated with high rates of miRNA diversification, although miRNAs serve as important phylogenetic markers because of their overall slow rate of evolution.

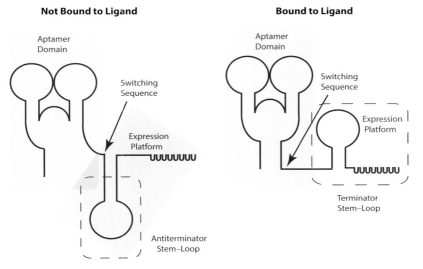

Figure B.4.1. Schematics of a riboswitch that is unbound (left) and bound (right) to its ligand.

temporally alternative physiological and reproductive states and respond adaptively to contingent environmental conditions by virtue of AS, IDP, and PTM. The conclusion is that AS, IDP, and PTM add to the "incompleteness theorem" when we attempt to reduce development to a single causal level. It is important to emphasize that rather than proposing AS, IDP, and PTM as a developmental mechanism in its own right, they collectively create an adaptive plasticity that significantly diminishes the strict determinism that is often attributed to GRNs.

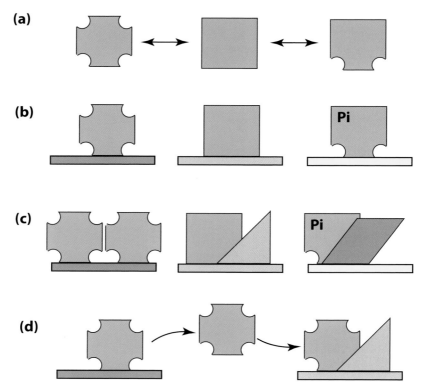

Figure 4.9. Schematic for the role of alternative splicing (AS), intrinsically disordered protein (IDP) domains, and post-translational modifications (PTMs) in cell-specific DNA target site selection by transcription factors (TFs). a. The structured DNA binding domain of a TF binds to a variety of DNA sequences with high affinity. However, most TF sequences are intrinsically disordered and can adopt a variety of conformations (represented by different green polygons), which can rapidly interconvert (as indicated by arrows). b. Specific spliceoforms and phosphoforms of a TF (denoted by Pi) are produced in different cell types or tissues. The variants can reinforce a subset of TF conformations or enable access to new conformations. Specific conformations may enhance or inhibit affinity for particular DNA sequences (denoted by different colored rectangles). c. When a TF binds to a "correct" DNA sequence, additional copies of the same protein, or additional other TFs (represented by different new polygons), may bind to both the TF and neighboring DNA binding sites, thus reinforcing TF-DNA binding. Alternately, the TF isoform may bind other proteins first, followed by DNA binding by the protein complex. d. When a TF binds a DNA sequence that is not appropriate for its variant form, an incorrect transcriptional readout may be transiently produced. Both the lower intrinsic binding affinity for this site and the lack of reinforcing interactions with other TFs eventually cause the protein to dissociate. The released TF protein then has an opportunity to bind to a higher affinity site to produce the appropriate response for the cell or tissue.

Alternative Splicing (AS) and Much More

Alternative splicing produces protein isoforms from the same precursor mRNA by retaining or excluding different exons to achieve differential translation. First observed in the infectious adenovirus cycle and subsequently in the transcripts of normal, endogenous genes, AS occurs in all eukaryotic lineages (and an analogue is known for prokaryotes) and becomes more prevalent as complexity, estimated by the number of different cell types, increases (see chapter 7). Although a number of scenarios have been advanced for the origin of AS, including a role in enabling the cell to filter out aberrant transcripts, there is ample evidence that AS is a molecular motif that facilitates evolutionary innovations.

Five basic types of alternative splicing exist: alternative 3′ acceptor site, alternative 5′ donor splice site, intron retention, mutually exclusive exon splicing, and exon skipping. The last is the most frequent. Regulation and selection of the splice sites are performed by trans-acting splicing activator and repressor proteins within an RNA–protein complex, the spliceosome, which is canonically composed of five small nuclear RNAs (i.e., U1, U2, U4–U6) and a range of assorted protein factors (fig. 4.10). Splicing is regulated by trans-acting repressor-activator proteins and their corresponding cis-acting regulatory silencers and enhancers on the pre-mRNA. The effects of splicing factors are often position-dependent. A splicing factor that functions as an activator when bound to an intronic enhancer element may function as a repressor when bound to its splicing element in the context of an exon.

The secondary structure of the pre-mRNA transcript also determines which exons and introns will be spliced—for example, by bringing together splicing elements or by masking a sequence that would otherwise serve as a binding element for a splicing factor. Consequently, activators, repressors, and secondary pre-mRNA structure constitute a splicing "code" that defines the protein isoforms produced under different cellular conditions. Additionally, the elements within

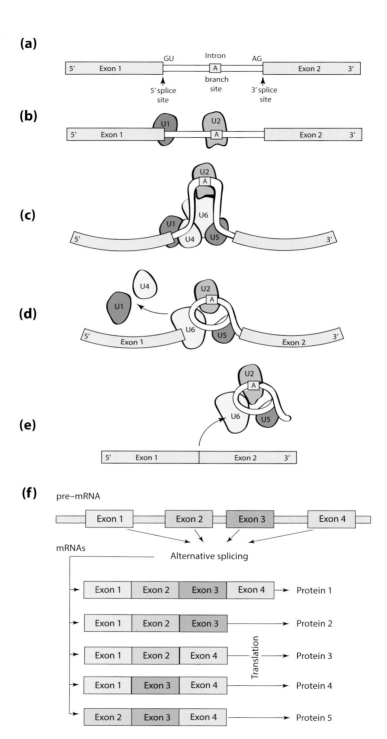

this code function interdependently in ways that are context dependent, both intracellularly and extracellularly. For example, cis-acting regulatory silencers and enhancers are influenced by the presence and relative position of other RNA sequence features, and the trans-acting context is affected by intracellular conditions that are in turn influenced by external conditions and other RNA sequence features. Furthermore, some cis-acting elements may reverse the effects on splicing if specific proteins are expressed in the cell. Indeed, the number of factors influencing AS is significantly large.

AS is adaptive and highly conserved. There is strong selection against mutations that alter splicing. For example, the AS mechanism for heat regulation among land plants is an ancestral condition as attested by the fact that the AS pattern for heat shock transcription factors is conserved in the moss *Physcomitrella patens* and the flowering plant *Arabidopsis thaliana*.

The phenotypic domain an individual organism can occupy without increasing its proteome size (as defined by the number of different amino acid sequences) is significantly expanded because AS produces a disproportionate number of intrinsically disordered protein (IDP) domains that lack equilibrium 3D structures and thus can assume multiple functional roles under normal physiological conditions. Additionally, the majority of eukaryotic transcription factors have IDP domains affected by AS. IDP domains are characterized as having little

Figure 4.10. (*opposite page*) Schematic of the structure and operation of a spliceosome to remove an intron flanked by exons on a pre-mRNA. a. The splicing process is guided by a highly conserved 5′ splice site GU sequence, an A branch site near a pyrimidine-rich region, and a 3′ splice site AG sequence. The spliceosome protein complex contains RNA and protein components (small nuclear ribonucleoprotein or snRNPs, designated as U1, U2, U4–U6) that recognize and bind to the pre-mRNA conserved sequences in a stepwise process. b. The process begins with U1, which binds to the 5′ splice site, and U2, which binds to the A branch site. c. U4, U6, and U5 subsequently bind the pre-mRNA transcript forming the mature spliceosome complex that configures the intron bringing the 5′ and 3′ splice sites into a loop. d. The mature spliceosome splices the 5′ first and the 5′ GU end second and creates a lariat by connecting the 5′ end to the A branch site. The U1 and U4 snRNPs are released and the 3′ splice site is cleaved. e. The intron, U2, and the U5–U6 ensemble are released, and the exons are attached. The intron will degrade and the snRNPs can be reused. f. Schematic of alternative splicing of a pre-mRNA with four exons that can yield five different proteins.

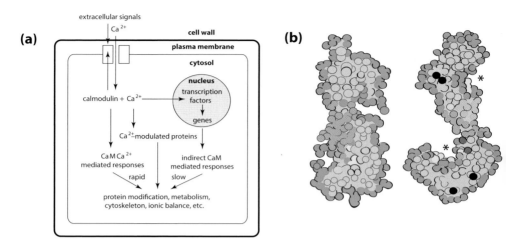

Figure 4.11. Schematic of slow and rapid calmodulin-Ca^{2+} and Ca^{2+} mediated responses to external and internal signals (a) and stereo-chemical diagram of calmodulin (right) and calmodulin bound to four Ca^{2+} ions (left) indicated by black circles (b). a. External signals (which must pass through a cell wall in the case of plants) registered by plasma membrane-bound receptors result in transient changes in intracellular Ca^{2+} ion concentrations in the cytosol and organelles, including the nucleus. Some of these ions become bound to calmodulin (CaM) and calmodulin-like (CaML) proteins, which interact with numerous target proteins interacting with metabolic processes and cytoskeletal structure. To date, over 440 target proteins for CaM are reported. CaM-Ca^{2+} also affects gene expression patterns directly by binding to transcription factors or indirectly by signal transduction pathways. Rapid changes in cell functionalities result from CaM-Ca^{2+} or CaML-Ca^{2+} protein binding to target proteins; slower responses result from CaM-Ca^{2+} mediated gene expression. b. When bound to Ca^{2+}, the CaM complex exposes non-polar surfaces (on EF-hand motifs bound by a linker, shown in light and dark blue, respectively) that then bind to non-polar regions on target proteins. When bound to Ca^{2+}, the non-polar amino acids form two grooves (shown with asterisks), which configure to bind with a target protein. Because these two regions can assume different configurations by virtue of their IDP domains, CaM acts as a versatile regulatory protein and its targets are not required to possess any specific amino acid sequence or structural binding motifs.

specificity and affinity and thus enhanced binding diversity. These proteins also have the ability to form large interaction surfaces, possess fast association and dissociation rates, have polymorphisms in the bound state, and have reduced intracellular lifetimes. These traits make IDP domains ideal signaling and regulatory molecules (as for example calmodulin and calmodulin-like proteins; fig. 4.11). Studies have identified IDP domains as enriched in non-constitutive exons, indicating that protein isoforms may display functional diversity due to the alteration of functional modules within these regions (as for example calmodulin EF-hand motifs and linker regions). IDP domains can exist as molten globules with defined secondary structure or as unfolded chains that can function through transitions among different folded states. Their functional conformations can change by binding to other proteins and nucleic acids. Post-translational modifications (as for example phosphorylation) can also alter IDP domain functionalities.

Examples of IDP domains involved in transcriptional regulation are well known. For example, the C-terminal activation domain of the bZIP proto-oncoprotein c-Fos, which effectively suppresses transcription in vitro, is intrinsically disordered and highly mobile, whereas the unbound N-terminal domains of the DELLA proteins, which are central to the integration of plant developmental and environmental signaling by interacting with other transcription factors as co-repressors or co-activators to regulate gibberellin (GA) response gene expression patterns, are intrinsically disordered and undergo disorder-order transitions upon binding to interacting proteins. The DELLA proteins are similar in their domain structures to the GRAS protein family whose N-domains are intrinsically disordered. Like the DELLAs, the GRAS proteins are extensively involved in plant signaling by virtue of their ability to undergo disorder-order transformations with a variety of molecular partners involved in root and shoot development, light signaling, nodulation, and auxin signaling and transcription regulation to biotic and abiotic stresses. Like the DELLA-GA signaling system in plants, metazoans carry out a significant fraction of their intercellular signaling via a collection of small molecules that bind to their cognate proteins called nuclear hormone receptors (NHRs). Following ligand binding, the NHRs translocate to the nucleus where they act as transcription factors. In addition to the structured ligand and DNA binding domains, like other transcription factors, these NHRs have flanking and linking IDP domains that use their flexibility to bind to large numbers of partners.

These examples illustrate that intrinsically disordered transcription factors play central roles in plant and animal development and homeostasis. They are not exceptional. Separate studies on transcription factors using different disorder predictors identify similar, very large amounts of disorder in eukaryotic transcription factors. For example, 82.6% to 93.1% of the transcription factors in three databases contain extended regions of intrinsic disorder, in contrast to 18.6% to 54.5% of the proteins in two control databases. Since protein-DNA recognition and protein-protein recognition are central transcription factor functionalities, the available data illustrate the extent to which eukaryotic transcription factors manifest extensive flexibility as a consequence of disorder-associated signaling and transcriptional regulation. This flexibility permits transcription factors with IDP domains to bind to a greater array of partners that in turn can induce conformational changes in bound protein and DNA substrates. An excellent example is the high mobility group A (HMGA) protein family associated with a large number of the transcription factors participating in intra- and extracellular signaling.

An additional and important molecular mechanism that influences developmental versatility is post-translational modifications (PTMs) that can regulate transcription at different levels such as chromatin structure and transcription factor interactions. For example, the multi-protein complex called Mediator, which is massively disordered and involved in RNA polymerase II-regulated transcription, is positively and negatively regulated by post-translational phosphorylation, which is a common PTM mechanism affecting different macromolecular recognition/binding events involved in cell- or tissue-specific protein-protein interactions. For example, DNA binding by the transcription factor ETS-1, which is allosterically coupled to a serine-rich region, is modulated by Ca^{2+} signaling that induces phosphorylation of this region. Phosphorylation of the intrinsically disordered PAGE4 protein as part of the stress-response pathway causes PAGE4 to release the transcription factor c-Jun, enabling its activity in transcription regulation. Phosphorylation can also increase interactions among cofactors. For example, the cytokines TNF and IL-1 induce phosphorylation of the p65 subunit of NF-κB, which in turn induces a conformational change that allows p65 ubiquitination and interaction with transcriptional cofactors. Association of Elk-1 and ETS domain transcription factors with Mediator and histone acetyltransferases is dependent on Elk-1 phosphorylation.

It is important to note that multiple PTM combinations greatly increase the diversity of signaling proteins. Consider a protein segment with three phosphorylation sites, each of which can be phosphorylated or unphosphorylated. This condition gives $2^3 = 8$ different PTM states, each of which can lead to different downstream consequences. This scenario is an oversimplification because PTM sites are typically IDP domains that are associated with target binding sites and because IDP domains are subject to alternative splicing. Consequently, an "AS, IDP, PTM code" exists that can greatly expand the functionalities of a cell's proteome (fig. 4.12).

Although AS, IDP, and PTM can operate independently of one another, they are more often co-localized to operate synergistically. The co-localization of AS, IDP, and PTM is apparent in many ways. For example, pre-mRNA segments undergoing AS are far more likely to code for IDP domains than for structured domains. These AS-associated IDP domains also frequently contain binding sites for protein or nucleic acid partners such that they operate together to "rewire" gene regulatory networks, often at the tissue-specific level, and are well conserved over evolutionary time.

IDP domains are also far more likely than structured regions to undergo PTMs, especially the phosphorylation of serines and threonines. These IDP-associated PTMs are often observed to alter partner choice for IDP-based protein-protein interactions, which can further rewire gene regulatory networks. In addition, different patterns of multiple PTMs in localized protein regions have been shown to signal different downstream results, leading to their designation as a histone or PTM "code." Finally, "constellations" of multiple PTMs generally occur in IDP regions, some examples of which have been shown to be further modified by AS.

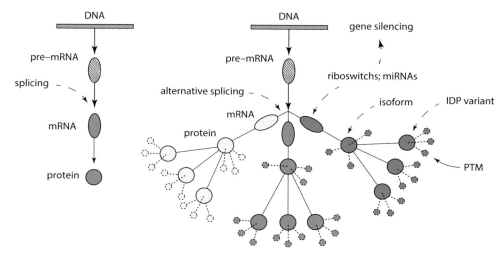

Figure 4.12. Schematic of a simple model for protein transcription-translation (left) and the combinatorial consequences of alternative splicing (AS), proteins with intrinsically disordered protein (IDP) domains, and post-translational modifications (PTMs) on protein functionalities (right). In the simplest case shown to the left (a "one gene-one enzyme" model), transcription produces a pre-messenger RNA (cross-hatched ellipse) that is spliced to yield a single mRNA (green ellipse) that is translated into a protein with one functionality (green circle). However, many signaling proteins and the majority of transcription factors are subjected to AS that yields multiple mRNAs and protein isoforms (three are shown here), each of which is subjected to PTMs (indicated by small circles with dashed lines). Assuming that each protein isoform has three IDP variant forms and that each variant has three PTM forms, a single pre-mRNA can obtain 27 proteins with different functionalities. The consequences of riboswitches (mRNAs containing regulatory segments) and microRNAs (that function in RNA silencing and post-transcriptional regulation of gene expression patterns) contribute to this combinatorial complexity (see box 4.1).

Less is known about the extent to which IDP and PTM have changed over evolutionary history. Quantitative measures of proteome intrinsic disorder are only recently becoming available. However, a positive relationship between a large number of proteins with intrinsically disordered domains and the extent to which species are evolutionarily derived has been noted. This relationship appears to be step-wise rather than continuous, which likely reflects major evolutionary transitions. For example, Xue and coworkers (2012) examined 3,484 viral, bacterial, and eukaryotic proteomes and found that the largest variance of intrinsically disordered content occurred among the viruses (between 7.3% and 77.3%), whereas only a weak correlation between complexity as gauged by the number of different cell types and overall IDP domain content was observed within the eukaryotes. These authors also report that the IDP domain content is generally independent of proteome size for both the prokaryotes and eukaryotes, but that it is significantly higher for eukaryotic compared to prokaryotic species and possibly correlated with the more elaborate signaling systems eukaryotes use to coordinate their intracellular functions. Schad and coworkers (2011) report that complexity (as gauged by the number of different cell types) and proteome size (measured as the total number of amino acids) correlate positively across diverse organisms. These authors also report that the fraction of IDP domains increases significantly from prokaryotes to eukaryotes, but does not increase further within the eukaryotes. A subsequent analysis at a finer taxonomic level indicates that the percentage of IDP domains in proteomes increases across plant and animal species (see chapter 7, fig. 7.23).

In this regard, using *Arabidopsis* and *Oryza* as respective eudicot and monocot model species, it is interesting to note that statistically significant and positive correlations are reported for chromosome regions that encode for intrinsically disordered domains (30+ amino acid sequences) and regions with a high frequency of crossing over and a large G + C content (Yruela and Contreras-Moreira 2015).

Revisiting Gene Regulatory Networks

Thus far, we have explored the hypothesis that gene regulatory networks (GRNs) are frequently, if not intrinsically, nondeterministic owing to the independent and synergistic consequences of alternative splicing (AS), intrinsically disordered protein (IDP) domains, and post-translational modifications (PTMs). This hypothesis does not assert that development is itself nondeterministic. It does claim that deterministic GRN dynamics are not a sufficient causal basis for developmental regularities. Although a GRN might provide a rough template for a cellular function (particularly if the GRN was established concurrently with the evolutionary origination of that function), remodeling of the GRN by AS, IDP, and PTMs will have rendered cell phenotype identity increasingly dependent on internal (physiological) and external (microenvironmental and extraorganismal) conditionalities beyond the GRN itself. This assertion is consistent with, if not confirmed by, somatic stem cell production and subsequent differentiation as well as examples of dedifferentiation.

The conservation of a useful cell function or morphological phenotype over the course of evolution accompanied by an unmooring from its originating GRN appears to be a common scenario in the history of multicellular plants and animals, reflected in what has been termed "developmental system drift." AS–IDP–PTM and its synergies also provide a context for understanding how the functionalities of ancient proteins and regulatory networks can be stably modified over the course of evolution to adapt to changing external conditions. Target sequence recognition and selectivity by a transcription factor are subtle properties of the latter's structure. It is well documented, for example, that novel relationships between protein structure and PTM induced by mutation can lead to altered protein-protein interactions resulting in dramatic changes in transcription factor function. However, synergy with AS and IDP provides an even greater multiplicity of functional states that can be explored ecologically and physiologically ahead of any mutational change.

Furthermore, nascent potentially adaptive mutations can be retained within (and subsequently integrated into) GRNs by virtue of AS, IDP, and PTMs modifications that can buffer GRNs from the immediate consequences of such mutations. In this scenario, a mutated GRN could survive by virtue of AS, IDP, and PTMs adaptive modifications that would permit the GRN time to adaptively reorganize. In this way, evolutionary changes would involve an interactive "genome ↔ AS, IDP, and PTMs" feedback loop.

The association of AS, IDP domains, and PTMs appears to be a core functional complex that mediates the modifications of protein functionalities required for context-dependent cell signaling, regulation, and differentiation. The combined effects of AS, IDP, and PTMs also likely buffer genomes from mutations (some of which can subsequently become adaptive to new conditions) and contribute to the evolvability of GRNs. AS, IDP, and PTMs are ancient and likely promoted variability and thus adaptive evolution to support more complex intracellular signaling processes coordinating the activities of functionally interdependent discretized organelles, cells, tissues, and organs. Unlike promoter activity, which primarily regulates the amount of transcripts, AS changes the structure of transcripts and their encoded proteins. The ability of IDP domains to assume different conformations expands the functional repertoire of proteins assembled by AS from pre-mRNAs to diversify the phenotypic domain that a single genome can provide. This repertoire is yet again increased by PTMs, which generate additional functionalities. Thus, AS, IDP, and PTMs can yield virtually limitless combinatorial possibilities, which can be adaptively sifted over the course of evolution.

Consequently, GRNs are inherently plastic and therefore adaptive. Moreover, they function in a noisy cellular milieu owing to the operation of AS, IDP, and PTMs in a multitude of other biochemical pathways as well as the effects of mutations and variations in gene and protein copy number. Note that this noisiness is over and above the described intrinsic indeterminacy. The evolution of cell differentiation may indeed have depended on such stochastic effects. However, heterogeneity at both the molecular and cell phenotypic levels must be suppressed for reliable development to occur. This is accomplished by a variety of "scaffolding" effects at multiple scales, including consistency of external cues from neighboring cells and the physical environment and the stabilizing effects of natural selection.

The multi-scale nature of developmental processes is increasingly acknowledged. In particular, tissue morphogenesis and cellular pattern formation involves the mobilization, by key gene products of the developmental "toolkit," of mechanical, electrical, and other physical phenomena external to the genome. It is therefore not surprising that

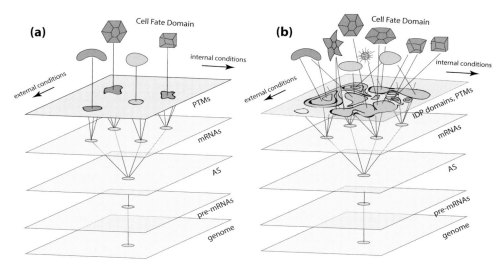

Figure 4.13. Schematics of cell fate specification viewed from a standard deterministic GRN perspective (a) and the non-deterministic GRN perspective on the consequences of AS, IDP, and PTMs (b). a. In the standard view, pre-mRNAs undergo alternative splicing (AS), and transcription factors specified by the variant mRNAs undergo post-translational modifications (PTMs) to form a cadre of proteins involved in GRN specification of cell fates (represented as irregular shapes) via their cis-acting targets. Discrete cell types result from the deterministic properties of these GRNs. b. In a non-deterministic perspective, transcription factors are generated by AS and PTM operating in the context of IDP domains and PTMs. Cell fate determination in this case (represented by interactions among components of variable, context-dependent identity and specificity) is a consequence of the time- and spatial-context dependency of each of the levels shown in the schematic, which depend on internal and external cellular conditions in a fashion that eludes deterministic description at the level of GRNs.

the determination of cell type identity does not reside at the single scale occupied by GRNs, but rather draws on factors at several causal levels, as described above, among the most important of which are the mechanical aspects of chromatin reorganization associated with changes in gene expression (fig. 4.13). Any biologically reasonable model regarding the operation of GRNs must acknowledge and integrate the ubiquitous effects of AS, IDP, and PTMs. Just as genes per se have long been rejected as the exclusive or privileged level of determination of the phenotype and evolutionary change, new understanding of the complexities of gene expression and the conditional identities of its protein products call into question a deterministic GRN-based reductionism in developmental and evolutionary biology.

Homology and the Ship of Theseus

The preceding discussion, which has revolved thus far entirely around the genomic and molecular levels of biological organization, establishes a context in which to explore the concept of homology and its empirical application at the organismic level of organization because evolution of form is inextricably linked to genomic evolution. Yet, as we shall see, the lineages between molecular and morphological evolution are neither simple nor clear.

The evolution of organic form involves the nonrandom, lineage-specific ordering and reordering of structural parts. Any legitimate evolutionary-developmental theory must account for the generation of structural parts, the fixation and combination of parts, and their subsequent modification (including their loss and reappearance). The neo-Darwinian theory of evolution accounts for some but not all of the necessary elements of a legitimate evo-devo theory. Although it accounts for the modification of parts (by means of natural selection), it fails to explain the appearance or reappearance of parts. Natural selection has no generative capabilities. It eliminates or maintains what exists. The relevance of the concept and application of homology becomes apparent when we consider that homology is the manifestation of structural organization and the processes from which it emerges and which maintain it.

Unfortunately, homology has been defined in a variety of ways, some ranging from idealistic (lacking a formal definition) to methodological (as in cladistic analyses). Some of these definitions emerged from the idealistic morphology of the eighteenth century (as for example, in Goethe's *Versuch die Metamorphose der Pflanzen zu erklaren*). What is arguably the most concise and precise definition for homology (as well as the first) was offered in 1843 by Sir Richard Owen (1804–1892): "the same organ in different animals under every variety of form and function." Note that this definition does not present criteria whereby homologous organs can be identified. More important perhaps, Owen's concept of homology deals with characters and their character states, such as leaves (a character) and simple leaves versus pinnately compound leaves (two character states). In the parlance of cladistics, all homologues are synapomorphies but the reverse is not true. (A synapomorphy is a character that is shared by two or more species sharing a last most recent common ancestor that is presumed to have possessed the character.) A character state shared by two or more species sharing a most recent common ancestor need not be a homologue. For example, the absence of an organ can be a synapomorphy, but it cannot be a homologue since the part does not exist. One of the difficulties of this approach to homology is that it has an element of circular reasoning in the sense that it requires us to identify a character and to infer that two or more different kinds of organisms possess it. For example, the leaves of lycopods and the leaves of ferns were called "leaves" by the earliest plant biologists. However, subsequent studies show that these structures evolved independently, just as roots have in different plant lineages. A more challenging example is multicellularity. In chapter 7, we will see that this presumed "functional trait" evolved independently in different lineages by means of similar developmental modules that are clearly not homologous.

The homology concept of Owen has been extended to other lower and higher levels of biological organization (fig. 4.14), often with considerable success. However, problems arise when homologies are drawn across different levels of organization. For example, a morphologically homologous character, such as a leaf or root, across different taxa does not provide prima facie evidence that the character is generated by the same developmental or genomic processes, nor do developmental or genomic homologies guarantee morphological homology. This caveat resonates with our previous discussion about logic circuits and the signal-activated subsystems they wire together. We saw that different logic circuits can process the same inputs to obtain the same outputs (see fig. 4.2). We also saw that each subsystem is

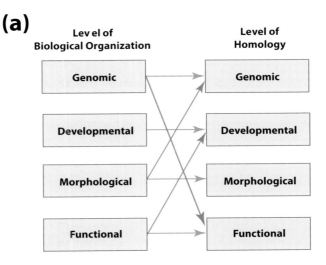

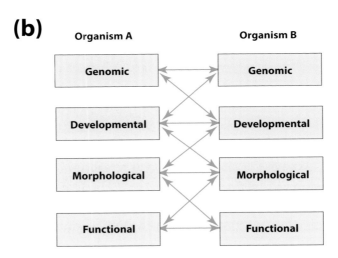

Figure 4.14. Valid and invalid comparisons of homology across four levels of biological organization (a) and an ideal scenario for the empirical determination of homology. a. Logically valid comparisons are those across equivalent levels of biological organization (blue arrows). Logically invalid or suspect comparisons are those across different levels of organization (red arrows). Comparisons across closely related levels of comparison (for example, genomic and developmental comparisons, or developmental and morphological comparisons) can be instructive when assessing homologies, but homologs can be asserted only across the same levels of comparison. b. A rigorous assessment of homology requires a multilevel understanding of how different levels of biological organization interact to produce similar character states. Ideally, this kind of assessment requires comparisons among all corresponding levels of biological organization (for example, genomic and genomic) and among all of the corresponding lower and upper levels of biological organization (for example, developmental and genomic, and developmental and morphological).

interconnected in one way or another with all of the other subsystems (see fig. 4.9). Although it is certainly true that the assessment of homology benefits inordinately the more levels of biological organization that are compared, the strongest evidence for homology comes from comparable level comparisons, which can be extended, albeit carefully, to cover all of the relevant levels of comparison (see fig. 4.14).

This caveat draws attention to the thought experiment known as the Ship of Theseus, discussed by ancient philosophers since the time of Heraclitus (535–c. 475 BPE) and Plato (c. 424–c. 347 BPE). The paradox deals with whether an object that has had all of its parts replaced is the same object. The ship of Theseus was so revered by the Athenians that it was preserved for decades as each rotten plank of wood was carefully replaced with planks identical in appearance. The philosophical conundrum is whether Theseus' ship remained the same "object." Aristotle (384–322 BPE) tried to solve the paradox by drawing a distinction between an object's formal cause (the functional design of an object's form; in our case, the ship of Theseus) and its material cause (the substances out of which the object is made; in our case, the wooden planks, masts, etc., out of which this ship is made). Aristotle concluded that the formal cause has precedence such that Theseus' ship was the same despite being refurbished. Other philosophers have drawn different conclusions on the basis of different definitions of "sameness." At issue here is whether an extensive rewiring of a GRN, or whether replacing some or all of its circuitry without affecting the morphology of an organism, results in homologous organisms.

The complexity of "sameness" is illustrated by studies of the MADS box gene called LEAFY (*LFY*), which is found in mosses, ferns, gymnosperms, and angiosperms. Among flowering plants, the single *LFY* gene product binds to sequences in the enhancers of several homeotic floral genes (as for example *APETALA1*). Among nonflowering plants, several *LFY* gene products control more general and manifold aspects of the life cycle. The *LFY* DNA binding domain is strongly conserved across all taxa. But the *LFY* protein as a whole has diverged in activity across taxa from mosses to angiosperms, which is indicated by the ability of *LFY* cDNAs (isolated from mosses, ferns, and various gymnosperms linked to the *Arabidopsis LFY* promoter) to progressively recover the *lfy* mutant of *Arabidopsis*. However, an important pattern of recovery emerges when the phyletic relationship among the different land plant lineages is considered. The ability to recover the mutant mirrors the phyletic distances of the mosses, ferns, and gymnosperms from the angiosperms—the success of recovery decreases the more distant the lineage used to construct the *LFY* cDNA is from *Arabidopsis*. Two possible explanations for this phenomenon exist. *LFY* either controls similar gene networks that have coevolved with target genes that have themselves become modified during plant diversification, or the function of *LFY* in early divergent plant lineages (mosses and ferns) and late divergent lineages (angiosperm) has changed as a result of the recruitment or intercalation of new target genes. In either case, it is clear that understanding phenotypic evolution requires thinking both within and outside the paradigm of transcription factors.

Another example, concerning plant apical meristems, is equally informative. Plants produce new organs as a result of the activities of apical meristematic populations of cells, one of which is the shoot apical meristem (SAM), which gives rise to the aboveground portions of the plant body. Despite its conserved functions during leaf, stem, and reproductive organ development, the structure of the SAM varies markedly across the land plant lineages (fig. 4.15). Seed plant SAMs can consist of hundreds of cells organized into a tunica, which gives rise to the epidermis, and the central and peripheral zones, which give rise to the other plant body tissues. In contrast, the SAMs of seedless land plants can consist of a single meristematic cell, called the apical cell (AC), which is subtended by a core region of meristematically active cells. As yet, the molecular genetic basis for the meristematic activity of the AC is problematic and is based largely on comparative molecular studies using homologs of angiosperm SAM related genes. For example, *CLASS I KNOTTED1-LIKE HOMEOBOX* (*KNOX*) genes are expressed in the SAM of some ferns, whereas the lycopod *Selaginella*

Apical Meristems

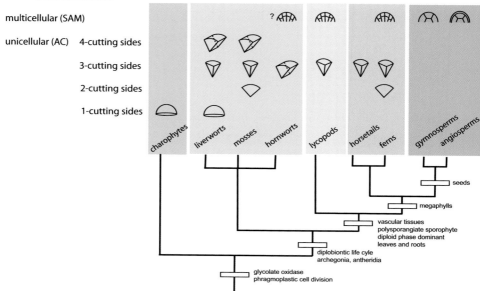

Figure 4.15. Schematics of the types of shoot apical meristem (SAM) configurations reported for the different land plant lineages and the charophycean algae (both groups are collectively referred to as the streptophytes). The phyletic relationships among the streptophyte lineages are shown in the form of a highly simplified cladogram based on a small number of informative character states. On the basis of this cladogram, unicellular apical meristems (AC = apical cell) such as those observed for the charophycean alga *Chara* are ancestral and common among the nonvascular land plants (liverworts, mosses, and hornworts) and some non-seed vascular plants (lycopods, horsetails, and some fern species), whereas multicellular meristems (with a tunica, and a central and peripheral zone configuration; see upper right hand diagrams) are the most derived. Less zonated multicellular shoot apical meristems occur among lycopods and ferns (and possibly some hornworts). See fig. 4.16 for photos of the real apical meristems of a horsetail (*Equisetum arvense*) and angiosperm (*Zea mays*).

kraussiana has transcript accumulations of the homologs of *CLASS I KNOX, ASYMMETRIC LEAVES1/ROUGH SHEATH2/ PHANTASTICA (ARP)*, and *HD-ZIP III* within the core cells beneath its AC. These and other observations indicate that the genes involved with angiosperm SAM meristematic regulation are present in the AC-type meristems of seedless vascular plants (fig. 4.16).

However, a recent study by Frank et al. (2015) indicates that the gene expression patterns observed for the AC-type meristems of

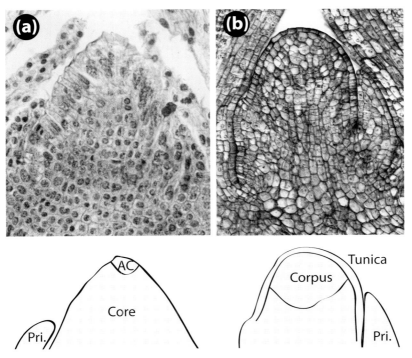

Figure 4.16. Longitudinal sections (top) and schematics (bottom) of the shoot apical meristems (SAM) of the horsetail *Equisetum arvense* (a) and corn *Zea mays* (b). The SAM of *E. arvense* consists of a single apical cell (AC) above a core region of meristematically active cells flanked in this picture by leaf primordia (Pri.). The SAM of *Z. mays* is comparatively massive and consists of a layer of cells (the tunica) that gives rise to cells that will differentiate into the epidermis. The tunica covers the corpus that gives rise to cells that will differentiate into the vascular and ground tissue systems. Fig. 4.16 b courtesy of Dr. Michael Scanlon (Cornell University).

Selaginella and the horsetail *Equisetum arvense* possess incongruous components despite sharing features in common with angiosperm SAMs. The data reported in this study demonstrate that the cellular subdomains of the apical meristems of *Selaginella* and *Equisetum* are transcriptionally distinct, and thus reminiscent of the molecular zonation patterns described for corn (*Zea mays*) angiosperm SAMs. However, the data also indicate that the gene expression patterns of developmental regulators in the subdomains of the *Selaginella* and *Equisetum* AC meristems differ (table 4.2). Two explanations for the

Table 4.2. Upregulated gene families for the shoot apical meristem of *Zea mays* and those shared with the apical cell– type meristematic subdomains of *Selaginella* (S) and *Equisetum arvense* (E)

Gene family	Apical cell	Core	Pri.
Ribosomal biogenesis	S, E		
Histone superfamily	S, E		
PROHIBIT 4	S, E		
4 Unknown proteins	S, E		
PHABULOSA	E	E	S, E
LEAFY	E	E	E
CHR11		E	
SPLAYED		E	
BARELY ANY MERISTEM 1	S	S	S
REPRESSOR OF GA1-3	S	S	
AUXIN RESPONSE FACTOR 3	S		S
ERECTA-like 2	S		
TOPLESS-related S			
AUXIN RESPONSE FACTOR 12			S
MERISTEMATIC RECEPTOR-like KINASE			S
AP2/B3 TF			S

Note. As reported by Frank et al. (2015). The locations of the apical cell, core meristematic cells, and the first leaf primordium (Pri.) are diagrammed in fig. 4.16.

unique gene expression patterns are obvious. Either the *Selaginella* and *Equisetum* ACs evolved independently in the two lineages represented by these taxa, or the lycopod and horsetail AC-type meristems evolved from a last common ancestor, but evolutionary diverged over the past 400 million years, such that their meristematic transcriptomic profiles no longer share strong similarities. The fossil record reveals that *Selaginella* and *Equisetum* belong to ancient lineages that shared a last common trimerophyte-like ancestor in the Middle Devonian. However, *Selaginella* is not an early divergent genus according to recent cladistic analyses of extant lycopods. Modern-day horsetails are also derived species. Consequently, both of these possibilities are plausible. However, the phyletic distribution of SAM configurations aligns best with the second explanation (see fig. 4.15). If true, the data reported by Frank and coworkers provide an excellent example of homology at the morphological level that is not exactly reflected in homology at the genomic level.

Our third example of the difficulty of attributing homology in the context of developmental patterns centers on the formation of the reproductive organs of the charophycean alga *Chara* and the modern representatives of early divergent vascular plant lineages such as *Equisetum*. As shown in fig. 4.15, the charophycean algae and the land plants comprise a monophyletic group of plants (the streptophytes). However, the charophycean algae have a life cycle in which the only multicellular organism is haploid, whereas the land plants have a life cycle involving two multicellular generations, a haploid gametophyte and a diploid sporophyte. Nevertheless, botanists have long noted a striking similarity between the morphologically complex multicellular egg- and sperm-bearing structures of the charophycean alga *Chara* and the corresponding structures of plants such as the horsetail *Equisetum* and the lycopods such as *Huperzia* (fig. 4.17). Indeed, the structural similarity shared by the reproductive structures of these plants was one reason why early botanists such as Eduard A. Strasburger (1844–1912) and Frederick O. Bower (1855–1948) speculated that these plants shared a last common ancestor.

Importantly, ontogenetic as well as structural similarities exist among the reproductive structures of *Chara* and the early land plant lineages (fig. 4.18). Each of the reproductive organs can be traced back ontogenetically to a single superficial meristematic cell that proliferates by means of a series of coordinated divisions to give rise to a sterile jacket of cells enveloping either a gametangenous or sporogenous mass of cells. The sterile jacket of cells provides physical protection and, in the case of the cells surrounding the egg, it also provides the developing egg with nutrients (matrotrophy, which is one of the hallmarks of the land plants, occurs in the Charales and Coleochaetales). Although a reasonable case for homology among the egg-, sperm-, and spore-bearing organs of the land plants can be made based on ontogenetic, structural, and physiological functional traits, the similarities between these organs and the reproductive organs are thus far super-

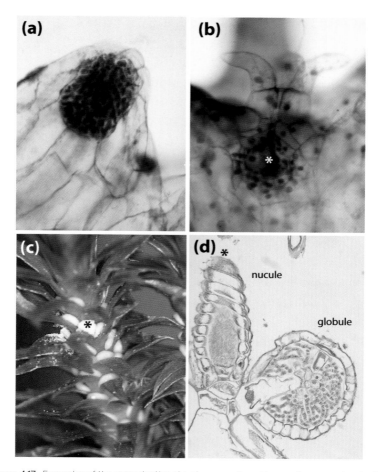

Figure 4.17. Examples of the reproductive structures produced by seedless vascular plants (a – c) and the charophycean alga *Chara* (d). a. The sperm-bearing structure of *Equisetum* (the antheridium) consists of a sterile jacket of cells surrounding a cluster of cells that will subsequently develop into sperm cells (the dark mass of cells). b. The egg-bearing structure of *Equisetum* (the archegonium) consists of a sterile jacket of cells surrounding a single egg cell (indicated by the asterisk). Motile sperm cells must swim to the egg and must pass through the neck of the archegonium to reach the egg (the neck consists of three flared rows of cells). c. A cluster of sporangia (one is indicated by the asterisk) of the lycopod *Huperzia*. Note the dusting of spores on the leaves to the left. d. The multicellular egg-bearing structure (the nucule) and the sperm-bearing structure (the globule) of the charophycean alga *Chara*. The nucule and the globule each consist of a sterile jacket of cells surrounding the gamete-producing cells. Motile sperm cells must enter an opening at the tip of the nucule (indicated by an asterisk) to reach the egg cell (the entrance is created at the tip of the nucule indicated by the asterisk).

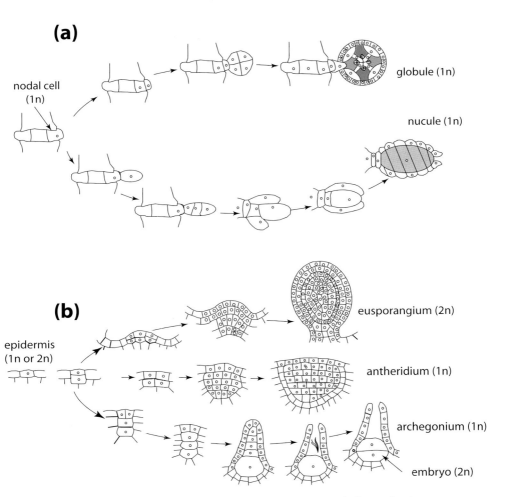

Figure 4.18. Schematics of the developmental patterns that give rise to the egg-bearing nucule and sperm-bearing globule of the green alga *Chara* (a) and the reproductive structures of land plants (b). All of these structures share three features. They are multicellular structures with a sterile external layer of cells and they originate from a single superficial cell.

ficial. Similar or identical genomic toolkits undoubtedly operate across all of the streptophytes, and molecular homologies will continue to be identified. However, the case for global structural homology among all these organs is weak for a number of reasons (for example, *Chara* has a branched filamentous body plan, whereas all land plants have a parenchymatous body plan; see chapter 7).

In summary, the preceding three examples (as well as many other examples that could have been presented) show that homology at one level of biological organization does not invariably translate into homology at other levels of biological organization. Each of the previous examples illustrates that homology is a manifestation of the maintenance and integration of a particular level of biological organization despite variation in the underlying mechanisms giving rise to it. The lesson that can be drawn from these examples, and that is reinforced by numerous examples of convergent evolution, is that natural selection typically acts on functional traits rather than on the mechanisms that generate them (*Many roads lead to Rome*) and that genome sequence homologies do not invariably translate into morphological homologies (*Rome isn't what it used to be*). Much as in the case of trying to produce a precise and universally acceptable definition of what is meant by *species*, the concept of what is meant by *homology* is perhaps best kept flexible depending on the research agenda being used.

The Transcription Factor Paradigm

A common ground between developmental and evolutionary biologists is the elucidation of the developmental mechanisms that influence the course of phenotypic evolution. This shared objective is the explicit impetus of the rapidly emerging evo-devo field. Curiously, the two worldviews of evo-devo have not been seamlessly welded together. As noted, the molecular regulation of development is traditionally described in terms of HOX and MADS genes whose products directly or indirectly control the transcription of gene products by binding to targeted DNA domains. These transcription-regulating factors influence development and morphology by activating or repressing downstream target genes according to the location of the developing body parts. HOX genes have been found in diverse lineages such annelids, insects, cnidarians, and vertebrates. Similar homeobox sequences have been reported for all eukaryotic lineages, including plants and fungi.

The phyletic ubiquity of this genomic machinery has fueled the gene regulatory network paradigm and a belief in the existence of a hegemonic mechanism for phenotypic evolution whose rules of operation shed light on some of the oldest theoretical problems in evolutionary biology. However, evolutionary biologists of every ilk are fully aware that the HOX and the MADS box paradigm is insufficient to explain how phenotypic evolution occurs for at least three reasons. First, as stressed in this chapter, the picture emerging from molecular developmental biology is not the evolutionary equivalent of the Beadle and Tatum "one gene–one enzyme" dictum for metabolism, but rather a picture of increasing complexity involving manifold gene-networks operating hierarchically and differentially in a nondeterministic manner that requires a continued conversation between a cell's genome and its cytoplasm. Second, phenomenological concepts such as individualization, dissociation, and heterochrony are inextricably linked to epigenetic processes that intervene between gene action and phenotypic expression. And, third, neither genomic nor epigenetic "rules" can account for the changes in allele frequencies that attend phenotypic evolution. The linkage between complex molecular genomic change and phenotypic evolutionary change thus will remain an open theoretical question until molecular genetic principles and the operation of natural selection, genetic drift, and other classical phenomena are conceptually integrated and pursued experimentally.

This claim does not in any way diminish the transcription factor paradigm, which is fundamental. The elucidation of global mechanisms often begins with the fastidious dissection of an individual model system. But when the model system reveals the paradigm is incomplete, the search for explication must be widened.

> "We shall not cease from exploration, and the end of all our exploring
> will be to arrive where we started and know the place for the first time."
> —T. S. ELIOT, *Little Gidding* (1942)

Suggested Readings

Dietrich, M. R. 2003. Richard Goldschmidt: hopeful monsters and other "heresies." *Nat. Rev. Genet.* 4: 68–74.

Frank, M. H., M. B. Edwards, E. R. Schultz, M. R. McKain, Z. Fei, I. Sørensen, J. K. C. Rose, and M. J. Scanlon. 2015. Dissecting the molecular signatures of apical cell-type shoot meristems from two ancient land plant lineages. *New Phytol.* 207: 893–904. doi: 10.1111/np4.13407.

Gilbert, S. F., J. M. Opitz, and R. A. Raff. 1995. Resynthesizing evolutionary and developmental biology. *Dev. Biol.* 173: 357–72.

Oldfield, C. J., and A. K. Dunker. 2014. Intrinsically disordered protein and intrinsically disordered protein regions. *Annu. Rev. Biochem.* 83: 553–84.

Schad, E., P. Tompa, and H. Hegyi. 2011. The relationship between proteome size, structural disorder and organism complexity. *Genome Biol.* 12: R120. doi: 10.1186/gb-2011-12-12-r120.

Wagner, G. P. 2014. *Homology, genes, and evolutionary innovations.* Princeton, NJ: Princeton University Press.

Woodward, J. B., D. Abeydeera, D. Paul, K. Phillips, M. Rapala-Kozik, M. Freeling, T. P. Begley, E. E. Ealick, P. McSteen, and M. J. Scanlon. 2010. A maize thiamine auxotroph is defective in shoot meristem maintenance. *Plant Cell* 51: 1627–37.

Xue, B., A. K. Dunker, and V. N. Uversky. 2012. Orderly order in protein intrinsic disorder distribution: disorder in 3500 proteomes from viruses and the three domains of life. *J. Biomol. Struct. Dyn.* 30: 137–49.

Yruela, I., and B. Contreras-Moreira. 2015. Genetic recombination is associated with intrinsic disorder in plant proteomes. *BMC Genomics* 14: 772.

Speciation and Microevolution

But the orderly sequence, historically viewed, appears to present, from
time to time, something genuinely new.
—C. LLOYD MORGAN, *Emergent Evolution* (1923)

"I know what kind of tree that is.—It is a chestnut."
"I know what kind of tree that is.—I know it is a chestnut."
The first statement sounds more natural than the second. One will
only say "I know" a second time if one wants especially to emphasize
certainty; perhaps to anticipate being contradicted.
—LUDWIG WITTGENSTEIN, *Über Gewißheit* (1950)

As we have seen in chapter 3, a central concept in population genetics
is that individuals can be grouped into populations and that similar
populations can be grouped into discrete biological units called species. The idea that species evolve to give rise to new species is the
fundamental precept of evolutionary biology, and, as we will discuss in
chapter 6, central to the concept of species selection and thus macroevolution. Nevertheless, many open-ended questions remain concerning how species preserve their biologically diagnostic features in spite
of heterogeneous and changing gene pools and environments, while

at the same time possessing the ability to evolve into new species. Other questions concern whether the mechanisms responsible for maintaining variation within a species are the same as those responsible for the origin of new species (speciation) or the origin of higher taxa ("genus-iation," "family-iation,", etc.).

In this chapter, we explore how the cohesiveness of genetically related populations can break down to give rise to new species within the context of four well-known speciation models. Two of these models emphasize the importance of geographic disjunctions among populations (vicariant speciation models). The other two emphasize how gene flow can break down within a population (non-vicariant speciation models). An exploration of the data relating to these models shows that the latter two models are particularly suitable when describing how plants evolve into new species owing to their propensity for polyploidy and hybridization, neither of which is common among animals. Nevertheless, all four models propose mechanisms that restrict gene flow within or among populations of conspecifics. Accordingly, these models incorporate elements of the biological species concept. At the end of this chapter, we will explore alternative species definitions as a prelude to the topic of macroevolution, which will be discussed at length in chapter 6.

The Biological Species Concept and Vicariant Speciation

According to the authors of the Modern Synthesis, a species consists of sets of individuals grouped into populations whose members can potentially interbreed and yield viable, fertile offspring but cannot breed successfully with members of other species. This definition is called the biological species concept. It has two components: (1) gene flow is maintained within and among populations of a species, and (2) gene flow is negligible or nonexistent between members of different species. According to this species concept, speciation can occur when gene flow between members of the same species is reduced or becomes impossible. In general, the biological species concept places greater emphasis on reproductive isolation compared to the capacity, whether real or potential, for interbreeding. This concept of what it means to be an extant species remains the most generally accepted species definition for organisms with biparental sexual reproduction. It can be applied to many plant species whose members are interfertile and reproductively isolated from other species by strong sterility barriers. For many plant species, however, emphasis must be placed more on the second of the two components of the biological species concept. Some plant species hybridize freely and so are not reproductively isolated from one another yet are geographically disjunct and have little opportunity to interbreed. In other cases, interbreeding is limited to immediate neighbors within populations. For these reasons and others, the biological species concept as it applies to plants has been richly criticized.

Nevertheless, with its emphasis on reproductive isolation, the biological species concept offers a valuable hypothesis for the evolutionary phenomenon of speciation. A new species evolves from a preexisting one as a consequence of processes or mechanisms that reduce or completely block gene flow between subpopulations originally sharing the same ancestral gene pool. The concept posits gene flow resulting from random mating as the glue that binds all the members of a species into a discrete biological entity and posits that species evolve from a preexisting one when this glue is eroded by any mechanism resulting in reproductive isolation (fig. 5.1).

This mode of speciation can be visualized in terms of Sewall Wright's fitness landscape metaphor discussed in chapter 3 wherein a cluster of fitness promontories identify genetically distinct subpopulations belonging to the same species. Because these subpopulations occupy different promontories, the average member of each subpopulation differs somewhat from the average member of the other subpopulations. As long as gene flow is maintained, all the subpopulations will share much of the same ancestral gene pool. If the subpopulations diverge far enough genetically or phenotypically owing to different selection regimes, barriers to gene flow between two or more

(a) 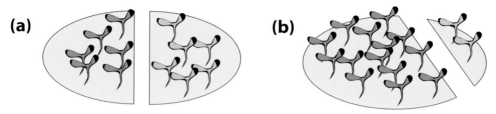 **(b)**

Figure 5.1. Schematics of two modes of vicariant speciation. Individuals sharing the same color depict members sharing the same gene pool. a. Allopatric speciation (or geographic speciation) occurs when members of a population become geographically isolated from one another and phenotypically or genotypically diverge as a consequence of genetic drift, mutation, or different selection pressures. b. Peripatric speciation occurs when a few individuals peripheral to a population become isolated. Because the isolated population is comparatively small, it is more susceptible to founder effects (the significant reduction of genetic variation resulting from a small number of individuals in a splintered population).

subpopulations may become established. If low gene migration rates persist, reproductive isolation will intensify and foster further genotypic or phenotypic divergence. If this persists, the subpopulations may eventually diverge until they warrant separate species designations.

The subpopulations of a species are expected to diverge over time because of the combined effects of the stochastic (random) processes of mutation and random changes in gene frequencies (genetic drift), and the deterministic (nonrandom) process of natural selection. In theory, the influence of random processes will be more pronounced in small subpopulations than in large groups of individuals. Nevertheless, natural selection will countermand random changes whenever the mean fitness of a subpopulation is reduced for long. Because subpopulations occupy different fitness promontories, which by definition can be reached only by genetic or phenotypic differentiation, selection acting on different subpopulations must differ in kind or degree. These different selection regimes have the potential to pull the entire population apart. In a randomly mating (panmictic) population, the fracturing of subpopulations is reduced because all members of the species have reproductive access to one another. The traditional view is that gene flow among subpopulations living in close proximity (sympatric populations) reduces the extent to which subpopulations genetically diverge. However, when subpopulations are geographically isolated from one another (allopatric populations), gene flow is reduced and different mutations or allele frequencies produced by chance may be fixed in populations. The paucity or absence of gene flow among allopatric subpopulations can result in speciation.

The portrait of speciation owing to vicariant geographic isolation is called allopatric speciation (fig. 5.1 a). It involves three sequential steps: (1) subpopulations become spatially isolated from the main range of their parent population, (2) subpopulations genetically diverge from one another, and (3) selection leads to genotypic or phenotypic differences that preclude subpopulations from interbreeding with one another. In sum, a physical or biological barrier isolates members of the same species so that they cannot interbreed, allowing them to diverge in appearance or behavior. At some point the divergence may be so great that a new species comes into existence.

Peripatric speciation is a variant of the allopatric speciation model. According to this model, one or more small peripheral subpopulations become separated from the main population of a species and undergo genotypic or phenotypic divergence (fig. 5.1 b). The principal difference between this model and the allopatric speciation model is that the splinter peripatric subpopulations are small and thus contain low genetic diversity. Under these circumstances, peripatric populations are subject to founder effects, selection bottlenecks, and genetic drift. As a consequence of reduced genetic variation, a peripatric population may not resemble its parent species (the founder effect) and thus may respond to different selection regimes resulting in phenotypic divergence (selection bottlenecks). Further, because the peripatric population is genetically unrepresentative of its parent population, there may be fewer copies of some alleles such that allele frequencies shift as offspring are produced (genetic drift) (fig. 5.2). These phenomena are expected to be prevalent in populations with no overlapping generations, as for example annual plants.

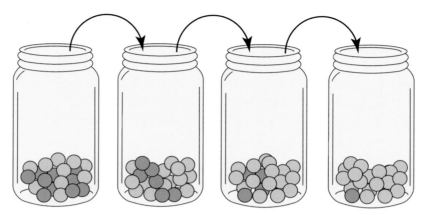

Figure 5.2. An illustration of genetic drift using marbles in a jar. The different colored marbles represent different alleles of the same gene. In the example shown here, ten red and ten green marbles are placed in the first jar (left) to represent a parapatric population. Marbles are then randomly drawn from the first jar and the color of each is used to represent the color of the marble placed in the second jar. The process is repeated to randomly select successive populations of parents. Over time, one color (allele) predominates. Indeed, because the selection process is random, it is possible that the first drawing will eliminate one color entirely.

Geographic Barriers and Vicariant Speciation

Geographic isolation, as emphasized in the allopatric and peripatric speciation models, is the most readily understood of all the barriers to gene flow. Perhaps the most dramatic examples of vicariant speciation are seen on island chains because they are physically isolated from the mainlands that supply colonizing plants and animals, because they tend to be colonized by only a few individuals (a small gene pool), and because ecological niches are initially unoccupied. These features are well illustrated by the Hawaiian archipelago, which is home to the largest number of endemic plant species in any one geographic area. An endemic species is a species that has evolved in a particular location and is found nowhere else in the world. A native species is a species that has arrived from a different location and established itself in a flora or fauna. The amount of endemism is particularly impressive in the Hawaiian flora: 89% of the 956 native flowering plants are endemic species, while 15% of the 32 native genera are endemic

Table 5.1. Summary of endemic and native Hawaiian flowering plant species, genera, and families

	Eudicots	Monocots	Total
Total species	822 (86%)	134 (14%)	956
Endemic species	759 (92%)	91 (68%)	850 (89%)
Total genera	165 (76%)	51 (24%)	216
Endemic genera	31 (19%)	1 (2%)	32 (15%)
Families	73 (84%)	14 (16%)	87
Total taxa	947 (87%)	147 (13%)	1,094
Total endemism	888 (94%)	107 (73%)	995 (91%)

Source: Wagner, Herbst, and Sohmer 1990.

(table 5.1). In terms of the number of colonizing species, it is estimated that the native flora is derived from as few as 272 successful
species introductions, migrating predominantly from the southwest
of the Hawaiian archipelago (most of the native flowering species have
a Malaysian taxonomic affinity). The rate of allopatric speciation has
been inferred on the basis of potassium-argon dating of Hawaiian volcanic rocks, which indicates that one of the oldest currently habitable
islands (Kauai) is only between 4 and 7 million years old.

It is tempting to believe that the taxa endemic to the Hawaiian
islands could not be older than 6 million years. However, the Hawaiian island chain extends far to the west to include formerly large and
much older volcanic islands that have been worn down by erosion and
are now reduced to low atolls and reefs barely breaking the ocean's surface. Species that once evolved on these much older western islands
may have colonized younger islands forming in the east, hopscotching
from older to younger islands. Thus, precise estimates of allopatric or
peripatric speciation rates for the Hawaiian island chain are problematic at best. Nevertheless, some species are endemic exclusively to the
younger Hawaiian islands and so could not possibly have evolved in a
time less than the age of these comparatively newly formed islands.
For example, 79 "single island" endemic species are known to grow
only on the Big Island, which is estimated to be about 1 million years
old. Likewise, 158 species are endemic only to Maui, which formed

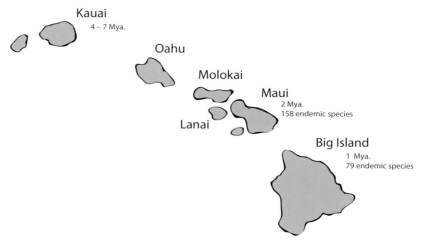

Figure 5.3. Schematic of the Hawaiian archipelago showing the estimated ages of the youngest island (the Big Island), Maui, and one of the oldest habitable islands (Kauai). The numbers of endemic plant species known only from the Big Island and Maui indicate that vicariant speciation can occur rapidly.

about 2 million years ago. These and other data indicate that the rates of vicariant speciation on geographically isolated islands can be very high indeed (fig. 5.3).

Other Barriers to Reproduction

Although vicariant speciation models emphasize geographic isolation, reproductive isolation can occur in several ways. Traditionally, the barriers to gene flow are grouped into two broad categories: those that prevent cross-mating or cross-fertilization and those that preclude reproductively viable hybrids (table 5.2). Because the zygote is the cell resulting from the fusion of egg and sperm, its formation is a discrete biological event in the sexual life cycle of all biparental species and thus conveniently allows us to distinguish mechanisms that preclude the formation of zygotes from the fusion of sperm and egg, each drawn from a different species (prezygotic barriers), from those that exclude the development of a zygote into a hybrid organism

Table 5.2. Barriers to sexual reproduction between species

A. Barriers to mating or fertilization of eggs (prezygotic barriers)

   1. Geographical isolation—physically separated populations

   2. Temporal isolation—reproduction occurs at different times

   3. Mechanical isolation—structural differences in reproductive organs

   4. Gametic incompatibility—sperm and egg fail to fuse

B. Barriers to the formation of viable hybrids (postzygotic barriers)

   1. Hybrid inviability—embryos fail to develop; hybrids fail to reach sexual maturity

   2. Hybrid sterility—hybrids fail to produce functional reproductive organs

   3. Hybrid breakdown—progeny of hybrids revert to parental types; progeny are
      inviable or sterile.

(postzygotic barriers). Here, we will briefly review a few examples of pre- and postzygotic barriers with a view to how the breakdown of these mechanisms can lead to speciation.

Experiments verify that reproductive isolation owing to temporal displacement can occur rapidly even within sympatric populations when selection is intense. For example, Paterniani (1969) planted a mixed population of two varieties of corn (*Zea mays* var. white flint and yellow sweet) and removed all hybrid seeds from successive populations over a period spanning four generations. This process is called truncated selection. The removal of all hybrids removes the genes that match the pollen shed with stigma receptivity. Even though the original mixed population had a comparatively high rate of intercrossing (35.8% for white flint and 46.7% for yellow sweet), the rate of intercrossing was dramatically reduced by the fourth generation (to 4.9% and 3.4%, respectively). The result of artificial disruptive selection for plants exclusively representing intra-varietal matings was a substantial alteration in the flowering times of the two corn varieties after only a few generations (fig. 5.4).

The displacement of morphological characters can provide an effective mechanical prezygotic barrier to reproduction. Character displacement is sometimes observed among closely related sympatric species that flower at approximately the same time of year. For ex-

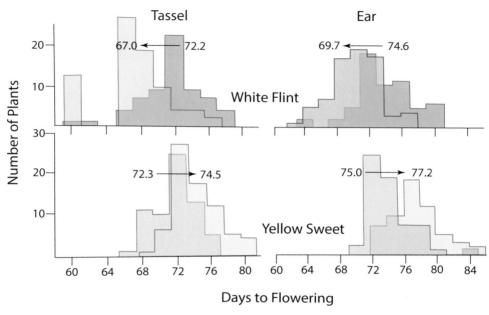

Figure 5.4. The displacement of flowering times for both the pollen-producing tassel and the seed-producing ear of two varieties of corn, *Zea mays*, white flint and yellow sweet (shown in blue and yellow, respectively) following four generations of intense selection to eliminate all naturally occurring hybrids between the two varieties. The darker shaded frequency distributions plot the data for the plants at the beginning of the experiment. The lighter shaded frequency distributions plot the data from the fourth generation of plants. The arrows above each pair of frequency distributions indicate the direction of the shift in the number of days to flowering; the numbers are the means of the days of flowering. Note that the two cultivars started out having nearly identical tassel and ear flowering times. However, disruptive selection resulted in statistically significant shifts in flowering times despite intra-varietal mating. Data taken from Paterniani (1969).

ample, the lengths of styles and anthers in the flowers of two night-shade species, *Solanum grayi* and *S. lumholzianum,* are similar for both species in allopatric populations in Arizona. However, when the two species coexist in close quarters, as in the Sonoran Desert, the anthers and styles of *S. grayi* are significantly shorter (63% and 73%) than those of *S. lumholzianum* (Levin 1978). The difference in style and anther length reduces the probability that the sympatric species will utilize the same pollinator and so reduces the likelihood that their reproductive processes will interfere. Gametic incompatibility involving genetically controlled pollen-stigma interactions is well known among flowering plant species. Space precludes a detailed discussion of these complex interactions here; interested readers are referred to the excellent treatments of this topic provided by June and Mikhail Nasrallah.

Postzygotic barriers to sexual reproduction also help to maintain species as discrete biological entities even when prezygotic barriers fail. Postzygotic barriers preclude the formation of viable hybrid zygotes or the survival of viable hybrids (see table 5.2). A hybrid can be aborted at any stage in its life or, if it survives to maturity, it may fail to perpetuate its own kind. The hybrid zygote may die immediately after sperm and egg fuse, or the developing plant may die before reaching maturity (hybrid inviability). Another counterintuitive possibility is that the largest and potentially most vigorous seeds are selectively eaten precisely because they are the largest and best. In flowering plants, the survival of the zygote and its successful development into an embryo typically depend on the formation of endosperm. The endosperm is a product of double fertilization, because one sperm fuses with the egg and the second sperm enters another cell that will normally develop into the endosperm. The abortion of hybrid zygotes or embryos is frequently the result of anomalous endosperm growth (as for example in the angel trumpet *Datura*, *Iris*, and cotton *Gossypium*). Even when hybrids develop into plants, they may be vegetatively weak and easily killed by competition with their neighboring parental forms for nutrients and space. Conversely, some hybrids exhibit "hybrid vigor"—that is, they grow more luxuriantly and are more competitive than either parent. But the hybrid may fail to develop functional reproductive organs (hybrid sterility) or may produce progeny that are either sterile or, if reproductively competent, genetically segregate back to the parental types (hybrid breakdown). In either case, the hybrid fails to perpetuate its own kind and so cannot establish a population of individuals. An escape route from this barrier is asexual reproduction, which gives the genome of a hybrid time to become reorganized and sexually competent.

The postzygotic barriers to hybridization embrace a wide variety of effects involving disharmony in gene and chromosome interaction that may be expressed in vegetative growth (somatic effects leading to poor growth) or during the formation of reproductive organs (gametic effects that result in hybrid sterility or breakdown). One classic example involving chromosomal disharmony must suffice. In 1927, Georgi D. Karpechenko (1899–1941) attempted to hybridize the common radish (*Raphanus sativus*) with cabbage (*Brassica oleracea*). Both genera belong to the mustard family (Brassicaceae), and both species have the same number of chromosomes in their diploid cells (2n = 18). Karpechenko's attempt to produce a hybrid was successful. However, the hybrid plants had nine pairs of chromosomes in their diploid cells, and the nine chromosomes from one parent failed to recognize their homologues from the second parent and so did not pair during meiosis. As a consequence, the division products of meiosis had varying numbers of chromosomes, mostly ranging between six and twelve per cell. Because normal sperm or eggs could not be formed in flowers, the hybrid was sterile as a consequence of chromosomal disharmony (the failure of homologous chromosomes to pair).

However, the experiments of Karpechenko also show how a new species can evolve from a sterile hybrid whose cells spontaneously double their chromosome numbers. Karpechenko found that some of his "sterile" hybrid plants produced a few viable seeds that grew into fertile plants whose diploid cells had thirty-six chromosomes. Apparently these viable seeds were produced when the chromosome number spontaneously doubled in the germinal tissue, giving rise to cells with two chromosomes of each kind from both parents. This doubling permitted the pairing of homologous chromosomes and the formation of gametes with eighteen chromosomes. Fusion of these gametes gave rise to fertile plants whose somatic cells contained thirty-six chromosomes. Karpechenko had "synthesized" a new species because it has no possibility of gene flow with either of its two parent species. He called this organism *Raphanobrassica*. Unfortunately, it had the roots of a cabbage and the leaves of a radish. Nonetheless, Karpechenko had witnessed the birth of a new species (or more precisely, a new genus) in the passage of three plant generations.

Distractions

As we have seen, the biological species concept states that gene flow in panmictic populations holds the members of a sympatric population together genetically, that species are maintained as discrete biological entities by reproductive isolation from other species, and that isolated subpopulations can undergo genetic and phenotypic differentiation as a prelude to speciation. However, several factors cast doubt on the general applicability of the biological species concept and on corollary vicariant models for speciation. Indeed, they indicate that the broader issue of the relations among genetic differentiation, phenotypic divergence, and the mechanisms that establish genetic incompatibilities leading to reproductive isolation are far more complex and diverse. For example, early studies of plant populations showed that gene flow may be spatially limited even among the members of a sympatric population. In these cases, gene flow may genetically bind together the individual plants composing subpopulations, but it cannot account for the features characterizing the entire sympatric population. Early studies also showed that plant species that are radically different morphologically may nevertheless produce viable hybrid swarms of such complexity and size that parental gene complexes become nearly or totally unrecognizable. By way of an example, consider the goldenrod species *Solidago rugosa* and *S. sempervirens*. The former species is a tall, strongly hairy plant with thin, dentate leaves, widely distributed in most of the eastern United States and Canada. In contrast, *S. sempervirens* is a short, glabrous plant with fleshy and entire leaves that grows along the Atlantic seaboard, penetrating inland only along tidal rivers. However, when populations of these two species make contact, they freely interbreed to produce fertile hybrids. In turn, these hybrids produce fertile later-generation backcross hybrids that differ so much

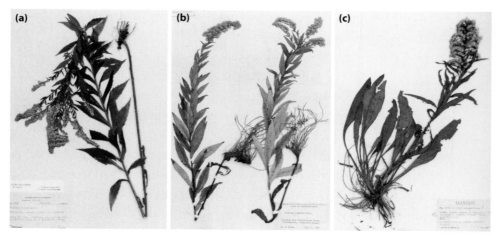

Figure 5.5. Herbarium specimens illustrating a hybrid between *Solidago rugosa* and *S. sempervirens*. a. *Solidago rugosa*. b. *Solidago × asperula*, the hybrid. c. *S. sempervirens*. Specimens from the Liberty Hyde Bailey Hortorium (Cornell University).

from the two parent species that they are accorded species status as *Solidago × asperula*. As a pair of entities, *S. rugosa* and *S. sempervirens* are so phenotypically and ecologically different that few taxonomists would attribute them to the same species. Nevertheless, the two species freely interbreed in nature and give rise to what authorities recognize as a third species (fig. 5.5).

Cryptic or sibling species also detract from the proposition that phenotypic divergence is a necessary consequence of reproductive isolation. Sibling species are genetically separated by strong barriers to sexual reproduction yet are barely or not at all distinguishable on morphological or anatomical grounds. Here, reproductive barriers may have set in suddenly or progressively, but their establishment has not (or has not yet) been supplemented by phenotypic characters sufficient to distinguish easily among sibling species. Accordingly, the mechanisms underpinning phenotypic and genotypic divergence are neither invariably identical nor correlated with one another. If one admits that the essence of speciation is the formation of genomic divergence and reproductive isolation, then sibling species show that phenotypic divergence is neither an attribute of species status nor invariably the end result of speciation, even though it is undeniable that in most cases, in both plants and animals, speciation has been accompanied by or has followed phenotypic differentiation.

Another difficulty with the biological species concept is that to be called a species organisms must pass the test of reproductive isolation. As we have seen, some sexually reproductive species fail this test because they can hybridize with other species (see fig. 5.5). By the same token, asexual life forms, many of which are formally recognized as bona fide species, also fail the test of reproductive isolation. The bacteria are perhaps the most familiar unicellular asexual life forms. Bacteria reproduce asexually by means of binary fission. Although genetic information can be transferred among different strains of bacteria (as for example by means of bacterial conjugation), genetically isolated lineages of bacteria have arisen through chance mutation amplified by asexual multiplication. Genetically different clones have clearly diverged through the combined effects of mutation and selection to evolve into different species.

Even if one argues that species definitions for prokaryotes must be based on criteria entirely distinct from those for eukaryotes, a substantial number of plants (from the algae to the flowering plants) reproduce exclusively or predominantly by asexual means.

For example, many flowering plants reproduce by apomixis (they produce seeds without the fusion of sperm and egg). The apomictic seed develops within the ovules within the ovary of the flower, as in sexually reproducing species, but the embryo within the seed originates by mitotic division of a somatic cell within the ovule. Except in special cases, the embryo is genetically identical to the parent plant, and so populations of apomicts are genetic clones. Apomixis has been described as an escape from sterility, but into a blind alley of evolution (but at least with an established mode of dispersal). The implication is that all apomicts are doomed to extinction. Nevertheless, apomixis has been reported in at least thirty families of flowering plants including more than four hundred species, some of which, like the dandelion, *Taraxacum officinale*, are highly invasive and ecologically success-

(a) **(b)**

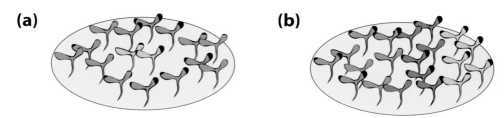

Figure 5.6. Schematics of two modes of non-vicariant speciation. Individuals sharing the same color depict members sharing the same gene pool. a. Sympatric speciation occurs when a new species evolves within the home range of one or both of its parental species. b. Parapatric speciation involves a continuous series of overlapping subpopulations with minimal contact zones resulting from either unequal dispersal, incomplete geographical barriers, or similar factors. The most common mechanisms for non-vicariant speciation among plants are disruptive selection, polyploidy, and hybridization.

ful organisms. Those who subscribe to the biological species concept may summarily dismiss uniparental organisms as legitimate species. Ironically, one may also argue that obligate asexual reproduction provides the ultimate reproductive barrier between species. Regardless, the dismissal of asexual life forms as bona fide morphospecies is impotent in the face of parthenocarpic animals and apomictic plants that are in every other respect real species, while the notion that asexual reproduction is a "barrier" to gene flow between species in the sense of the biological species concept is logical legerdemain.

There is no doubt that allopatric and peripatric speciation occur. The diversity and abundance of endemic "single island" species in the Hawaiian islands makes this abundantly clear (see table 5.1 and fig. 5.3). However, from our foregoing discussions, it should be clear that the vicariant speciation models that emerged from an unwavering adherence to the biological species concept are *necessary but insufficient* to cover the diversity of species and the processes that give rise to them. To fill this gap, we now turn attention to two additional models, the sympatric and the parapatric speciation models (fig. 5.6). The sympatric speciation model posits that a new species can evolve within the home range of its ancestral species despite high initial gene flow and without any special spatial segregation. Parapatric speciation involves the evolution of reproductive isolation between populations that have

a continuous biogeographic distribution and the potential for significant gene flow. The mechanisms by which gene flow can cease even in
panmictic populations include disruptive selection, autopolyploidy,
and hybridization. We have touched on these mechanisms, albeit very
briefly (see, for example, figs. 5.4 and 5.5), but not in sufficient detail.

Non-vicariant Speciation

Karpechenko's experiments showed that a new species can evolve
through polyploidy. The cells of a polyploid individual contain extra
sets of chromosomes. There are two ways a polyploid species can
evolve. The first and simpler to understand is autopolyploidy, which
results from the spontaneous doubling of chromosome number in
germ line cells to produce tetraploid germ line cells. A species could
arise through autopolyploidy when the sperm cells and eggs of members of a population with accidentally doubled chromosome numbers
interbreed and produce tetraploid offspring. When these offspring
mate with one another, they can produce fertile tetraploid progeny
that are genetically isolated from the parental diploid population.
Thus, in a single generation, autopolyploidy can establish a barrier to
gene flow between a fledgling species and its parent species.

Autopolyploidy can result from either meiotic nondisjunction or
mitotic nondisjunction (fig. 5.7). Meiotic nondisjunction is the failure of homologous chromosomes to move apart during meiotic cell
division. Mitotic nondisjunction is the failure of sister chromatids
within chromosomes to move apart during mitotic cell division. In
the case of meiotic nondisjunction, an autopolyploid species may
evolve when normal diploid germ line cells begin to divide meiotically
but homologous chromosomes fail to separate. The resulting diploid
gametes then fuse to produce a tetraploid zygote that may develop
into a self-fertile tetraploid plant. Likewise, speciation through autopolyploidy could result when somatic cells within a diploid species
become tetraploid because sister chromatids fail to separate during
mitosis (mitotic nondisjunction) in cells leading to the germ line. After

Meiotic nondisjunction

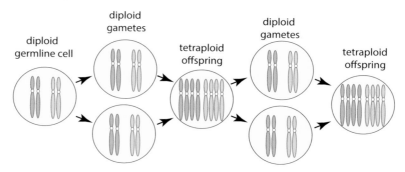

Mitotic nondisjunction

Figure 5.7. Schematics of two ways autopolyploidy (the spontaneous doubling of chromosome number that yields a tetraploid cell) can occur and possibly give rise to a new species. The ploidy of a "normal" diploid cell is depicted as having two pairs of homologous chromosomes (shown by different colors). Meiotic disjunction occurs when homologs fail to pair and separate (upper panel). This results in diploid gametes that, if viable, can produce a tetraploid offspring. The tetraploid offspring can produce diploid gametes that in turn can produce offspring like its parents. Mitotic nondisjunction occurs when the chromosome number of a "normal" somatic cell spontaneously doubles and gives rise to a tetraploid germline cell that subsequently undergoes meiosis to produce diploid gametes. If viable, the gametes can produce a tetraploid offspring that can produce diploid gametes that in turn can produce offspring like its parents.

meiosis, the resulting diploid gametes can fuse to produce a tetraploid zygote that develops into a self-fertile tetraploid. Speciation by means of autopolyploidy was first discovered by Hugo de Vries (1848–1935) while he was examining the genetics of the evening primrose. The parent species, *Oenothera lamarckiana*, has fourteen chromosomes; the autopolyploid, *O. gigas*, which cannot backcross successfully

with *O. lamarckiana*, has twenty-eight. Many commercially important plants are autotetraploids (*Zea mays* and horticultural strains of
snapdragon).

The second way a sympatric species can evolve through polyploidy is by allopolyploidy, which has occurred frequently among the
angiosperms (such as *Galeopsis* and *Iris*) and the ferns and the "fern
allies" (such as the fern genus *Asplenium* and the lycopod *Selaginella*)
(box 5.1). An allopolyploid results when two species successfully
mate to produce offspring containing the chromosomes of both parents (fig. 5.8). These interspecific offspring would be sterile whenever
the haploid set of chromosomes from one species excludes the possibility that meiosis can produce functional gametes, which can happen
whenever the tetraploid has an odd number of chromosomes (such
that not all chromosomes have homologues) or when homologues do
not exist among even numbers of chromosomes.

The second way allopolyploidy can occur, which may be more
common than the first, is when meiotic nondisjunction in one of
two parental diploid species produces an unreduced "diploid" gamete
that fuses with a normal gamete from the second parent species (see
fig. 5.7). Even though the resulting hybrid is sterile, meiotic nondisjunction can produce unreduced triploid gametes that, when backcrossed with one of the parent species, produce viable offspring. The
allopolyploid species resulting from these offspring has a chromosome
number equal to the sum of the chromosome numbers of the two parent species. It is estimated that between 25% and 50% of all plant species have evolved from allopolyploidy. Many of these polyploid species
are commercially important crops such as oats, cotton, and potatoes.

Two mechanisms permit the evolution of a viable hybrid species. The first possibility is that mitotic nondisjunction occurs in the
germ line tissues of the sterile diploid hybrid to produce unreduced
polyploid gametes. This mechanism was the route to the new species
observed by Karpechenko. Another example is seen in triticale, a
polyploid organism resulting from a cross between wheat (*Triticum*,
2n = 42) and rye (*Secale*, 2n = 14) combining the high yields of wheat

Box 5.1. Reticulate Evolution and the Appalachian *Asplenium* Species Complex

Unlike vertebrate speciation, allopolyploidy (the hybridization of two species to produce a new species) is not uncommon among plants. Indeed, it may be one of the most prevalent modes of plant speciation. In contrast to the genetic divergence among diploids that precedes speciation among animals, allopolyploidy can result in immediate reproductive isolation from progenitors and fixes a hybrid's heterozygosity at many loci owing to the nonsegregation of nonhomologous chromosomes. Additionally, spontaneous doubling of chromosomes in hybrids allows for the hybridization among hybrid species, which can produce a reticulate swarm of species differing in their ploidy levels.

The phenomenon of reticulate evolution is nowhere better seen than in what has been referred to the Appalachian *Asplenium* species complex (fig. B.5.1), which has been studied intensely by numerous pteridologists. This complex consists of three diploid species (*A. platyneuron*, *A. rhizophyllum*, and *A. montanum*) and six hybrid species, three of which are hybrids between hybrid species (*A. kentuckiense*, *A. gravesii*, and *A. trudellii*). Two noteworthy attributes of the hybrid species in this swarm is that many are fertile and manifest phenotypes that combine widely different morphologies observed in the parental species. A third feature of note is that some hybrids have extended geographic ranges beyond their parental species (as for example *A. bradleyi* and *A. pinnatifidum*). Finally, although some hybrids in this complex are typically sterile (for example, *A. ebenoides*), most of the species are fertile and all are capable of asexual reproduction by means of rhizome fragmentation, or the production of adventitious plantlets. These four features dispel the notion that hybrids are invariably less vigorous than their parental species and the notion that they are invariably sterile.

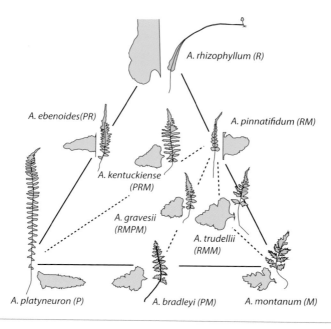

A. rhizophyllum (R)

A. ebenoides(PR)

A. pinnatifidum (RM)

A. kentuckiense (PRM)

A. gravesii (RMPM)

A. trudellii (RMM)

A. platyneuron (P) A. bradleyi (PM) A. montanum (M)

Figure B.5.1. Schematic of species of the Appalachian *Asplenium* species complex resulting from hybridization and allopolyploidy. Each species is depicted by a representative frond and leaflet (pinnule). Solid lines denote hybrids between the three diploid species; dashed lines denote hybrids among hybrids. Redrawn from Wagner (1954).

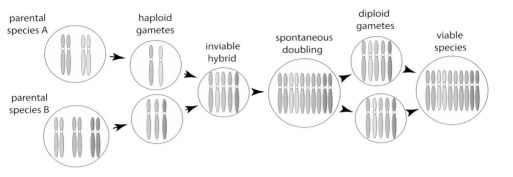

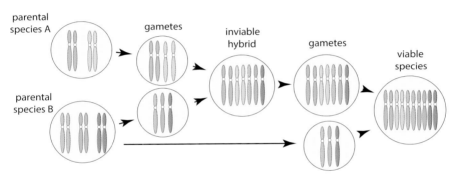

Figure 5.8. Schematics illustrating how allopolyploidy resulting from the hybridization of two species (species A and species B) can give rise to a new viable species. The ploidy of a "normal" diploid cell is depicted as having two pairs of homologous chromosomes (shown by different colors).

with the hardiness of rye. Commercially important species have been produced in this way from three *Brassica* species. Known as the *Brassica* "triangle of U," the allopolyploids and their parent species are as follows: *B. oleraceae* (n = 9) and *B. nigra* (n = 8), which give rise to *B. carinata* (n = 17, the Abyssinian mustard); *B. oleraceae* (n = 9) and *B. campestris* (previously called *B. rapa*) (n = 10), which give rise to *B. napus* (n = 19, the oil rape and the rutabaga); and *B. campestris* (n = 10) and *B. nigra* (n = 8), which give rise to *B. juncea* (n = 18, the leaf mustard) (fig. 5.9).

The birth of a new species through polyploidy does not require a

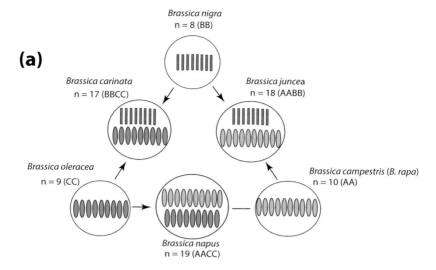

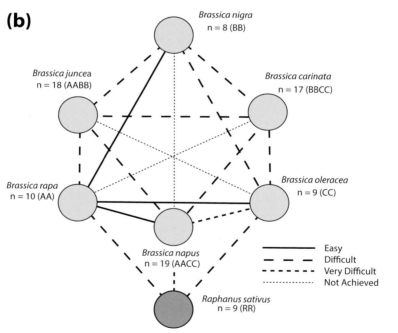

Figure 5.9. Schematics of the intergeneric hybrids among the species within the genus *Brassica* (a) and the radish *Raphanus sativus* (b). a. The "triangle of U" for the various species in *Brassica* showing the chromosome number (n) and genomes of each species. b. The *Brassica* "species triangle" (modified to include radish, *Raphanus sativus*). The haploid chromosome numbers (n) and genomes for each species are indicated parenthetically (A, B, C, or R). Different lines indicate the ease with which hybrids can be created between species (see insert).

geographic barrier, so it can occur within the home range of the parent species. Thus, speciation by means of polyploidy provides an example of how sympatric speciation may occur. Polyploidy is comparatively rare among animals, but comparatively common among plants. This explains perhaps why some zoologists believe allopatric speciation is the most prevalent mode of speciation, and why many botanists believe the reverse is true. Although both modes of speciation are equally theoretically viable, from a practical point of view the issue revolves around the "relative frequency" of the two modes. This can be judged only within the context of the type of organism being studied. That plants differ from animals in many important respects is an undeniable fact of life. That plants likely have also evolved and speciated differently from most animals may also be true.

Hybridization and Genetic Distance

Numerous studies of plant populations verify that hybridization occurs naturally and that hybrid zones can evolve into stabilized, genetically distinctive populations and even species. Hybridization may also give rise to new adaptations or combinations of adaptations by transferring and combining phenotypic features among species. It is profitable, therefore, to briefly review some examples of species hybrids, emphasizing the genetic distance among the progressively higher ranks in the taxonomic hierarchy (species, genera, and so forth).

Although all the members of a species are closely related genetically and presumably are reproductively compatible, no two subpopulations of the same outcrossing species are identical in every respect. Taxonomic categories below the species rank, called infraspecific units, are frequently established to deal with recognized differences between subpopulations in one or more heritable traits. The most widely used infraspecific unit is the subspecies. Subspecies inhabit different geographical areas or habitats within the range of their species, and they often presage parapatric speciation events (see fig. 5.6 b). Nevertheless, they can interbreed and so may genetically and phenotypically intergrade, particularly where their ranges normally make contact or when they are artificially brought together. Experiments verify that the ability of subspecies to interbreed reflects differences in divergence times between subspecies. In this respect subspecies may be viewed as biological entities that have taken the first steps toward the genetic and phenotypic divergence traditionally posited to preface the evolution of a new species, even though gene flow typically is sufficient to hold subspecies together genetically and phenotypically so that they are taxonomically recognizable as belonging to the same species.

Infrageneric taxonomic groups are also recognizable. Infrageneric taxa are groups of closely related species assigned to the same genus. The most commonly recognized is the species group called the section. The species assigned to one section are presumably more closely related to one another than to species assigned to another section within the same genus. Thus the extent to which species within and across sections of the same genus can interbreed illustrates simultaneously the genetic and phenotypic divergence attending the evolution of species and higher taxa. Numerous studies verify that the potential for gene flow between species in the same section varies over a continuum but is generally greater than the potential for gene flow between species belonging to different sections within the same genus. A good example is seen in the North American genus *Ceanothus*, which ranges from southern Canada to Guatemala and is principally found along the Pacific coast. The genus consists of two sections, *Cerastrus* and *Ceanothus* (which was formerly called the section *Euceanothus*). Species within each section can interbreed and produce vigorous hybrids that may extend into the second generation. However, interbreeding between species belonging to different sections typically yields either sublethal hybrids or hybrids that fail to set seeds even under otherwise favorable garden conditions (Nobs 1963). Presumably, species in the two sections have genetically and phenotypically diverged so far that hybrids between them are inviable, whereas vigorous hybrids are possible between more closely related species in the same section.

The situation in the genus *Ceanothus* is far from unique. Many intrageneric hybrids are known, and they are sometimes so genetically complex and phenotypically distinctive that the boundaries between species nested in the same genus may be blurred out of existence and hybrids may properly be accorded species status. This has happened for the two species *Clarkia nitens* and *C. speciosa*. Naturally occurring intrageneric hybrids of oak are also well known. Intrageneric hybrids may result when species that formerly were geographically isolated are brought into artificial contact. For example, the sycamore trees of the eastern United States (*Platanus occidentalis*) and the eastern Mediterranean (*P. orientalis*) are geographically separated and so do not normally interbreed. But when planted together they can produce fertile progeny that have been designated as a separate species (*P. acerifolia*). Habitat disturbance also can result in hybridization. Perhaps the classic example of the "hybridization of habitat" is seen in species of *Iris* inhabiting the Mississippi Delta. Before the human disturbance of their native habitats, two distinct sympatric species, *Iris fulva* and *I. hexagona* var. *giganticaerulea* existed, respectively, in well-drained forests along the edges of rivers and streams and in poorly drained swamps. The advent of deforestation and the draining of swamps produced habitats combining the features of both parental habitats, resulting in an enormous diversity of hybrids and hybrid derivatives between the two species.

Hybrid populations are not accorded species status unless they maintain their unique biological identity in successive generations. The evolution of a new species from a hybrid has been called hybrid speciation, which occurs when a small proportion of the descendants of a partly fertile species hybrid become stabilized and give rise to a population within which free gene exchange is possible, though it is not possible between either parental type. Among flowering plants, one way this can happen is when the pollination syndrome of the hybrid diverges from that of either of its parents. This has occurred for *Delphinium gypsophilium*, a species resulting from the hybridization of *D. hesperium* and *D. recurvatum*, and for *Penstemon spectabilis*, which evolved from a hybrid between *P. centranthifolius* and *P. grinnellii*.

The ability of species to hybridize has received close attention because interspecific crosses are one way to introduce new traits into commercially important crops. Plant breeding programs therefore shed considerable light on the genetic barriers among related species and genera. The genus *Solanum* (formerly *Lycopersicon*) consists of two major species complexes, the *lycopersicon* complex (to which the commercial tomato *S. lycopersicum* belongs) and the *Eriopersicon* complex. A major barrier to hybridization separates these two complexes, although hybrids between *Eriopersicon* and *Lycopersicon* species have been reported on very rare occasions (fig. 5.10). This barrier is more easily circumvented by obtaining intermediate hybrids with two sibling species in the *Arcanum* species complex. Highly involved programs of hybridization and embryo culturing techniques have introduced numerous genes conferring disease resistance and stress tolerance isolated from the *Eriopersicon* complex into commercial *Lycopersicon* tomato lines. Intrageneric hybrids have also been achieved between the tomato and *S. lycopersicoides*, a species related to the potato *S. tuberosum*. These hybrids show that the genetic barriers between species are not absolute, although they frequently result in extreme hybrid sterility.

There is little doubt that species in different genera are genetically and phenotypically more divergent than species within the same genus. Thus, viable intergeneric hybrids tend to be rare. Nevertheless, intergeneric hybrids do occur, and there is good evidence that some can evolve into new species as a consequence of introgression, or backcrossing with subsequent fixation of backcross types. One example of a stabilized introgressant occurs between the two shrub species *Purshia tridentata* and *Cowania stansburiana*. Hybrids between these genera apparently backcrossed with their parent species, and the backcrossed derivatives became genetically and phenotypically fixed in the form of the new species *P. glandulosa*, which has a wide geographic distribution.

It was formerly believed that plant hybrids typically are morphologically intermediate and less fit than either of their parent species.

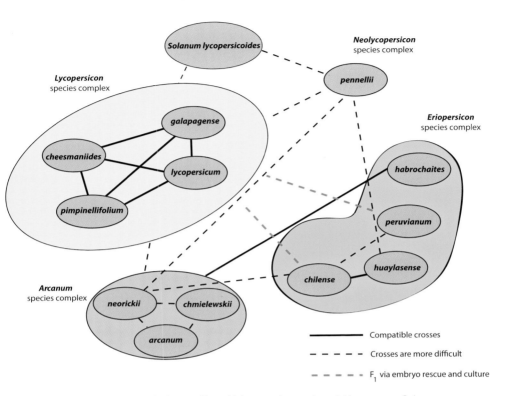

Figure 5.10. Summary of interspecific and intergeneric crossing relations among *Solanum* species and *Solanum lycopersicoides* (a relative of the domestic potato). The genus *Solanum* (formerly *Lycopersicon*) consists of two large species complexes, the *lycopersicum* complex and the *Eriopersicon* complex, and a small complex, the *Arcanum*, which consists of two sibling species and is recognized by some authorities as distinct from but related to the *lycopersicum* complex. Hybrids between the two large complexes are difficult to achieve (one viable seed was produced from among hundreds of seeds set by a hybrid between *S. lycopersicum* and *S. peruvianum*). Most can be achieved only through rescuing hybrid embryos by growing them in culture. Hybrids between the two large species complexes are more easily achieved by making intermediate hybrids with species in the *Arcanum* complex. Intrageneric hybrids among these species indicate that the genetic barriers among species are not absolute.

It was also believed that the parental characteristics of hybrids were likely to remain associated or correlated in segregating hybrid progenies. However, these beliefs have been challenged by mounting evidence that hybridization can act as a catalyst for speciation. Currently, few doubt that some species have evolved from hybrids. What remains unanswered, however, is the relative frequency of speciation by this means. By providing an expanding arsenal of analytical techniques that permit high-resolution descriptions of the distributions of molecular markers in hybrid populations from which the genetic dynamics of populations may be inferred, molecular techniques may eventually resolve this issue. One of the difficulties in drawing a general portrait of the birth of a species is that speciation can involve a succession of events that can span considerable time. Another difficulty is that what looks like incipient speciation may not ultimately conclude with the appearance of a new species. In general, speciation is not directly observed but merely adduced after the fact. Nevertheless, many of the individual events believed to immediately prefigure or follow the origin of a species have been observed in nature or simulated in the laboratory.

One of the more stunning "laboratory" investigations of speciation is that of Loren Rieseberg and coworkers, who reproduced the genetic changes leading to the formation of a naturally occurring species of sunflower—*Helianthus anomalus,* an outcrossing diploid restricted to swales and sand dunes in Arizona and Utah. Under laboratory conditions these changes are repeatable across independent experiments. *H. anomalus* appears to have arisen by recombinatoral speciation (a process in which two species hybridize and the mixed genome of the hybrid becomes a third species that is reproductively isolated from both of its ancestral species). The ancestral species of *H. anomalus* are *H. annuus* and *H. petiolaris* (fig. 5.11). All three species are self-incompatible annuals. The two ancestral species often produce large hybrid swarms whose members are typically semisterile because the species differ by fixed chromosome arrangements that cause meiotic nondisjunction. However, over several generations these arrangements can sort out into a new genome that is fertile but incompatible with its ancestral genomes. Rieseberg et al. (1996) hybridized *H. annuus* and *H. petiolaris* to produce three independent hybrid lines that were subjected to different sib-mating and backcrossing regimes. With the aid of molecular analyses, they found that plants from all

Figure 5.11. The hybrid between *Helianthus petiolaris* and *H. annuus*. a. *H. petiolaris*. b. *H. annuus*. c. The hybrid *H. anomalus*. Courtesy of Dr. Loren Rieseberg (University of British Columbia).

three lines converged to nearly identical gene combinations including parallel changes in the nonrearranged portions of chromosomes that must represent gene fitness effects. Although many features of this experiment still need to be understood, it is clear that a complex network of genetic interactions was involved in this laboratory example of microevolution and that despite this complexity, the path of evolutionary change was repeatable in ways suggesting that selection rather than chance governs the genomic composition of hybrids between *H. annuus* and *H. petiolaris*.

Molecular techniques may also shed light on the genetic mechanisms and ecological circumstances responsible for transspecific evolution, the origin of higher taxa. The differences between species drawn from very different taxonomic groups are so impressive that some investigators have suggested that higher taxa arise through mechanisms distinct from the microevolutionary processes traditionally viewed as responsible for speciation. The argument made is that the mechanisms responsible for variation within a species *are not the same* as those leading to the appearance of higher taxa. This proposition has led to some of the major controversies in contemporary evolutionary theory. Nevertheless, the examples of species hybrids reviewed here cast doubt on the need to invoke unknown (and unspecified) genetic phenomena for transspecific evolution.

Although islands provide clear opportunities for allopatric speciation, a single species may also consist of a graded spectrum, or cline, of subpopulations that differ genetically and phenotypically along a continuous geographic range in accord with the parapatric speciation model (see fig. 5.6). Neighboring subpopulations will be interconnected by gene flow, but interbreeding diminishes as the distance between subpopulations increases. The failure of some subpopulations to interbreed even when given the opportunity in laboratory settings indicates that subspecies may evolve along clines. Differences in soil conditions or in the availability of water may provide selection pressures favoring different genotypes in subpopulations along a cline or in patchy microgeographic settings. For example, soil moisture has been correlated with different allele frequencies for a total of six loci coding for enzymes and seed color in subpopulations of the wild oat (*Avena barbata*). Races of various grasses (*Agrostis*, *Festuca*, and *Anthoxanthum*) that are tolerant and intolerant to heavy metal soil toxins are reported to grow in very close proximity. Similarly, different species of *Ceanothus* occupy neighboring sites differing in soil acidity and moisture (Nobs 1963). Differences in the physical environment may also evoke temporal reproductive isolation. Soil moisture and temperature affect the time when reproductive organs develop, mature, and shed pollen. A good example of temporal isolation is seen in *Pinus radiata* and *P. muricata* (Stebbins 1950). Similarly, staggering the time of year that flowers are receptive to pollen can also prevent or greatly reduce gene flow among closely related sympatric species (Gentry 1974; Levin 1978).

Finally, however tempting it might be to argue that all of the foregoing can be swept aside in light of a cladistic approach and the use of the phylogenetic species concept (see table 6.1), it should be noted that the algorithms that sort out phylogenetic relationships are incapable of distinguishing one breeding population from another. In other words, cladistics is neither taxonomically nor nomenclaturally literate. Cladistics is also incapable of dealing with reticulate evolution.

Portraits of Speciation

Thus far, we have seen that the evolutionary process of speciation spans a spectrum of possibilities with two extreme ends. The first is by the gradual accumulation of selectively favorable or neutral genetic variation in spatially isolated populations, leading to genetically incompatible groups of organisms that, although derived from the same original gene pool, are distinguishable as separate species (fig. 5.12 a). The second possibility is by a rapid evolution inspired more by the random processes of mutation and genetic drift than by natural selection. The first portrait of the tempo and mode of speciation, which posits the gradual divergence between ancestor and descendant species owing to selection pressures, is called *phyletic gradualism*. This is the traditional Darwinian view. The alternative, which is shown in fig. 5.12 b, depicts speciation as a rapid transition between ancestor and descendant owing to the random processes of mutation and genetic drift. When cast in the context of a genetic model, this view of speciation has been called *genetic revolution*. In the context of paleontological patterns of speciation, it may result in what is called *punctuated equilibrium*.

The genetic model for phyletic gradualism depicts a slow transition of an ancestral phenotype to a new form as a series of steps propelled by the combined effects of random and nonrandom evolutionary pro-

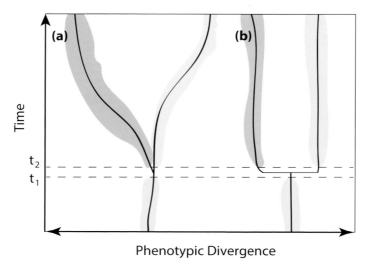

Figure 5.12. Schematic contrast between the gradual phenotypic divergence of two populations leading to speciation, called phyletic gradualism (a), and the abrupt phenotypic divergence of two populations leading to sudden speciation, called punctuated equilibrium (b), between t_1 and t_2. Note that phyletic gradualism views speciation as a more or less continuous process that nevertheless culminates in phenotypically different kinds of organisms, whereas punctuated equilibrium emphasizes phenotypic status for long periods of time after a speciation event occurs.

Time (y-axis), **Phenotypic Divergence** (x-axis)

cesses operating in a gradually changing environment. Nonetheless, the consequences of selection are balanced by mutation and other random genetic changes. Natural selection will tend to move the evolving population toward fitness peaks—that is, to regions of optimum fitness at a given time and place. Random genetic changes may by chance also increase the mean fitness of the evolving population, or they may by chance reduce it, causing the population to descend to a region of lower mean fitness. The model of phyletic gradualism sees natural selection as the principal evolutionary force directing the walk over the fitness landscape. As the gene pool of the population adaptively changes, the population may diverge into new genotypic and phenotypic entities, each associated with its own fitness peak. Isolated populations would continue to differentiate as a result of the selection pressures acting on each population in its particular environment. Even if the populations could interbreed at their borders, the progeny would likely be less fit than either parent population and thus would fail to become established, because these borders would occur in regions of lower mean fitness. Thus the evolving populations would continue to differentiate genetically and phenotypically owing to sexual or geographic isolation and might assume the status of separate species.

Using the metaphor of a landscape discussed earlier, Sewall Wright showed that as a consequence of genetic drift a small population may diverge into two or more populations occupying different fitness peaks in the same fitness landscape. Because the movement down an adaptive peak was considered an arduous process for a very large population (owing to the "struggle" against natural selection and the negligible effects of genetic drift), genetic revolutions were thought to be one reasonable vehicle for rapid speciation in very large populations. But mathematical models derived from conventional population genetics show that even a moderately large population may move quickly from one fitness peak to another. These models show that the decrease in the mean fitness attending the downhill movement may be so rapid that natural selection has no opportunity to act on the population. And as the population reaches the foothills of a fitness peak, selection pressures can quickly carry the population uphill, where it increases in size. Because large populations are expected to remain phenotypically stable for long periods owing to the inability of random genetic changes to overcome the pull of natural selection, the rapid transit of a moderate-sized population from one fitness peak to another is expected to be followed by little or no phenotypic change unless the environment changes.

The portrait of speciation by means of a genetic revolution posits that genic or chromosomal variations can be rapidly fixed in very small populations by the combined effects of natural selection and random fluctuations in gene frequencies, and that the low fitness of heterozygotes prevents significant gene flow among such small populations. In small populations, random genetic changes may become rapidly fixed provided by chance they confer higher fitness. The result is that a small population may abruptly shift phenotypically. The isolation of subpopulations occupying different habitats would favor phenotypic divergence that in turn could produce one or more new species. Notice that this scenario conforms to the peripatric speciation model (see fig. 5.1 b). Small isolated subpopulations may be founded by a few individuals at the periphery of the main population of a species. These small subpopulations undergo rapid peak shifts on the fitness landscape owing to random changes in gene frequencies. Because each small subpopulation occupies a geographic area subject to selection pressures that differ from those acting on other subpopulations or the main population, extensive and diverse modifications of the subpopulation gene pools may occur. These may lead to genetic incompatibilities between subpopulations and the main population that could preface speciation.

As noted, peripatric speciation is sometimes referred to as the "founder effect" because it emphasizes the role played by a small number of individuals, which can lay the genetic foundation for a new species. First, genetic drift destabilizes adaptive gene complexes in small, geographically peripheral subpopulations, and selection pressures subsequently stabilize randomly inspired genetic differences to accord with different environmental conditions. As originally proposed, the peripatric speciation model presupposes a dramatic loss of genetic variance in founders. Because isolated breeding pairs would contain only a small fraction of the total genetic variance of their main population, small populations of founders are assumed to be at risk. However, some mathematical models indicate that the loss of variance in a founder population is equal to one-half the effective population size, that is, $1/(2N_e)$. Thus, with as few as four pairs of founders ($N_e = 8$), as much as 94% of the original variance would be retained. Alternatively, other models indicate that small founder populations can rapidly accumulate slightly deleterious mutations that may eventually cause extinction. It is worth noting that the founder effect can also occur in very small sympatric populations. Thus, peripatric speciation and the founder effect are not synonymous, although they share many of the same conceptual elements.

The peripatric speciation model also assumes that the initial changes in the gene frequencies of a founder population would lead to simultaneous changes in many loci, with cascading effects on the phenotype. More recent speciation models based on genetic revolutions have abandoned this assumption. For example, changes in the allele frequencies of a few founders resulting from random genetic drift may affect only one or at most a few loci that have major phenotypic effects. Once these gene frequencies become stabilized in small subpopulations, natural selection is expected to act on other alleles of interactive loci, modifying phenotypic traits and adapting them to local environmental conditions. Thus the mutation of one or two genes may produce reproductive barriers between ancestral and descendant populations provided the phenotypic effects of the mutation are stabilized by modifier alleles. If true, cycles of severe reduction and rapid increase in population size would result in population "bottlenecks and flushes." Some blocks of strongly interacting genes will resist change even if other loci change readily as a result of natural selection. Because genetic recombinants of the alleles of these "selection resistant" loci will have lower fitness, the ongoing population is predicted to retain its adaptive gene frequencies. During a bottleneck and flush cycle, however, the selection-resistant blocks of loci may become destabilized under conditions of relaxed selection pressure, and even recombinants with low fitness may significantly increase in number. With the onset of another bottleneck, intense selection would act on these variants, picking out those best adapted to the new environmental conditions. The cycles of population flushes and bottlenecks leading to the genetic and phenotypic differentiation of successive subpopulations engendered by the ancestral gene pool has the potential to evoke new species.

The requisite conditions for genetic revolution are frequently found on active volcanic islands, where repeated lava flows can fragment large populations into many smaller ones, some of them essentially biotic islands in a sea of cooling lava. The surviving individuals on each of these biotic islands may subsequently differentiate genetically as their population size increases and as they colonize unoccupied territories until new lava flows reiterate the cycle of population size "bottleneck and flush." On the Big Island of Hawaii, major lava flows occur roughly every 250 years—an extremely short interval in geological or evolutionary terms. Clearly, another factor contributing to genetic revolutions is the time it takes an organism to complete its life cycle, which in turn is related to body size. As noted, the rate at which allelic frequencies change in a population is influenced by the generation time, which tends to decrease with decreasing body size. The Hawaiian islands are home to many small plant and animal species that complete their life cycles comparatively quickly. The suggestion that speciation occurs rapidly for isolated populations of small, rapidly growing organisms living in chronically disturbed or very stressful environments is supported by the fact that the Hawaiian islands have the highest number of endemic plant and animal species known for any geographic area on Earth (see table 5.1). Indeed, the diversification of Hawaiian island invaders is far more stunning than that of Darwin's finches (fig. 5.13).

Monogenic Speciation

The evidence for simple one-step mutations that engender significant mechanical or temporal differences in reproductive organs is overwhelming, particularly for flowering plants (Hilu 1983; Gottlieb 1984). Among the outstanding examples are simple one-step mutations that alter the number, position, symmetry, and fusion of floral parts and even the identity of floral parts. Because systematists often use such features to characterize taxonomic groups above the level of the genus, these mutations could conceivably account for the sudden appearance of floral structures so novel that taxonomists would be

Figure 5.13. Molecular and morphological analyses reveal that the genera collectively assigned to the Hawaiian silversword alliance (examples of which are shown in b–e) are all descendants of an ancestor belonging to the California tarweeds (shown in a) that arrived on the Hawaiian archipelago hundreds of thousands of years ago. a. *Carlquistia muirii*. b. *Dubautia latifolia*. c. *Dubautia scabra*. d. *Argyroxiphium sandwicense*. e. *Wilkesia gymnoxiphium* (by G. D. Carr). Figures 5.13 a–c, courtesy of Dr. Bruce Baldwin (University of California, Berkeley). Figure 5.13 d, courtesy of Mr. Justin Kondrat (US Botanic Garden Conservatory, Washington, DC). Figure 5.13 e, courtesy of Dr. Gerald D. Garr (Oregon State University, Corvallis).

compelled to assign them to new genera or even higher taxa (families, orders, and so forth).

For example, flowers lacking petals (apetalous flowers) are typically wind pollinated or self-pollinated, while flowers with numerous petals and other floral parts (polypetalous) are generally pollinated by animals. Single-gene mutations resulting in apetalous, fertile flowers have been reported for mountain laurel (*Kalmia latifolia*), evening primrose (*Oenothera parodiana*), tobacco (*Nicotiana tabacum*), and a variety of annual chrysanthemum species, all of which normally have large-petaled flowers that are insect pollinated. Conversely, monogenic mutations resulting in flowers with supernumerary petals occur in the mountain laurel, geranium (*Pelargonium hortorum*), soybean (*Glycine max*), gloxinia (*Sinningia speciosa*), garden nasturtium (*Tropaeolum majus*), and petunia (*Petunia hybrid*), whereas an X ray–induced mutation of soybean resulted in flowers with two or more carpels instead of the single carpel typical of the bean family (Leguminosae). Flowers with supernumerary carpels, which developed normal fruits, were the result of a single recessive gene. Although atypical for the bean family, the flowers of *Swartzia ignifolia* and *S. littlei* typically have two or three ovaries. In theory, the single-gene mutation observed for soybean could explain the multi-carpellate condition of these two "anomalous" species.

The number of flowers in an inflorescence, which can distinguish related species and genera, can also be influenced by single-gene mutations. The corn inflorescence (*Zea mays* subsp. *mays*) differs from that of its wild ancestor teosinte (*Z. mays* subsp. *mexicana*) in a number of ways. But one of the most important differences is that the female spikelets of corn are paired, whereas those of teosinte are single. This phenotypic trait is controlled by a single gene whose mutation could account for the difference between these two subspecies.

Monogenic mutations may establish mechanical isolation among subpopulations by changing floral symmetry, thereby potentially altering the mode of pollination. A simple mutation alters the bilaterally symmetrical flower of the insect-pollinated snapdragon (*Antirrhinum*) into an almost radially symmetrical flower similar in appearance to that of the mullein (*Verbascum*), which is in the same family, while another mutation in the snapdragon confers a spurred petal similar to that on the flowers of the butter-and-egg plant (*Linaria vulgaris*), also in the same family (Hilu 1983). Species within the family Asteraceae are distinguished in part by whether their inflorescences contain radially symmetrical "disk" flowers, bilaterally symmetrical "ray" flowers, or both. Yet, by performing artificial crosses between two composite species, *Haplopappus aureus* and *H. venetus* subsp. *venetus,* which have rayed and rayless florets, respectively, research has shown that the presence and absence of ray florets is controlled by a single gene. Thus a single mutation may determine the difference between these two species.

Evidence also suggests that single-gene mutations can establish or remove temporal or mechanical barriers to sexual reproduction. For some flowering plant species, a temporal barrier to hybridization can be created by a change in day length or temperature. Examples of monogenically inherited effects on the time of flowering are the difference between day-neutral and day-sensitive forms of *Salvia splendens,* the formation of flowering versus vegetative branches in the peanut (*Arachis hypogaea*), and early flowering in the sweet pea (*Lathyrus odoratus*). In terms of mechanical barriers to hybridization, research has shown that gibberellin-induced heterostyly in tomatoes is governed by a single gene. (Heterostyly is the formation of flowers differing in the lengths of their styles so that flowers with one style length are predominantly pollinated with pollen from a flower with a different style length.) Another mechanical barrier to hybridization is the formation of flowers that never open and therefore are self-pollinated (cleistogamy). In the cucumber, a single recessive gene mutation results in cleistogamy, as it does in a mutant of the soybean. Finally, genetic analyses of the sexual expression of flowers and the sex ratios of populations indicate that these phenomena are often controlled by one or only a few genes, although their effects are modulated by other modifier genes and environmental factors such as light intensity, day length, temperature, soil conditions, and plant health.

Mutations affecting floral development have received a great deal of attention because alterations in the pattern formation of reproductive organs provide insights into the evolution of new species from old ones. A bewildering array of mutations is now known to alter the pattern of flower formation. Nevertheless, relatively few genes appear to specify the type of floral organ that will normally develop. Mutations in most of these genes cause homeotic transformations. Homeotic loci are believed to contain the genetic information required to shunt developing cells along a particular pathway. Thus homeotic mutations switch the development of cells from one fate to another and may replace one type of organ with another that is perfectly normal in every respect except its location in the overall body plan. Some of the best understood floral homeotic mutations occur in the mouse-ear cress, *Arabidopsis,* and the snapdragon, *Antirrhinum.* Like many other angiosperms, these two plants have four concentric whorls of floral organs, of which the outermost develops into sepals and the innermost develops into carpels. Homeotic mutations result in dramatic departures from the normal pattern of flower development. Most of these mutations alter the type rather than the number of organs formed, suggesting that the pattern is regulated at several levels, with some genes regulating organ identity and others organ number.

For example, mutations of the *AP3* or *PI* genes of *Arabidopsis* or the *DEF* gene of snapdragon cause petals to be replaced by sepals, and stamens by carpels. Mutations in the *AG* gene of *Arabidopsis* and the *PLENI* gene of the snapdragon appear to convert stamens into petals and carpels into sepals (careful examination of developing flowers reveals that the initiation of stamens proceeds normally at first and that the homeotic organ is really not a true petal). Because gene flow within populations of flowering plant species may be maintained by the ability of flowers to attract specific pollinators, changes in the appearance of flowers may alter the kinds of pollinator species visiting them. Thus, because they have the potential to establish reproductive barriers, homeotic mutations may serve as a mechanism for character displacement, genetic divergence, and subsequent speciation.

Naturally, a balanced view must be maintained regarding the role of monogenic mutations. Although these mutations have the *potential* to engender novelties that could evolve into new species, there is little evidence that the monogenic mutations previously reviewed *did*, whereas there is ample evidence that floral characters can change, sometimes very rapidly, in response to environmental changes. For example, populations of the alpine wildflower *Polemonium viscosum* rapidly adapt to abrupt changes in pollinator assemblages. The data indicate that the broadly flared flowers of the bumblebee-pollinated *P. viscosum* could have evolved from narrower ones in a single generation because corolla flare increased by 12% from populations pollinated by a wide assemblage of insect pollinators to those pollinated only by bumblebees. This appears to be another example of truncation selection—if bumblebees pollinate the flowers with the widest corollas, the next generation of plants will have allele frequencies biased for a broader corolla flare.

Regulatory Gene Networks and Heterochrony

The development of plants as well as animals depends not only on structural genes encoding for proteins, but also on regulatory genes that coordinate the activities of hundreds of structural genes. Recall from chapter 4 that regulatory genes encode for proteins that either activate or repress the ability of other genes to express themselves. Regulatory genes are usually constitutive—that is, they produce repressor proteins continuously but at comparatively slow rates. Even though these repressor proteins are always present, they do not always block the transcription of their targeted genes, because the protein's ability to function depends on the chemical environment within the cell, which in turn is indirectly self-regulated by the activity of the genome as a whole as well as subject to changes in the external environment. Regulatory and structural genes therefore operate in vast, complex feedback systems that receive and respond to physiological cues generated from within and without the organism.

As discussed in chapter 4, networks of regulatory genes participate in developmental coordination and indirectly influence cell metabolism, growth, and development. Their role may be likened to the fingers of a violinist. Virtually every piece of music ever written can be played on a violin, even though this instrument can play only a limited number of musical notes. The versatility of the violin comes from the musician's ability to vary the sequence, duration, and volume of the notes played. In like manner, the genome of any organism consists of a finite number of genes, each encoding for a very particular gene product. By varying the sequence, duration, and activity of individual genes, regulatory genes control the combinations of gene products simultaneously existing in a cell. The same gene product thus can play many different roles depending on its physiological or developmental context. Naturally, regulatory genes are themselves influenced by this context, and so the music of the cell is a fugue that turns in upon itself in complex and strangely beautiful ways.

Advances in molecular genetics have provided some insight into how some regulatory gene networks influence growth and development. One network has been postulated to control the time of transition from vegetative to reproductive growth and the development of floral organs in the mouse-ear cress *Arabidopsis thaliana* (see for example Yang, Chen, and Sung 1995). Two genes, called the *EMBRYONIC FLOWERING* or *EMF* loci, figure prominently in this network (fig. 5.14). In wild-type plants, the activity of *EMF* genes gradually declines in the normal course of vegetative development. Once the activity level drops below a critical threshold, the shoot apical meristem undergoes a transition from vegetative to reproductive growth. Other genes, such as the *EARLY FLOWERING* and *CONSTANS* (*ELF* and *CO*) genes, are believed to regulate the rate at which the *EMF*

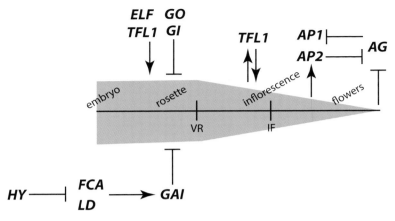

Figure 5.14. A proposed gene network influencing the transition time from vegetative to reproductive growth (VR) and the time of initiation of flower development (IF) in *Arabidopsis thaliana*. The developmental stages for the wild type are shown from the time the embryo germinates to form a rosette to the formation of mature flowers. The decrease in the activity level of *EMF* gene loci is indicated by the thickness of the wedge-shaped pentagon. Gene loci repressing other genes are indicated by ⊣; gene loci promoting other genes are indicated by →. External environmental factors known to influence gene activity are not indicated. Gene loci names are as follows: *EMF* = EMBRYONIC FLOWERING, *ELF* = EARLY FLOWERING, *TFL* = TERMINAL FLOWERING, *CO* = CONSTANS, *GI* = GIGANTEA, *AP* = APETALLA, *AG* = AGAMOUS, *GAI* = GIBBERELLIC ACID INSENSITIVE, *FCA* = FLOWERING, *LD* = LUMINIDEPENDENS, *HY* = LONG HYPOCOTYL.

products decline, either extending or reducing the transition time from vegetative to reproductive growth. These genes, together with *EMF* genes, provide a feedback loop.

As we have seen in chapter 4, rendering gene networks as shown in fig. 5.14 is misleading. The linear sequence of "cause and effect" with focus on the *EMF* genes obscures the fact that no gene in a regulatory network of interacting genes can take precedence because the role and influence of each gene is defined only in the context of the whole system. Likewise, the number of genes involved in the transition from vegetative to reproductive growth and the formation of flowers in *Arabidopsis* or any other flowering plant is undoubtedly much greater. Indeed, the genes involved in the transition from vegetative to reproductive growth and the ultimate development of flowers most likely encompasses the entire genome.

Nevertheless, someday our understanding of gene regulatory networks and the consequences of their mutation, such as those briefly described for *Arabidopsis thaliana*, may help us uncover the genetic mechanisms resulting in the phenomenon called heterochrony, which has been implicated theoretically in the evolution of some species and even higher taxa. Heterochrony is broadly defined as any evolutionary change resulting from any alteration in the sequence, duration, or timing of developmental events. Unfortunately, this cavalier definition tends to invite unbridled speculation about the ability of heterochrony to engender phenotypic novelties that may preface transspecific evolution. Strictly speaking, heterochrony refers only to the morphological expression of adult form with respect to the presumed ancestral adult condition. Therefore it is a purely descriptive term lacking precise reference to the genetic or developmental mechanisms responsible for a phenotypic novelty. Consider that there are two general classes of heterochrony: (1) paedomorphosis (from the Greek *pais, paidos*, "child," and *morphe*, "form, shape"), which is the retention of juvenile (ancestral) features by adults of the species, and (2) peramorphosis (from the Greek *pera*, "beyond, across," or *peraiteros*, "farther"), which occurs when adult size or shape is developmentally extended beyond that of the ancestral condition. However, each of these two classes of morphological expression can be achieved by a number of categories of developmental modifications (table 5.3). For example, paedomorphosis can result from neoteny or progenesis. Neoteny occurs when somatic growth is delayed or retarded so that the reproductively mature adult retains juvenile (ancestral) features. Progenesis achieves the same result but by accelerating the rate of sexual development with respect to somatic (vegetative) development.

The *emf* mutants of *Arabidopsis thaliana* may show how progenesis produces paedomorphosis because they result in the formation of flower-like structures on embryos. Before seed germination, *emf* mutant embryos either produce a shoot apical meristem whose features are similar to those of the apex that gives rise to the inflorescence

Table 5.3. Two classes of the morphological expression of heterochrony and their developmental categories

Class of Morphological Expression	Developmental Category	Alteration		Growth Rate Onset
		Sexual	Somatic	
Paedomorphosis				
	Neoteny	Same	Delayed	
	Progenesis	Accelerated	Same	
	Postdisplacement			Same Later
Peramorphosis				
	Acceleration	Same	Accelerated	
	Hypermorphosis	Delayed	Same	
	Predisplacement			Same Earlier

Note. Each category is described in terms referring to the presumed ancestral condition.

in mature wild-type plants, or they produce a short-lived vegetative shoot apex that quickly changes to a reproductive shoot apex. Experiments verify that *emf* mutations can suppress the expression of a number of other regulatory genes such as the *AG* homeotic gene that converts stamens into petals in *Arabidopsis thaliana*. Because *emf* mutants are sexually mature when they are somatically (vegetatively) embryonic, they result in paedomorphosis via the developmental route of progenesis.

It is very tempting to speculate that gene mutations like *emf* may produce developmental "monsters," like flower-bearing embryos, that are nevertheless "hopeful" in the sense that they may have some selective advantage over their normal counterparts and so may ultimately evolve into a new form of life. Indeed, neoteny and progenesis have been implicated in the evolution of various plant and animal lineages. In some cases the evidence for the roles of these and other heterochronic phenomena is compelling (Takhtajan 1958, 1959; Gould 1977; Hébant 1977), especially in light of single gene mutations that can evoke dramatic changes in morphology (box 5.2). As a general rule, however, "developmental monsters" are anything but "hopeful"—

Box 5.2. Hopeful Monsters and the Evolution of the Land Plant Sporophyte

A major transformation in the history of the land plants was the evolution of plants with life cycles that involved a branched sporophyte bearing multiple sporangia (a poly-sporangiate diploid generation). Although the fossil record provides limited information about how this occurred, it is likely that this kind of organism evolved from an embryophyte ancestor with a life cycle that involved an unbranched sporophyte bearing a single sporangium (a monosporangiate diploid generation) that was similar to the sporophytes of modern day mosses, liverworts, and hornworts (collectively called bryophytes). One of the advantages conferred by developing a branched sporophyte is the production of more sporangia (and thus more spores) from a single fertilized egg. In a terrestrial environment, fertilization events can be rare owing to the failure of sperm cells to reach eggs. The sporophytes of modern-day bryophytes reach a certain size, produce a single sporangium with a developmentally proscribed number of spores, and die after the spores are released. Even a singly branched sporophyte would confer an advantage, at least in theory, because it would produce twice as many spores as its monosporangiate counterpart. Although rare, some moss sporophytes are branched and bear two sporangia instead of one (fig. B.5.2 a). These bisporangiate plants have the general appearance of some of the most ancient vascular land plants, like the fossil sporophyte called *Cooksonia*.

The developmental mechanisms responsible for this kind of anomaly in mosses have remained obscure. However, recent studies have shown that disruption of PIN-mediated polar auxin (IAA) transport in the sporophytes of the moss *Physcomitrella* can induce a single-branched sporophyte (fig. B.5.2 b-c) that is morphologically intermediate between modern-day moss sporophytes and some of the most ancient vascular plant sporophytes (Fujita et al. 2009; Bennett et al. 2014; Coudert et al. 2015). (The PIN proteins are active cell membrane transporters that facilitate the influx and efflux of IAA from one cell to another [see chapter 4].) The induction of this "hopeful monster" is likely complex because the formation of lateral branching in the corresponding moss gametophyte generation involves the participation of two other regulatory hormones (cytokinin and strigolactone).

Although there is sufficient evidence to conclude that evolutionary modification of IAA transport was involved in the transformation from mono- to polysporangiate sporophytes, the "moss model" may not be apropos because the apex of the developing

---

most are incapable of surviving, let alone reproducing—and, for biparental species, two hopeful monsters of the same kind are required to produce a population. Although *Arabidopsis emf* mutants have sexual characteristics when in the embryonic condition, they do not survive because they lack functional foliage leaves and their flowers are completely sterile.

sporophyte is developmentally programmed to develop into a sporangium, whereas the apex of polysporangiate sporophytes are programmed to develop vegetatively. Thus, the apices of mosses and polysporangiate plants differ in more ways than "one doesn't branch" versus "the other branches" before producing a sporangium. As mentioned earlier in the text, the mosses, liverworts, and hornworts are lineages that likely evolved the sporophyte generation independently. It is possible that the polysporangiate lineage was yet another evolutionary experiment that nevertheless may have used developmental toolkits shared among all of the nonvascular embryophyte lineages.

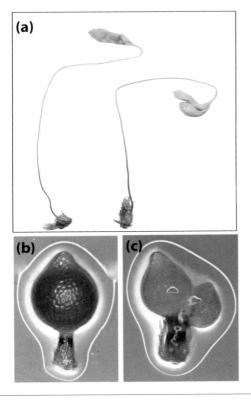

Figure B.5.2. Examples of normal and abnormal moss sporophytes. a. Examples of a normal monosporangiate *Funaria* sporophyte (left) and a naturally occurring bisporangiate sporophyte. Courtesy of Dr. Yoan Coudert (University of Cambridge). b–c. Examples of a normal and a *pin3 Physcomitrella* sporophyte. The PIN proteins are efflux proteins that facilitate the intercellular transport of auxin (IAA). Courtesy of Dr. C. Jill Harrison (Bristol University).

Alternatives to the Biological Species Concept

As valuable as the biological species concept has been to the study of taxonomy and evolutionary biology, its central tenets continue to be challenged by various biological phenomena such as limited gene flow in sympatric populations, the formation of complex and vast hybrid swarms, sibling species, and asexual life forms. It should not come as a surprise, therefore, that alternative species definitions have been advanced. We will return to these alternative definitions in chapter 6 when we turn to the topic of macroevolution. As a preamble to the next chapter, we will briefly review the strengths and weaknesses of some of these species definitions.

The mate recognition concept defines a species as the most inclusive population of a sexually reproductive organism that has a common fertilization system. This definition emphasizes reproductive adaptations that have maximized successful mating among members of an inclusive population rather than reproductive isolation barriers (Paterson and Macnamara 1984; Paterson 1985). Here the view is that the sexual isolation that underlies the biological species concept is a derivative feature of reproductive adaptations to recognizing mates. In this respect it asserts the opposite side of the coin. Recall that the biological species concept defines a species as "groups of actually or potentially interbreeding populations, which are reproductively isolated from other groups" (see Mayr 1963, 19). The mate recognition definition focuses on the reproductive mechanisms cloistering a species into a separate and internally coherent biological unit. By doing so, it draws attention to traits presumably directly acted on by natural selection during the process of speciation. This emphasis successfully deals with the problem of hybrid species. Provided organisms share the same fertilization mechanism, they belong to the same species regardless of the pathology or success of their hybrids. However, this assumes that species are biparental organisms, just as does the biological species concept, and excludes a number of kinds of organisms that are asexual or uniparental life forms (apomictics or self-fertilizing plants).

An alternative definition is the morphospecies concept, according to which a species is any kind of organism identifiable by one or more unique and heritable phenotypic traits (Cain 1963). This concept successfully copes with asexual life forms as well as species hybrids, but it is conceptually challenged by the biological reality of sibling species and may fail to discern whether a collection of morphologically overlapping organisms is a set of subtly different species or a single species with extreme phenotypic variation resulting from a broad norm of reaction.

Yet another definition of species is a single lineage of ancestral-descendant populations of organisms that maintains its identity from other such lineages and has its own evolutionary tendencies and historical fate (Simpson 1961; Wiley 1978). This definition, called the evolutionary species concept, emphasizes evolutionary modification through descent. It benefits and suffers from many of the same theoretical and practical reasons as the morphospecies concept.

The advent of cladistic analyses has led to the phylogenetic species concept, according to which a species is the smallest aggregation of populations (of a sexual organism) or lineages (of an asexual organism) diagnosable on the basis of a unique combination of character states found in all members of the aggregate (Eldredge and Cracraft 1980; Cracraft 1983; Nixon and Wheeler 1990). This species concept relies on genetically fixed molecular or phenotypic characters that identify each terminal branch of a cladogram as the smallest discernibly different aggregate. Because each terminal branch is a group divergent from all others in the cladogram by virtue of its combination of diagnostic characters, it constitutes a unique phylogenetic entity or "species." Much like the biological species concept, the phylogenetic species definition emphasizes genetically fixed character states reflecting limited gene flow among divergent population systems. But because it focuses on how every discernible difference, even those resulting from a single molecular or morphological marker, distinguishes a phylogenetic species, this concept amplifies the subtleties in the pattern of divergence and so tends to identify many more entities as species. In this sense the phylogenetic species concept is a more modern version of the morphospecies concept. The strength of this approach is its ability to bring to the forefront the intricate genetic diversification patterns that may preface the origin of biological species. Its drawback is its potential to partition diverging population systems into evolutionarily meaningless biological entities. Applying the phylogenetic species concept would separate broccoli, brussels sprouts, cabbage, and kohlrabi within the context of *Brassica oleracea*, but it is dubious that each warrants individual species recognition (see fig. 5.9). Put differently, the phylogenetic species definition renders the smallest discernible taxonomic units (aggregates of populations or lineages), but well-reasoned debate may fail to reach a consensus on whether these smallest "units" are in fact "species."

Space does not permit a full discussion of all the species definitions so far advanced. The quest for a universally applicable species definition, one that will fit all manner of different fossil and living plants and animals, continues to remain something of a Holy Grail in systematic biology. Virtually every definition has its merits, but the principal difficulty with each is the attempt to reduce the tremendous diversity of life and its multifarious biological entities to a single, uniformly applicable verbal formula. Ultimately each species can be diagnosed by a basic set of genes that all its members share. In many cases but not all, these shared genes have translated into shared morphological features as a consequence of reproductive barriers. Thus genetic and phenotypic markers and reproductive barriers are common elements in many species definitions. But reproductive barriers and morphological divergence are lacking or weakly developed in what many consider to be good species. For this reason, reproductive isolation and phenotypic differentiation are best seen as secondary rather than principal attributes of species. Each may or may not accompany or follow the process of speciation. Seen in this light, it may be unrealistic to expect all kinds of organisms to fit neatly into a single species definition and much more practical to employ the definition that is best suited to understanding the biology of the particular group of organisms of interest and to think of species as occupying a multidimensional matrix of criteria in which different criteria take precedence depending on circumstances (see chapter 6 for an extension of this point of view).

Plants Are Not Animals

Most attempts to define what is meant by species are based on animal biology, and many have neglected the conspicuous differences between plants and animals. Five of these differences suggest that species definitions (and the models for speciation they evoke) suffer whenever plant biology is neglected. First, plants tend to be developmentally simpler than animals. The integration and delicate balance between the various organs and organ systems required for the motility, sense perception, and coordinated behavior in animals are far greater than anything existing in even the most morphologically and anatomically complex plants. Thus, it is possible that different developmental patterns may be more easily integrated in plant species hybrids than in animal hybrids (Grant 1971). Second, the motility, sense perception, and coordinated behavior of animals provide opportunities to quickly establish barriers to species interbreeding. In contrast, many groups of plants promiscuously cross-pollinate, or interbreed in other ways, such that reproductive isolation may be more difficult to achieve or maintain (Baker 1959). Third, many plants have an open system of growth based on the retention of perpetually embryonic tissues (meristems) that, in addition to amplifying the number of body parts and overall size, makes possible great individual longevity and asexual (vegetative) propagation. Given the capacity for continued growth in size and vegetative propagation, a single mutant or hybrid plant has the potential to establish a large clonal population without sexual reproduction. Thus, in contrast to most animal species, low sexual fecundity or sexual sterility does not preclude the establishment of a plant population, nor does it afford an intrinsic obstacle to speciation. The open system of growth by means of apical meristems also provides the opportunity for genetic mosaics through random mutation of meristematic cells. An individual plant could generate a population of genetically different descendants. Fourth, sterile species hybrids may become fertile when the chromosome number in their germinal cells spontaneously doubles (polyploidy). By this mechanism, a sterile hybrid can give rise to fertile progeny that may serve as the founding members of a new species. And fifth, many plant species are capable of self-fertilization. For these hermaphroditic species, uniparental rather than biparental reproduction is the normal method of sexual reproduction.

A sixth difference is also worth noting, although it has less to do with plant biology and more to do with computational biology—phylogenetic analyses have considerable difficulty when dealing with reticulate evolution resulting from hybridization or extensive lateral gene transfer, which are far more common among plants than among animals. Reticulate evolution refers to the anastomosing from time to time of formerly separate branches on the evolutionary tree. Separate branches can graft together when species successfully hybridize and the resulting hybrids evolve into new species. Speciation may occur immediately by virtue of polyploidy or later when populations of sterile organisms persist for long periods (apomictic populations) and subsequently acquire the capacity for sexual reproduction (Babcock and Stebbins 1938). In either case, the evolutionary tree will bear a number of interweaving and merging branches in addition to conventional dichotomous branches. Certainly, the evolution of the eukaryotes and the persistence of endosymbiotic events attest to the continued evolutionary importance of reticulate evolution.

Suggested Readings

Babcock, E. B., and G. L. Stebbins. 1938. *The American species of* Crepis. Carnegie Institution Publication no. 504. Washington, DC: Carnegie Institution.

Baker, H. G. 1959. Reproductive methods as factors in speciation in flowering plants. *Cold Spring Harbor Symp. Quant. Biol.* 24: 177–91.

Bennett, T. A., M. M. Liu, T. Aoyama, N. M. Bierfreund, M. Braun, Y. Coudert, R. J. Dennis, D. O'Connor, X. Y. Wang, C. D. White, E. L. Decker, R. Reski, and C. J. Harrison. 2014. Plasma membrane-targeted PIN proteins drive shoot development in a moss. *Current Biol.* 24: 2776–85.

Cain, A. J. 1963. *Animal species and their evolution.* Rev. ed. London: Hutchinson
    University Library.

Coudert, Y., W. Palubicki, K. Ljung, O. Novak, O. Leyser, and C. J. Harrison. 2015.
    Three ancient hormonal cues co-ordinate shoot branching in a moss. *eLife* doi:
    10.7554/eLife.06808.

Cracraft, J. 1983. Species concepts and speciation analysis. *Current Ornithol.* 1:
    159–87.

Eldredge, N., and J. Cracraft. 1980. *Phylogenetic patterns and the evolutionary process.*
    New York: Columbia University Press.

Fujita, T., and M. Hasebe. 2009. Convergences and divergences in polar auxin
    transport and shoot development in land plant evolution. *Plant Signal. & Behav.*
    4: 313–15.

Gentry, A. H. 1974. Flowering phenology and diversity in tropical Bignoniaceae.
    Biotropica 6: 64–68.

Gottlieb, L. D. 1984. Genetic and morphological evolution in plants. *Amer. Nat.* 123:
    681–709.

Gould, S. J. 1977. *Ontogeny and phylogeny.* Cambridge, MA: Harvard University
    Press.

Grant, V. 1971. *Plant speciation.* 2d ed. New York: Columbia University Press.

Hébant, C. 1977. *The conducting tissues of bryophytes.* A. R. Gantner Verlag Komman-
    ditgesellschaft. Vaduz, Ger.: J. Cramer.

Hennig, W. 1966. *Phylogenetic systematics.* Urbana: University of Illinois Press.

Hilu, K. W. 1983. The role of single-gene mutations in the evolution of flowering
    plants. In *Evolutionary biology,* vol. 16, ed. M. K. Hecht, B. Wallace, and G. T.
    Prance, 97–128. New York: Plenum.

Levin, S. A. 1978. On the evolution of ecological parameters. In *Ecological genetics:
    The interface,* ed. P. F. Brussard, 3–26. New York: Springer-Verlag.

Mayr, E. 1963. *Animal species and evolution.* Cambridge, MA: Harvard University
    Press.

Nasrallah, J. B., and M. E. Nasrallah. 1993. Pollen-stigma signaling in the sporo-
    phytic self-incompatibility response. *Plant Cell* 5: 1325–35.

Nixon, K. C., and Q. D. Wheeler. 1990. An amplification of the phylogenetic species
    concept. *Cladistics* 6: 211–23.

Nobs, M. A. 1963. *Experimental studies on species relationships in* Ceanothus. Carnegie
    Institution Publication no. 623. Washington, DC: Carnegie Institution.

Paterniani, E. 1969. Selection for reproductive isolation between two populations of
    maize, *Zea mays. Evolution* 23: 534–47.

Paterson, H. E. H. 1985. The recognition concept of species. In *Species and speciation,*
    ed. E. S. Vrba, 21–29. Transvaal Museum Monograph 4. Pretoria: Transvaal
    Museum.

Paterson, H. E. H., and M. Macnamara. 1984. The recognition concept of species.
    S. Afr. J. Sci. 80: 312–18.

Raven, P. H. 1980. Hybridization and the nature of species in higher plants.
    Canadian Bot. Assoc. Bull. 13 (suppl.): 3–10.

Rieseberg, L. H., B. Sinervo, C. R. Linder, M. C. Ungerer, and D. M. Arias. 1996. Role of gene interactions in hybrid speciation: Evidence from ancient and experimental hybrids. *Science* 272: 741–44.

Simpson, G. G. 1961. *Principles of animal taxonomy.* New York: Columbia University Press.

Stebbins, G. L. 1950. *Variation and evolution in plants.* New York: Columbia University Press.

Takhtajan, A. L. 1958. *Origins of angiospermous plants.* [Trans. from the original 1954 Russian version.] Washington, DC: American Institute of Biological Sciences.

———. 1959. *Essays on the evolutionary morphology of plants.* [Trans. from the original 1954 Russian version.] Washington, DC: American Institute of Biological Sciences.

Wagner, W. H., Jr. 1954. Reticulate evolution in the Appalachian Aspleniums. *Evolution* 8: 103–18.

Wagner, W. L., D. R. Herbst, and S. H. Sohmer. 1990. *Manual of the flowering plants of Hawai'i.* Vol. 1. Honolulu: University of Hawaii Press.

Wiley, E. O. 1978. The evolutionary species concept reconsidered. *Syst. Zool.* 27: 17–26.

Yang, C. H., L. C. Chen, and Z. R. Sung. 1995. Genetic regulation of shoot development in *Arabidopsis*: Role of the EMF genes. *Dev. Biol.* 169: 421–35.

6

Macroevolution

> And the result of this examination is: we see a complicated network of
> similarities overlapping and criss-crossing: sometimes overall similar-
> ities.
>
> −LUDWIG WITTGENSTEIN, *Philosophical Investigations* (1953)

In the previous chapter, we saw that microevolution is traditionally
defined as evolution below the species level (microevolution deals with
the evolutionary divergence of related populations and the processes
leading up to speciation). Logically, therefore, macroevolution can be
defined as evolution at and above the species level. Conceptually, this
definition includes a broad spectrum of biological phenomena including long-term historical trends (such as the succession of major floras
and faunas), the appearance of developmental innovations (such as
the vascular tissues, the seed, or the flower), and coevolution (such
as the evolutionary dynamics between plants, animals, and fungi).
Clearly the range of topics that could be discussed under the rubric of
macroevolution is too broad to be covered in one chapter. Indeed, it
would require a book of substantial length to do the topic justice. Consequently, the goal of this chapter is to narrow our focus onto a few
examples of macroevolution that nevertheless have broad applicability

to understanding evolution in general. One of these topics is species selection—the notion that a species is an "individual" or unit that is subject to natural selection as a discrete entity. This topic continues to evoke heated debate because it rests on the concept of multilevel (or group) selection theory, which argues that natural selection can act at the level of a group instead of at the more conventional level of an individual organism. Another topic to be covered is the evolution of sexual reproduction (or, more precisely, the evolution of meiosis). Although interesting on its own merits, this topic emerges as an essential ingredient when treating species selection if sexual reproduction is viewed as a species-level adaptation (and therefore seen as evidence for species selection). A far less controversial topic to be covered is the evolution of the flowering plants. The flowering plants are the most species-rich clade of land plants, and we have a reasonably good knowledge of their evolutionary history. Consequently, they provide an excellent clade to review species selection, sexual reproduction, co-evolution, and macroevolution in general.

About the Tempo of Speciation

Before entering into discussions about species selection, sexual reproduction, and the evolution of the angiosperms, we need to address a problematic topic, namely, the tempo of speciation. Unfortunately, most definitions of complex processes draw upon other concepts to give them meaning, and these definitions can change subtly depending on how these ancillary concepts are interpreted or defined. In the case of macroevolution, much depends on the species concept and on how the speciation process is conceptualized. Recall from the Introduction that one of the major tenets of the Modern Synthesis was that the processes giving rise to the differences among individuals within the same species and that ultimately give rise to new species are the same processes giving rise to the differences between species and higher taxa. Accordingly, the appearance of new species and the appearance of higher taxa were assumed to be gradualistic. Another important tenet of the Modern Synthesis is that species evolve as a consequence of barriers to gene flow among neighboring populations. Many such reproductive barriers exist, either by precluding sexual reproduction (prezygotic barriers such as ecological isolation) or by negating the consequences of sexual reproduction (postzygotic barriers such as hybrid sterility or the spontaneous abortion of embryos). Even though, in some instances, reproductive isolation can vary from little to complete, the neo-Darwinists of the Modern Synthesis assumed that, like the evolution of any other functional trait, reproductive isolation emerges by the gradual substitution of alleles in populations.

This gradualistic perspective did not go unchallenged. The geneticist and developmental biologist Richard B. Goldschmidt (1878–1958) argued strenuously and at times disagreeably that the processes giving rise to a new species cannot be understood in terms of the processes that underwrite the genetic differences within or between species. Goldschmidt coined the phrase "hopeful monster" to describe the sudden appearance of novel traits resulting from macromutations that are sufficient to engender dramatically new phenotypes so different from their predecessors as to require these organisms to be classified as new species. Goldschmidt did not reject the Darwinian view of the gradual accumulation of small mutations as the source of intraspecific variation. He veered from the conventional wisdom only by arguing that microevolutionary gradualism cannot account for the processes giving rise to new species or higher taxa. Today, this radical perspective resonates with the discovery of homeotic mutations (mutations that cause one structure to develop in a place where another structure would normally develop) and with a number of other developmental phenomena capable of isolating a subset of organisms from their populations, or dramatically changing phenotypes in one generation. Indeed, one of his lasting contributions to evolutionary theory was the attempt to link the physiological and developmental effects of genes to the phenotype. Whereas homeotic mutations were thought to be caused by alterations of genes, Goldschmidt argued that homeotic mutations are part of a genetic hierarchy of structural and functional units. He therefore departed from the philosophy of the neo-Darwinists by emphasizing the importance of the genetic regulation of development and the importance of developmental transformations in evolutionary history. Perhaps his greatest failure was his inability to see the importance of population genetics and dynamics when explaining the evolutionary role of developmental processes. Changes in gene regulatory networks that affect development must spread in a population if they are to affect subsequent evolutionary patterns. Therefore, to effect evolutionary change, a hopeful monster has to (1) find another of similar kind to sexually reproduce, (2) be sexually self-compatible, or (3) be capable of asexual reproduction. Indeed, the disconnect between traditional population genetics and the role of developmental biology in evolution continues to be one of the major challenges of what is known today as evolutionary-developmental (evo-devo) biology, which we discussed in chapter 4.

How to Define a "Species"

Setting aside the debate about whether speciation is a gradual or a sudden process, we now turn our attention to the species concept itself. In this context, it is always a good idea to bear in mind that our perceptions of organisms, including ourselves, are filtered through neural networks that are not without biases and preconceptions. One phenomenon that emerges repeatedly is called apophenia—seeing a scripted pattern or connection where none exists. Another feature of human behavior is a desire to classify things into related sets of objects. As yet, we are unaware of organisms other than ourselves that manifest these behaviors. However, a very basic evolutionary fact is that a species is an ancestor-descendent "related set" of organisms that recognize each other in one or more ways, since there is ample evidence that most organisms recognize members belonging to their "related set." Consequently, the concept we call species has a biological reality regardless of how we define what is meant by "species."

Nonetheless, as we saw in the previous chapter, deciding on how

Table 6.1. Examples of how species have been defined or conceptualized (in alphabetical order)

Species Concept	Explanation / Definition
Biological Species	A species is a group of individuals in which all individuals have the potential to interbreed and that are reproductively isolated from other groups of similar organisms.
Ecological Species	A species is a lineage or set of closely related lineages occupying an adaptive zone demonstrably different from any other lineage that evolves separately from the other lineages outside of its range.
Evolutionary Species	A species is a single ancestor-descendant sequence that retains its identity from other lineages (and thus manifests unique evolutionary tendencies and a historical fate).
Morphospecies (Phenotypic Species)	A species of a group of individuals sharing the same phenotypic traits.
Phylogenetic Species	A species is the smallest monophyletic group sharing a common ancestry.
Recognition Species	A species is the most inclusive population of biparental organisms sharing a common fertilization system.

Note. With the exception of the phenotypic species concept, most definitions implicitly assume that the organisms in question are capable of sexual reproduction. Each also rests on one or more concepts of "individuality."

to define "species" has challenged biologists and philosophers for hundreds of years, and consensus on one definition is not within sight. Although it is certainly true that biologists continue to describe new species and to resolve taxonomic and phylogenetic relationships quite well without reaching consensus on how to define species, a formal definition continues to be sought (as if it could emerge from first principles). Yet, the advocacy of disparate definitions persists for practical as well as philosophical reasons. Table 6.1 lists just a few of the alternative definitions discussed in chapter 5. As noted, the best known of these is the biological species concept advocated by Ernst Mayr (1904–2005). Largely based on his lifelong studies of birds, Mayr's

Figure 6.1. Herbarium specimens illustrating a hybrid between the bear oak *Quercus ilicifo-lia* and the eastern black oak *Q. velutina*. a. *Quercus ilicifolia*. b. The *Q. ilicifolia* × *Q. velutina* hybrid. c. *Q. velutina*.

definition emphasizes sexual reproductive compatibility among the members of "related sets" (populations) and posits that an absence of gene flow between populations fosters speciation as a consequence of genetic divergence between once related populations. Although it is mentioned in every textbook dealing with biology and evolution, there are a number of problems with this definition. Recall from chapter 5 that many organisms reproduce largely or entirely by means of asexual reproduction (the unicellular green alga *Chlorella* and the gametophytes of fern species assigned to *Vittaria* and *Trichomanes*), many organisms that are recognized as separate species are capable of hybridization (fig. 6.1), and the vast majority of organisms are extinct and yet assigned species status despite our complete ignorance as to whether morphologically similar fossils were capable of sexually reproducing with one another (the early land plant gametophytes of *Cooksonia pertonia* and *C. cambrensis*). Another problem revolves around the phrase "all individuals have the potential to interbreed." What does "potential" really mean? If a population of interbreeding organisms is suddenly separated by a physical barrier that cannot be breached, are we obliged to consider the two populations as separate species because the potential for interbreeding is lost? Likewise, if a population of plants or animals is neutered, are they a new species? It is also worth mentioning that the vast majority of species has never been identified on the basis of the criteria established by the biological species definition. Indeed, most species have been identified and classified largely or exclusively on the basis of easily observed phenotypic traits, which include reproductive organs but by no means in an operational context.

The reason for reintroducing the biological species concept here is to draw attention to a feature that is shared by other species definitions—the notion of cohesiveness. In the case of species that reproduce sexually, cohesiveness is in large part the result of gene flow within and among populations. In the case of other species, cohesiveness can be judged on the basis of phenotypic or ecological criteria as well as the propinquity of descent from a last common ancestor. Indeed, many of the current definitions for species rely on or emphasize one or more of these features (see table 6.1). Collectively these and other definitions draw attention to the work of Ludwig Wittgenstein (1889–1951), who concluded that sets of objects are perceived as such because of their *Ähnlichkeit* (similarity), or more pointedly their *Familienähnlichkeit* (family resemblance). Wittgenstein noted that some things can be identified as related on the basis of overlapping similarities rather than on a global and discrete verisimilitude. In this sense, we can think of organisms as occupying a multidimensional space defined by a minimum of four axes: (1) reproductive isolation, (2) phenotypic dissimilarity, (3) ecological differentiation, and (4) phyletic distance. The position of two or more groups of organisms can be used to ascertain whether they constitute a species, if they manifest incipient speciation, or if they belong to two species (fig. 6.2). Each of these four axes establishes a set of criteria for assessing whether organisms belong to the same species, but none has a legitimate biological priority over any other. Each contributes to assessing a species and species differences.

It must be noted that this conceptualization of what constitutes

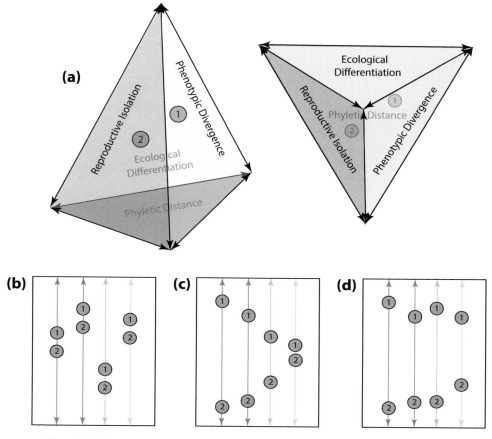

Figure 6.2. Schematic of a multidimensional space that identifies the degree to which two groups of organisms (denoted by circles labeled 1 and 2) are related to one another. The multidimensional space has four axes: (1) reproductive isolation, (2) phenotypic divergence, (3) ecological differentiation, and (4) phyletic distance. Additional axes are possible, as for example one defined by the fossil record of the two groups of organisms. a. One lateral and one polar view of the four-dimensional space. None of the four axes has priority in defining the relatedness between the two organisms. Depending on the type of organisms, some of these axes may be irrelevant or problematic as for example reproductive isolation in the context of fossil organisms, or of extant organisms that reproduce exclusively by asexual methods. The four axes are interrelated. For example, phyletically divergent organisms are likely to be ecologically or phenotypically divergent, or both. b–d. The locations of two or more organisms in the four-dimensional space can be displayed graphically by aligning all four axes and plotting the similarity or dissimilarity as a function of each axis. Organisms aligning near one another on each axis potentially qualify as members of the same species (b), whereas divergent locations can serve as evidence for incipient speciation (c), or as evidence that they belong to different species (d).

a species is not a formal definition of what is meant by "species" because different sets of criteria necessarily come into play when dealing
with diverse organisms such as bacteria and mosses, or algae and land
plants. Rather than advancing one definition based on one set of criteria to evaluate relatedness, it presents a pluralistic yet quantifiable
perspective on assessing what constitutes a species. Note that it also
accommodates the fact that there are different routes to becoming a
species, as for example allopatric and sympatric speciation (see chapter 5), by virtue of giving equal attention to niche specialization and
propinquity to a last common ancestor.

Nevertheless, it is also important to note that some of the axes
defining the location of two or more groups of organisms may be
unknowable or problematic, as for example in the case of fossil organisms, or organisms that reproduce extensively but not exclusively
asexually (as for example, the dandelion *Taraxacum officinale*). Further, two or more of the axes defining the locations of organisms in
any multidimensional space are likely to be interrelated (and in some
cases autocorrelated). Phyletically closely organisms are more likely
to be phenotypically or ecologically similar than organisms sharing
a very distant last common ancestor, and organisms sharing a recent
last common ancestor but adapted to very different habitats are more
likely to be more phenotypically different than organisms living in
the same habitat but evolving from a more distant last common ancestor. These concerns resonate with what Charles Darwin referred
to as "unity of type" and "conditions of existence"—respectively, the
extent to which organisms share the same body plan and the extent
to which selection under different environmental conditions affects
the appearance of a body plan (see chapter 7). For now, it is sufficient
to point out that judgment always enters into determining whether
two things are the same, or whether they are different enough to be
designated as belonging to different groups.

Finally, nothing in the perspective offered here explicitly dictates
how "sameness" or "differentness" can or should be measured. How
sameness is measured depends on the set of criteria delineating the

different axes in the multidimensional space. For example, molecular analyses combined with cladistics can be used to estimate phyletic distance. Genomic markers can be used to assess reproductive isolation. And morphometrics and comparative anatomy can be used to evaluate phenotypic divergence. The common denominator to all of these methodologies is that we are comparing heritable traits that contribute to (or that have contributed to) fitness.

Are Species Individuals?

The conceptualization of "species" discussed in the previous section involves another question with an elusive answer—is a species an individual, and if so, does natural selection act upon it in a manner analogous to its action on a population? The concept of species selection was first proposed by Steven M. Stanley who conceived of species as cohesive units that evolve because they are each subject to selection. It is important to not confuse species selection with group selection. The latter concerns itself with the interactions among groups of individuals within a population on the order of time scales that are of the same order as those affecting selection among individuals. Species selection concerns itself with the evolution of a species on the order of geological time scales.

At its core, species selection perceives species as possessing relative fitness as reflected by differential speciation and extinction rates. Thus, a species is conceived of as an *evolutionary unit or individual*. To understand this concept, we return to some of the concepts explored in chapter 3, wherein we saw that an evolutionary unit such as a population must fulfill three conditions: (1) it must manifest phenotypic variation (trait z must take on different values), (2) the phenotypic variation must correlate with relative fitness (variation in z correlates with variation in fitness w), and (3) fitness must be heritable. It therefore follows that an evolutionary unit (1) has a birth and a death, (2) is subject as a unit to selection, (3) is capable of interacting with its environment in ways that obtain differential replication, and (4) thus manifests adaptation. Although there is no question that each species has a "birth" and will have a "death" (an origination event and an extinction event in the fossil record), the meaning of "individuality" and "fitness" as they pertain to a species remains hotly debated. One issue is whether the traits that define a species and that confer its differential fitness are emergent and adaptive properties of the species. (The use of the word "adaptive" in this context includes aptation—a term that includes traits that evolved directly by means of natural selection [adaptions] and traits whose functionalities are co-opted for other functions [exaptations].) For example, consider the biogeographic range of a species. Is this a species-level emergent and adaptive property, or the statistical summation of the abilities of individual organisms to persist within the range of the species? Even if the biogeographic range of a species expands or contracts over geological time scales, are these fluctuations not the consequences of millions of individual organisms surviving or perishing according to their differential fitness? It should not escape notice that individual organisms do not evolve. Populations of organisms evolve. Consequently, the term "individual" in the context of selection theory has multiple meanings. If an individual organism has a relative fitness, cannot an individual population have a relative fitness? Indeed, geneticists frequently calculate and compare the fitness of different populations.

Perhaps the most relevant feature in this debate is intraspecific genetic variation, since this is a measure of a species' ability to evolve just as genetic variation is a measure of the ability of a population to evolve. Arguably the best evidence for species selection would be evidence for species-level adaptations that increase "evolvability" by means of increasing intraspecific genetic variation, which would presumably increase the ability of a species to survive and produce new species. One such adaptation is sexual reproduction, which has evolved independently in every major eukaryotic group of organisms (plants, fungi, and animals). As we will see, however, the argument that sexual reproduction is a species-level adaption is not clear-cut.

Species Selection and Modes of Reproduction

One view in evolutionary biology is that sexual reproduction increases population-level genetic variation and thus confers an advantage because it increases the ability of organisms to adapt to changing conditions. Although this standard explanation has been used to support the notion of species selection, it is confronted by two uncomfortable facts. First, many very ecologically successful species reproduce either predominantly or exclusively asexually, and, second, it ignores the two-fold cost of sexual reproduction.

The proposition that sexual reproduction confers an evolutionary advantage can be traced back in part to the seminal work of August Weismann (1834–1914), who postulated that cellular differentiation denies somatic cells the opportunity to contribute heritable information to the next generation. This concept, which is called "Weismann's doctrine" or "Weismann's barrier," played an important role in the formulation of the Modern Synthesis during the early and middle parts of the twentieth century because it helped to establish the organism as the indivisible unit of biological organization. Although a reasonable, but not unchallengeable, argument can be made for the predetermination of the egg cell during the early ontogeny of the megagametophytes of angiosperms, the preponderance of evidence shows that Weismann's doctrine does not hold true for the land plants since germ cells are not sequestered from vegetative cells owing to meristematic activity. Indeed, most organisms lack a clearly defined germ line. Likewise, although it is generally true that the somatic cells of many animals are denied the ability to contribute to inheritance, it is also true that even differentiated cells can de-differentiate and become germ cells, as demonstrated by conversion of choanocytes into sperm cells in the sponge *Hipposponga* and the formation of gonadal cells from parientopleural cells in some annelids such as *Lumbricillus*. Like some zoologists, Weismann often meant "mammal" when he said "animal," even though his studies of *Hydra* and Diptera are brilliant and classic.

Likewise, numerous successful species abstain from sexual reproduction in part or entirely, as for example the crustacean *Daphnia pulex* and the green alga *Chlorella vulgaris*. The number of such species is difficult to calculate accurately. However, it is not difficult to show that sexual reproduction is an unnecessary occupation for many species. For example, a simple matrix can be constructed using four functional traits: (1) the presence or absence of rigid cell walls (to distinguish plants and fungi from metazoans), (2) the ploidy levels of the different phases in a life cycle (such as the diploid sporophyte and the haploid gametophyte), (3) the type of body plans in a life cycle (such as a unicellular versus a multicellular phase), and (4) the presence or absence of sexual and asexual reproductive phases. This matrix contains a total of 72 theoretically possible life cycle combinations (fig. 6.3). Eight of these are biologically impossible because they combine "asexual reproduction" with an alternation of a diploid with a haploid individual. Some combinations describe the life cycles of many species, while others occur in only a few species. For example, most land plants have life cycles described by "rigid cell walls" + "an alternation of generations" + "multicellular body plan" + "both asexual and sexual reproduction" (*Rubus* and *Equisetum*). Likewise, many aquatic species have a life cycle consisting of "rigid cell walls" + "zygotic meiosis" + "unicellular body plan" + "sexual reproduction" (*Chlamydomonas*). To be sure, this matrix is a polite fiction because it neglects many important functional traits such as the degree to which cells or tissues are differentiated, or whether germ line cells are sequestered during the early stages of ontogeny. Nevertheless, the fact that the life cycles of numerous organisms occupy almost every conceivable "niche" in this matrix is testimony against the supposition that sexual reproduction is a necessity for evolutionarily or ecological success.

The literature reveals that many organisms can clone themselves in ways that provide for the coexistence of genetically different individuals produced by a population of sexually reproductive genotypes (table 6.2). Such clones may be adapted to different seasonal variations of the environment, or they may be adapted to different biotic or abiotic "subniches" in a stable but heterogeneous environment. Likewise, the ability to reproduce asexually permits a sexually sterile organism

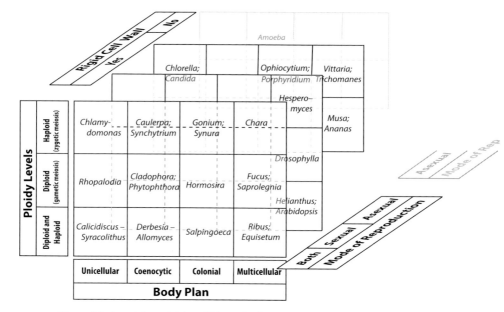

Figure 6.3. A matrix containing all theoretically possible life cycles based on the co-occurrence of four functional traits: (1) the presence or absence of a rigid cell wall, (2) the type of body plan (unicellular, coenocytic, colonial, or multicellular), (3) the ploidy level of the body plan(s) in the life cycle, and (4) the mode(s) of reproduction (asexual, sexual, or both). There are 72 theoretical possibilities, but eight are biologically impossible (for example, an organism possessing an alternation of generations that only reproduces asexually). For clarity, two panels of the matrix are not shown (those for organisms lacking rigid cell walls that can reproduce sexually, or that can reproduce asexually and sexually). Representative taxa are included for some of the biologically possible permutations. The absence of representative organisms in some panels does not imply that these kinds of organisms do not exist, although some biologically possible permutations may not.

Table 6.2. Developmental mechanisms of plant growth (horizontal axis) and growth stimuli (traumatic versus programmed reiteration) that can result in different modes of asexual reproduction

	Developmental Mechanisms			
	Apical meristems	Axillary meristems	Lateral meristems	Callus
Reiteration due to damage	Rhizome fragmentation	Branching	Root Suckering	Foliar embryos
Reiteration due to programmed development	Rhizomes Stolons	Budding Bulbils	Root Suckering Adventitious Roots	Gemmae Bulbils

to escape death and to capitalize on subsequent genomic changes that may confer fertility. Karpechenko's success at producing a new genus derived from an originally sterile organism is proof of that.

If the capacity for asexual and sexual reproductive modes is wedded to a rapid growth cycle, new clones can be produced continuously and rapidly by a sexually reproductive population, which permits the population to remain adapted to its local environment by retaining genetically identical clones, while simultaneously exploring the possibility of colonizing different habitats by means of dispersing genetically different propagules. This convergence of functional traits is common among land plants and invertebrates. But it is nowhere better expressed than among angiosperms, ferns, and mosses, which are the three most species-rich groups of embryophytes.

The second argument against the claim that sexual reproduction is an adaptation emerging from species selection is the two-fold cost of sex. All other factors being the same, an asexual species would reproduce twice as fast as a sexual species owing to the demographic cost of males. A related problem is that in many sexually reproducing species every gene has a two-fold disadvantage due to meiosis. For example, among mammals and most flowering plants, every gene in a primary oocyte or in an egg cell has a 50% chance of being in the functional megaspore, as a result of nuclear breakdown or fusion during the formation of the egg-bearing gametophyte (fig. 6.4 a). There is a way of bypassing this cost as illustrated by some land plants such as the lycopod *Selaginella*, since the formation of haploid spores does not require some to abort (fig. 6.4 b).

However, the issue before us is not whether asexually reproducing species exist (they most certainly do), or whether there are costs to sexual reproduction (there most certainly are), but whether these features refute the notion of species selection. Species that rely exclusively on asexual reproduction are comparatively rare, and molecular studies indicate that most are of recent evolutionary origin. The rarity of asexual reproduction is highlighted when we examine the matrix of life cycles diagrammed in fig. 6.3, which reveals that the majority

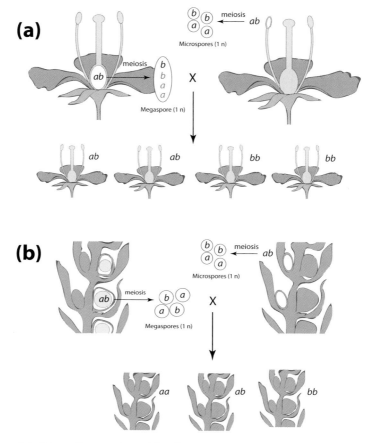

Figure 6.4. Schematic comparison of the life cycle of a stereotypical flowering plant (a), which illustrates one of the costs of sex, and a non-seed vascular plant with unisexual gametophytes such as the lycopod *Selaginella* (b), which does not experience the same cost of sex. In both cases, the diploid plant (the sporophyte) is identified as heterozygous recessive at one locus (*ab*). a. The formation of the haploid spore that will develop into the egg-producing gametophyte (the megagametophyte) involves the disintegration of three haploid nuclei such that either *a* or *b* is lost as a consequence of meiosis. The formation of microspores (pollen grains) does not eliminate *a* or *b*. However, in this example, the frequency of *a* is reduced by 50% when pollination and fertilization are successful. This schematic illustrates monosporic megagametogenesis (only one of the four haploid nuclei survives to produce the megagametophyte). An alternative form of megagametogenesis, called tetrasporic megagametogenesis (in which all four haploid nuclei participate in the formation of the gametophyte), also results in the elimination of either *a* or *b*, since three of the haploid nuclei fuse together and do not participate in the formation of the egg. b. The formation of haploid megaspores, which give rise to megagametophytes, and haploid microspores, which give rise to sperm-bearing gametophytes (microgametophytes) does not result in the loss of *a* or *b*. Assuming random mating among all mega- and microgametophytes, the frequencies of *a* and *b* remain unchanged (meiosis does not result in a 50% cost of sex).

of life cycles involves some form of meiosis in every major lineage of plants, fungi, and animals. Mathematical models also predict that sexual reproduction will be maintained provided that the fitness advantage of maintaining this form of reproduction is greater than the mutation rate to revert to asexual reproduction. Finally, the short-term advantages to sexual reproduction in unpredictable and rapidly changing environments are rather obvious, since sexual reproduction obtains genetic variation that can be adaptive to new conditions. It must also not escape attention that this genetic variation is well suited for colonizing new habitats. A species that is capable of asexual and sexual reproduction is well suited "to stay put" (by asexual cloning) and "to get out there" (by dispersing genetic variants of itself).

Finally, as we will see in the next section, there are valid reasons to consider sexual reproduction as a functional trait that evolved to function in one way that was subsequently co-opted to function differently (an exaptation) in different lineages.

Meiosis, Sexual Reproduction, and Species Selection

As noted, inspection of the diversity of life cycles reveals that meiosis and the fusion of haploid gametes (syngamy) to form diploid zygotes occur in almost every variant (see fig. 6.3). However, no model for the evolution of meiosis or explanation for its functions has been proposed without being criticized, sometimes even by its creator. For example, John Maynard Smith (1920–2004) proposed one of the best-known scenarios for the evolution of meiosis and sexual reproduction. In this scenario, two genetically different but related haploid cells fuse together to produce a diploid heterokaryon with separate spindles that subsequently evolved into a diploid cell with one spindle and later evolved chiasmata (the pairing of homologous non-sister chromosomes and the exchange of genetic material), meiosis, and syngamy (fig. 6.5). Although this model is plausible, Maynard Smith rejected it because he rejected the notion that it first asserts that organisms "give up genetic variability for hybrid vigour; then abandon hybrid

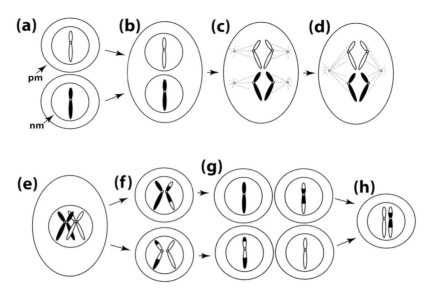

Figure 6.5. A scenario for the evolution of meiosis and syngamy developed by John May-nard Smith (1978). For simplicity the ancestral unicellular condition (a) is depicted to have only one chromosome (pm = plasma membrane, nm = nuclear membrane, and different genetic materials are depicted in white and black). Two genetically different cells (a) fuse to form a dikaryon with separate spindles (b-c) that becomes a diploid cell with a single spindle (d) with the capacity for chiasmata and meiosis I and meiosis II (e-g). The subsequent evolution of syngamy results in a diploid cell (h). A simpler scenario is shown in fig. 6.6.

vigour for the benefits of variability; [and] finally, regain hybrid vigour through reduction division and syngamy" (Maynard Smith 1978).

Arguably, a far more parsimonious scenario emerges from the proposition that meiosis evolved as a mechanism to correct for spon-taneous whole-genome duplication (autopolyploidy or endoduplica-tion), an idea that has been independently proposed by more than one author (fig. 6.6). This scenario removes the necessity of evolv-ing syngamy before meiosis and it removes the necessity of reducing dikaryotic spindles to one, while permitting the subsequent sequence of events proposed by Maynard Smith (see fig. 6.5).

An even more simplified scenario asserts that syngamy, genomic variation, and chiasmata evolved much later, perhaps in response to the problems arising from polyvalency, gene overexpression, an-euploidy, and other negative consequences of polyploidy (fig. 6.7).

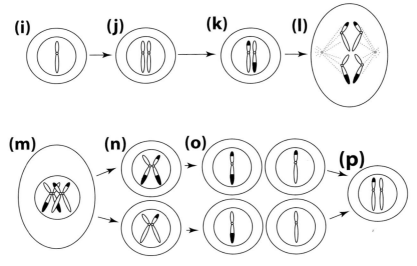

Figure 6.6. An alternative scenario in which meiosis evolved. For simplicity the ancestral unicellular condition is depicted to have only one chromosome (different genetic materials are depicted in white and black). In this scenario, meiosis evolved as a mechanism to correct for autopolyploidy as a haploid unicellular organism (i) undergoes spontaneous whole-genome doubling (j) followed by gene divergence (k) indicated by black chromosome segments. Using the single ancestral spindle (l), the diploid cell evolves the capacity for chiasmata, meiosis I and II, and syngamy (m–p) as depicted in fig. 6.5.

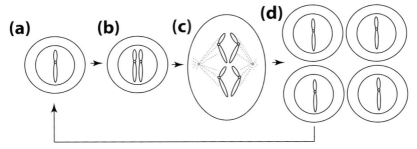

Figure 6.7. A modified version of fig. 6.6 in which meiosis evolves before syngamy, genomic divergence, and chiasmata. In this version, autopolyploidy (a–b) is followed by a one-step early form of meiosis (c–d) that immediately re-establishes the ancestral ploidy level (see arrow).

This scenario, in which meiosis increases fecundity, (1) evades the negative aspects of the cost of segregational genetic load (the fitness cost of breaking down genetic correlations and linkage disequilibria established previously by natural selection), (2) avoids the need to invoke multilevel selection theory to explain the evolution of meiosis (because the benefits of meiosis are conferred immediately during cell division), (3) is consistent with the existence of different recombination-independent mechanisms in homologous chromosome pairing (because the scenario predicts some form of "proto-meiosis" before ancient eukaryotes diversified into the current extant lineages), and (4) is compatible with the prevalence of unicellular haploid (as opposed to unicellular diploid) species in diverse protists.

None of the aforementioned scenarios for the evolution of meiosis provides a mechanistic explanation for the evolution of the unique features of meiosis, namely the pairing of homologs, the suppression of centromere splitting, and the absence of the S phase. However, it should not escape attention that a variety of dinoflagellates, sporozoa, archaezoans, and parabasalia undergo one-step meiosis—that is, chromosome segregation occurs before duplication. This phenomenon indicates that the critical evolutionary innovation was the acquisition of mechanisms capable of juxtaposing homologs to form bivalents (synapse), and their concomitant separation. This proposition is reasonable because the capacity for recombination occurs among prokaryotes and therefore predates meiosis. How this happened remains unknown, although research has shown that chromosomes that fail to undergo chromatin remodeling lose their competency to pair. For example, chromatin reorganization in the leptotene-to-zygotene transition is required for normal chromosome dynamics in *Zea mays*, whereas telomere and centromere interactions in early meiosis evoke chromatin conformational changes required for pairing interactions. Nevertheless, the molecular mechanisms underlying meiotic chromatin reorganization and homolog pairing are unclear, and other factors are undoubtedly important.

Much about the evolution of meiosis as we know it today remains hidden. However, we can ask whether it has any current adaptive value, or, more precisely, we can ask why sexual reproduction is retained in so many species. Many workers have attempted to provide a single canonical answer, but no one has provided an explanation that has satisfied everyone. The view taken here is that meiosis does more than just provide for additive genetic variation. For example, it is involved in the repair of double strand DNA damage, the maintenance of euploidy, and the resetting of epigenetic mechanisms. It also provides a mechanism for purging deleterious genomes (as when three haploid nuclei disintegrate following meiotic cell division in monosporic megagametogenesis or after zygotic meiosis in algae such as *Chara*). It is reasonable to suppose that selection has acted on each of these functionalities with different degrees of intensity over evolutionary time, and that sexual reproduction has been retained in different organisms for different reasons. This perspective may not be satisfying to those who desire a single reason for the evolution of a specified functional trait. However, the current available information is consistent with these three propositions. Although the evolution of asexually reproducing species within very different lineages of plants, animals, and fungi attests, at the very least, to the short-term gains of asexuality, no known asexual variant has been observed to outcompete and displace its sexually reproducing conspecific, even in stable environments.

Also, as noted, there are important immediate short-term advantages to sexual reproduction. For example, sexual reproduction generates variation among siblings, which can reduce sibling competition (also known as sibling rivalry). A sexually reproducing organism will have fewer surviving progeny if there is competition among siblings, and competition will likely increase as the genetic similarity of siblings increases. Meiosis and syngamy reduce the genetic similarity among progeny and thus reduce competition. The cost of sex might be counterbalanced by an increased potential for progeny to survive. This might be the case in particular when the potential for dispersal is low (when "the apple doesn't fall far from the tree").

The evolution and persistence of sexual reproduction provides one line of evidence supporting the notion of species selection provided that we accept the argument that sexual reproduction is a functional trait that emerges either as an exaptation or as an adaptation. Intuitively, we might surmise that sexual reproduction is advantageous under conditions favoring the extinction of species, or that it provides a phyletic competitive advantage in terms of fostering speciation. In either case, fitness defined at the species level is not an intrinsic component of microevolutionary dynamics. In this sense, sexual evolution fosters "evolvability" because it allows species to respond to both favorable and unfavorable conditions, and to explore new possibilities.

One counter-argument to this perspective is that sexual reproduction can slow down the rate of evolution because of the reassortment of loci and thus maintains genetic diversity at the expense of retaining currently nonadvantageous gene loci. Although it is true that this is a short-term cost, it can have long-term benefits, since currently "bad" or neutral loci might increase fitness under future conditions. It seems reasonable to think about the short-term costs and the long-term gains of sexual reproduction in terms of bet hedging in different time scales. Depending on the time scale, the units of selection can shift from the population-level to the species-level.

Large-Scale Species Selection

Regardless of whether we see species as units or "individuals" subject to selection at a macroevolutionary level, the fossil record shows clearly that some clades have been more successful in terms of spawning new species than others and that vast changes in the composition of Earth's flora and fauna have occurred over the course of the past 420 million years (fig. 6.8). Compilations of the first and last appearances of species in the geological column and statistical analyses of these "births and deaths" verify that the intensities of species origination and extinction have varied widely over a continuum of low to high. Episodes of high origination rates are often but not invariably

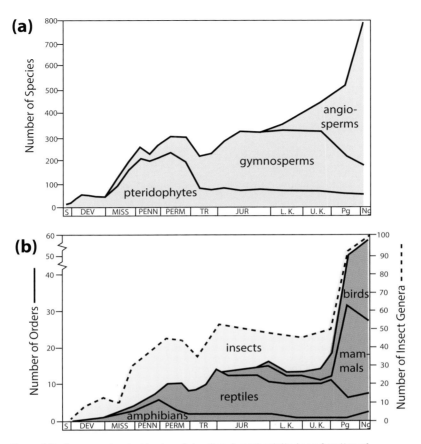

Figure 6.8. Coarse-grained estimates of standing diversity plotted as a function of geological time from the end of the Silurian to the end of the Neogene (from 420 Mya to 2.6 Mya). The data used to construct these plots were drawn from the primary literature published in the late nineteenth century and the twentieth century reporting the first and last occurrences of fossil taxa, largely from North America (Niklas, Tiffney, and Knoll 1980). New discoveries and taxonomic revisions have altered the plots shown here, both qualitatively and quantitatively. However, the data set used here has three advantages: (1) it was assembled by the same authors (and thus reflects a uniformly applied set of criteria for accuracy), (2) it covers a specific but large area and it was assembled over a specified time period, and (3) the data are at the species level. Despite not being at the global level, the data are consistent with the longstanding and general agreement regarding the temporal sequence in the appearances of the major groups of organisms shown in these plots. a. At the species level, the oldest paleo-floras were dominated by non-seed-bearing vascular plants (pteridophytes), which were the ancestors of modern day lycopods, horsetails, and ferns. These floras were replaced by flowerless seed-bearing plants (gymnosperms) that were subsequently replaced by paleo-floras dominated by flowering plants (angiosperms). b. Ordinal- and generic-level compilations (note different axes as indicated by solid and dashed lines) indicate turnover in the composition of paleo-faunas that coincide in some respects with those observed for plants. The oldest terrestrial fauna were dominated by amphibians; these were replaced by those dominated by reptiles and later by mammals.

identified with taxonomic radiations, while episodes of extinction of 70% or more of all standing diversity ("mass extinctions") are rare but nevertheless catastrophic. These intense episodes differ profoundly in magnitude from "background" origination and extinction rates, which taken at face value reflect the normal course of evolutionary affairs. Paleontologists have learned a great deal about the timing, magnitude, and taxonomic biases of previous episodes of rapid diversification and intense extinction, but the implications of the patterns of origination and extinction are still not fully understood. Much less is known about the patterns of plant origination and extinction, although the fossil record for terrestrial vascular plants shows that the pattern of intense episodes of plant extinctions does not correspond well temporally with the pattern of mass extinctions among either marine invertebrates or terrestrial vertebrates. The apparent discrepancy between the patterns of plant and animal extinction indicates that our understanding of the overall history and proximate causes of mass extinction is far from complete.

With all its potential pitfalls, uncertainties, and limitations, estimates of the origination and extinction rates of vascular plants (tracheophytes) show that the rates at which new species entered and departed the fossil record varied widely over the past 420 million years and that the angiosperms stand out as different from the rest (fig. 6.9). The highest origination rates occurred during Devonian and Mississippian times, when vascular plants taxonomically diversified into all but the angiosperm clade. The Devonian may be a special case, however, because origination and extinction rates become inflated whenever the standing diversity is low, as is the case for this time period (the Devonian extinction rates may be artifacts). But episodes of high origination also occurred during Permian, Triassic, and Tertiary times, when standing diversity was comparatively high. A recurrent pattern of initially high and variable rates of origination followed by declining and less variable origination rates is typically seen when tracheophytes are sorted into the three grades of reproduction (pteridophytes, gymnosperms, and angiosperms). Lyrically speaking, this

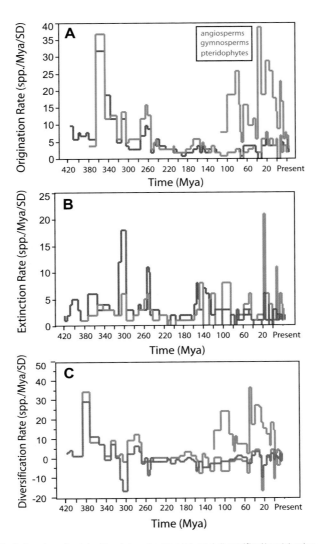

Figure 6.9. Estimates of origination (a), extinction (b), and diversification (c) rates of non-seed-bearing vascular plants (collectively called pteridophytes), flowerless seed-bearing plants (gymnosperms), and angiosperms plotted as function of time from the present (the present is at the right of the abscissa). Note that the metric used to calculate origination and extinction rates is the number of species appearing or disappearing per geological time interval divided by standing species diversity per geological stage (SD). Other metrics are possible, but this method is generally good demographic practice because it "normalizes" origination and extinction rates with respect to the number of species that can potentially give rise to new species or that can go extinct. Inspection of these graphs draws attention to the high origination and low extinction rates of the angiosperms compared to the gymnosperms and pteridophytes, which accounts for the high diversification rates of the flowering plants. Note further that each of the three groups of plants exhibits comparatively high origination rates early in its evolutionary career.

pattern begins with a vibrant *crescendo* whose reproductive theme is attenuated as a *diminuendo* ending in a *basso profundo*, the fundamental harmonic. The mechanisms responsible for this pattern remain unknown.

In theory, each of the three reproductive grades could have evolved as a result of a genetic revolution that literally created a new, totally unoccupied niche containing many opportunities for reproductive isolation and subsequent speciation. As the new niche became occupied, the subsequent genotypic changes required either to depose older species, or to create new opportunities would become increasingly more unlikely (or difficult to achieve), and so the rate of origination would steadily decline. The baseline for the origination rate may reflect the normally slow replacement of old species by new ones either by means of interspecific rivalry, or through random species extinction due to changes in the physical environment. Alternatively, the pattern may simply be a taxonomic artifact resulting from an inflated view of the real number of species during the early phase of the evolution of each of the three grades. All that can be said with any confidence is that the pattern is an observation awaiting explanation.

Patterns of Extinction

Clearly, one of the factors that may be responsible for patterns of origination is the pattern of extinction. The death of formerly successful species affords opportunities for other species to fill vacated niches. Recall that the traditional Darwinian view is that extinction is a gradual process driven by the struggle for existence among species for the same resources and space. Darwin believed that less well-adapted species would be slowly but inevitably pushed to extinction by better-adapted ones competing for the same resources. According to this view, extinction is active in the sense that rivalry among species is expected to gradually reduce the population size of less well-adapted species. At some point in its lifetime the inclusive population of a less well-adapted species will drop below a critical mass, and then effects like inbreeding depression may supply the final nudge toward extinction. Darwin was well aware of sudden disappearances of species from the fossil record, but he was convinced that these abrupt departures were "artifacts" due to unrecognized gaps in the geological column. Like his mentor the geologist Charles Lyell, Darwin was rather contemptuous of the notion that extinctions were the result of environmental catastrophes. He firmly believed that the death of species was due to biological, not physical, causes.

Darwin's view may explain background rates of extinctions—the comparatively low extinction rates that characterize much of the history of life. But the gradual replacement of old species by new does not comply with the lag in geological time between the extinction of major groups and their subsequent ecological replacement by others. Nor does it comply with mass extinction. If the cause of extinction is invariably interspecific rivalry, as Darwin believed, then no significant span of geological time should separate the extinction of a group of organisms from the filling of their ecological roles by competitors. In this sense interspecific competition abhors an ecological vacuum. Yet, the fossil record shows that significant intervals often separate episodes of intense extinction and the filling of vacated niches by other groups of organisms. Ichthyosaurs became extinct millions of years before their role in the biosphere was replaced by marine mammals. Bats and modern birds diversified millions of years after the extinction of pterodactyls. And dinosaurs died out well before terrestrial mammals assumed similar roles in the terrestrial landscape. These and other examples show that extinction is not invariably the result of interspecific competition as Darwin believed. Rivalry over the same resources and space necessarily involves coexisting species. Logically, interspecific competition must have immediate and direct effects on the number of competing individuals.

A reasonable explanation for the geological time lag between extinction and ecological replacement is that the proximate cause of mass extinctions is an abrupt and radical departure from previous environmental conditions. Once a formerly successful group is by chance extinguished, another kind of organism, perhaps even a formerly less successful one, has the opportunity to adaptively radiate and fill the vacated ecological space. Because adaptive modifications take time to evolve, the replacement of one group by another will not be immediate, even in geological terms, and a temporal lag will be evident in the fossil record of taxonomic succession. Note that Darwin's theory of natural selection permits the physical environment to sort among species as ruthlessly as does biological competition for limited resources and space. At a fundamental level, Darwin's theory of natural selection is indifferent to whether competition among species or physical emergency drives the sorting of species or higher taxa. In this sense the salient difference between the two modes of extinction (those caused by physical stimuli versus those resulting from species rivalry) is not whether natural selection occurs but the time organisms are given to adapt to new conditions. During mass extinctions, the time scale is too brief to give even formerly well-adapted organisms the opportunity to cope with events. During background extinctions the time scale is long enough for organisms to adaptively respond to changing conditions if it is within their genetic capacity to do so.

The fossil record for the mass extinctions of marine invertebrates and terrestrial vertebrates is relatively clear. Marine invertebrates have experienced at least five mass extinctions. Terrestrial vertebrates have experienced two mass extinctions and a number of "minor" mass extinctions. During each mass extinction, most animal species are extirpated in a geologically "brief" period. Many of these species were formerly very successful in terms of both geographical distribution and prior longevity in the fossil record. Attempts to find shared traits among the animals affected have garnered many rewards. For marine invertebrates, physiological, morphological, and ecological threads appear to bind together the animals that died out, such as the ability to deal with hypercapnia during the Permo-Triassic extinction event. In contrast to mass extinction events, background extinctions often involve great selectivity for life history traits. Animals that are geographically widespread appear to be spared more often than those that have narrow geographic coverage. The current view, therefore, is that mass and background animal extinctions differ in both tempo and mode. Mass extinctions occur quickly and appear to be due to sudden and severe physical environmental crises well beyond the normal ability of most animals to adapt. Background animal extinctions appear to reflect the more normal state of affairs involving gradually applied environmental stresses, of either a biological or a physical nature, that have normally been experienced by animals.

As noted, the fossil records for mass extinctions of vascular land plants and animals differ. The history of tracheophytes is characterized by nine single or clustered episodes of intense species extinction, and only two of these coincide with the "minor" mass extinctions reported for terrestrial vertebrates. Because the mass extinctions of land plants and animals do not correspond temporally, it is possible that either the proximal cause of extinction or the mode of response to the cause is frequently different in land plants than in marine and land animals. Also, unlike the animal mass extinctions, which show no evidence of selectivity for specific life history traits, intense episodes of plant extinction appear to favor some reproductive grades over others.

Regardless of the apparent differences between plants and animals to major extinction events, well over 90% of all the species that ever lived are now dead. Thus, the history of life is as much a saga of extirpation as of adaptive tenacity and success. However, the fossil record shows that extinction has not taken the upper hand. The tremendous morphological and taxonomic diversity seen today reflects a slight surplus of species birth and survival over death accumulated over hundreds of millions of years (see fig. 6.8). The terrestrial landscape was colonized by perhaps only a few small plant species existing in isolated populations. Today the surface of the Earth is draped in green, and the number of land plants is on the order of hundreds of thousands of species. The balance between speciation and extinction has shifted over time, and consequently the taxonomic composition of Earth's floras has dramatically changed over the past 420 million years. At one time Earth's forests were dominated by giant tree lycopods, horsetails, and ferns. As these forests passed into history, they left behind their fossil remains in the form of the great coal beds that fuel the fires of intellectual curiosity as well as those of commerce and industry. The early pteridophytic forests were replaced by forests dominated by cycad, maidenhair, and dawn redwood trees, which in turn gave way to the current stands of oak, maple, birch, and poplar. Evolution has no foresight—it is an unconscious process of descent with modification directed by the combined effects of random and nonrandom biological and physical forces. Thus the taxonomic composition and physical appearance of future forests cannot be predicted. What can be said with certainty is that some of today's descendants will become tomorrow's ancestors, despite the worst human efforts to extinguish many of evolution's fascinating organic products.

Darwin's Abominable Mystery

The rapid diversification and ecological dominance of the flowering plants raises a question that puzzled Charles Darwin, namely, why are there so many angiosperm species and why are they so ecologically diverse and successful? In a letter to the great botanist Joseph D. Hooker (1817–1911), dated July 22, 1879, Darwin wrote, "Saporta believes that there was an astonishingly rapid development of the high plants [angiosperms], "as soon [as] flower frequenting insects were developed." (Gaston de Saporta [1823–1895] was a paleobotanist.) Darwin's interest in the angiosperms had less to do with the origins of the "higher plants" than with their "sudden appearance" and rapid diversification, neither of which complied with his adherence to the notion of evolutionary gradualism. A number of equally plausible hypotheses have been advanced to resolve Darwin's "abominable mystery," among which the most widely accepted supports Saporta's speculation that angiosperm-animal coevolution fostered the rapid diversification of the flowering plants, particularly the coevolution with animal vectors for pollination and dispersal (box 6.1). Nevertheless, consensus acknowledges that there are many other attributes unique to or characteristic of the flowering plants. In addition, the remarkable coevolution of the angiosperms and animals could be an effect of the intrinsic adaptability of the flowering plants rather than a primary cause of their success, suggesting that the search for underlying causes should focus on an exploration of the genetic and epigenetic mechanisms that might facilitate adaptive evolution and speciation.

Here, we will explore angiosperm attributes that might promote diversification in their general form and draw particular attention to those that, either individually or collectively, have been shown empirically to favor high speciation rates, low extinction rates, or broad ecological tolerances. Among these are the annual growth form, homeotic gene effects, asexual/sexual reproduction, a propensity for hybrid polyploidy, hybrid necrosis, and an apparent "resistance" to extinction. A brief survey of these contending hypotheses fails to identify a single vegetative, reproductive, or ecological feature that taken in isolation can account for the evolutionary success of the angiosperms. Rather, it seems as if the answer to Darwin's abominable mystery lies in a confluence of features that collectively make the angiosperms unique among the land plants.

Most hypotheses about angiosperm diversification fall into one of three camps: (1) those that have focused on vegetative attributes, which can confer ecological and thus evolutionary advantages, such as diverse organographic morphology and anatomy, rapid growth rates coupled with high hydraulic conductivity, and the capacity for extensive phenotypic plasticity, (2) those that draw attention to the importance of reproductive features, such as floral display and morphology, pollination syndromes employing nondestructive biotic as well as abiotic vectors, double fertilization, endosperm formation and rapid embryogenesis, and the benefits of broadcasting seeds by means of edible fruits, and (3) those that take a more pluralistic perspective by pointing out that the flowering plants manifest a constellation of genomic, chemical, and structural attributes, which collectively confer vegetative and reproductive advantages as well as considerable plasticity.

Box 6.1. The "Harmony" of Plant-Animal Pollination Syndromes

A long-standing hypothesis for the rapid and successful diversification of the flowering plants rests on the coevolution of animal pollinators and the angiosperms. Until the evolution of the flowering plants, there is little fossil evidence that plants and animals evolved many positive feedback loops whereas the signature of herbivory is abundant. Plants and animals continued in their warfare, escalating chemical defenses and the means with which to defeat them. However, with the appearance of the flowering plants in the fossil record, a partnership evolved between some species of plants and some species of animals that resonates with the "idea of harmony between [*pollination*] visitor and blossom" (Fægri and van der Pijl 1979).

Among the hypotheses for the rapid rise of the angiosperms, one variant posits that animals with the same body plan share the same behavioral traits and thus tend to garner their nourishment similarly. If they obtain some or all of their nourishment from flowers (in the form of nectar, pollen, supernumerary floral parts, etc.), they will tend to visit similar flowers, and thus over time they will coevolve with the flowers that provide them with consistent nourishment. Evolutionary modifications of either the animal or the flowering plants upon which they rely would in theory establish a reciprocal selection regime in which both plant and animal would benefit. This selection regime would also provide a mechanism for speciation and macroevolution, because any sudden or dramatic phenotypic departure from the established norm in one participant would require a reciprocal adjustment in the other. If one or the other partner fails to adjust to the change, either or both would either suffer or adjust to a different relationship.

Whether this hypothesis explains the rapid diversification of the flowering plants shortly after their appearance in the fossil record remains uncertain. The flowering plants as a whole manifest a constellation of functional traits that give them a competitive edge (as for example rapid vegetative growth and short reproductive cycles). However, on the basis of empirical observations and experimentation, there is evidence that some animal pollinators visit certain kinds of flowers more frequently than other kinds of flowers, and ecologists have attempted to codify some perceived preferences as "pollination syndromes" (fig. B.6.1). Many of these syndromes are not hard and fast. For example, carrion beetles visit flowers of many different hues (brown, purple, and green); bees visit some red flowers because some red flowers possess UV cues (bees see in the UV and not in the visible spectrum). Many exceptions exist, and some long-standing "classic" syndromes have been discredited. Nevertheless, a certain harmony between the behavioral traits of pollinators and the blossoms they visit most certainly exists.

As noted, while no single hypothesis advanced to account for angiosperm success has met with universal acceptance, it is fair to say that the most influential, intensively investigated, and widely accepted are those that have followed Saporta's conjecture and point to the strong and often very complex biological interconnections that

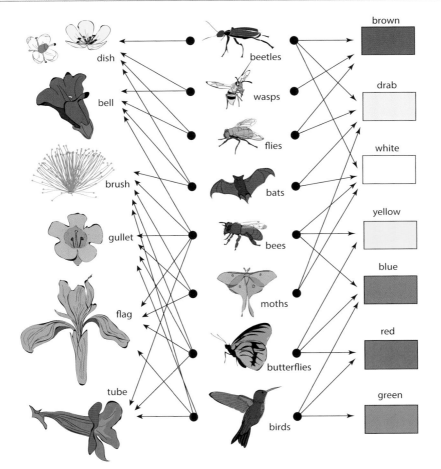

Figure B.6.1. Schematic of some of the relationships among floral display (left), animal pollinator (middle), and flower color (right). Considerable variation exists among these three pollination components, and many other factors enter into animal pollination (such as fragrance, the chemistry of the pollinator reward, the phenology of flower development). Adapted from Fægri and van der Pijl (1979; see their table 4).

have evolved between the flowering plants and animal species that facilitate pollination or seed/fruit dispersal (box 6.2). This focus is entirely justified given the large body of neontological studies that have documented numerous chemical, anatomical, morphological, and phenological plant adaptations that attract specific animal vectors

Box 6.2. The Titan Arum

One of the more remarkable flowering plants is popularly known as the Titan Arum, or more properly *Amorphophallus titanum*, which produces the largest known unbranched inflorescence (fig. B.6.2). The plant is endemic to Sumatra and grows in rainforest openings on limestone hills. The species was first described by the Italian botanist Odoardo Beccare (1843–1920) in 1878. The plants rarely produce flowers and are more often seen in the wild by virtue of their distinctive leaves. The inflorescence can reach over 3 meters (10 ft) in height and consists of a modified leaf, called a spathe, that wraps around a central hollow stem bearing flowers (a spadix). Ovule-producing flowers, which will develop seeds when pollinated, develop near the bottom of the spadix, and pollen-producing flowers develop along the mid-to-upper portions. When fully mature, the spathe unfurls to expose its upper surface, which is red and mottled, and the flowers to pollinators, which are attracted to the plant's odor (which smells to me like rotting fish due to the presence of trimethylamine).

Owing to the appearance and fragrance of the inflorescence, which mimics a decomposing animal, the Titan Arum is typically pollinated by carrion-eating beetles and flesh-eating flies. The flowers that will eventually produce seeds develop first; the pollen-producing flowers shed pollen one to two days later. This sequence of maturation reduces self-pollination. When mature, the seeds are red and likely attract birds as a dispersal agent. After seeds are produced, the inflorescence collapses onto itself and withers. Either the underground corm becomes dormant, or it produces a single large leaf.

It is worth noting that this elaborate pollination life-history is "adaptive" in the sense that many functional traits had to evolve to achieve this mode of pollination, but "maladaptive" in the sense that it is remarkably ineffective (the species is rare and declining in numbers in its native habitat).

Figure B.6.2. (*opposite page*) Reproduction of the Titan Arum (*Amorphophallus titanum*). a. The spathe and spadix before the release of volatiles to attract pollinators (mostly flies and beetles). b. The spathe unfurls and releases volatiles, and the carpellate flowers at the base of the inflorescence are receptive to pollination. c. Carpellate flowers (bottom) and staminate flowers (top) occur on the same inflorescence, but pollen is not released until after the styles on carpellate flowers begin to wither. d. Pollen is released in the form of long sticky strands. e–f. Fruits are fleshy and red to attract birds.

for pollination and long-distance disseminule dispersal and a paucity of insect-pollinated extant gymnosperm species, such as the cycads. Likewise, paleontological studies indicate that plant-animal relationships predating the appearance of the flowering plants were, with few exceptions (generally inferred on the basis of pollen size and structure or evidence of phytophagy), either neutral or mutually antagonistic.

Importantly, the ability of angiosperms to establish beneficial relationships with animals involves the capacity to evolutionarily modify
many characters that are either absent or rare in each of the many
other plant lineages. For example, many pollination syndromes involve adaptive chemical, anatomical, and morphological modifications
of vegetative as well as reproductive organs.

Seen in this light, the remarkable coevolutionary history of the angiosperms and their animal compatriots can be interpreted as more of a manifestation of the intrinsic adaptability of the flowering plants than as a mechanistic explanation for the success of the angiosperms. If true, the focus of the search for underlying reasons should shift to an exploration of genetic or epigenetic mechanisms that might facilitate adaptive flexibility, which may explain why the angiosperms are so successful. This agenda is particularly attractive in light of the suggestion that multicellular species may be largely unaffected by intense natural selection because of their propensity for gene duplication and divergence, pleiotrophy, and epistasis. If true, rapid diversification within a clade is not a priori evidence for rapid *adaptive* diversification. It may simply reflect an evolutionary episode of genomic restructuring among related taxa that affords opportunities to occupy new niches, some of which may involve coevolution among plant-animal interactions. Whether these niches remain occupied is a consequence of natural selection and can only be judged in light of the fossil record of extinction patterns, or after careful cladistic analyses of coevolutionary relationships. However, nonadaptive models provide no explanation for the high diversity of angiosperms *relative* to other groups, because the phenomenology they predict should apply to all or at least most multicellular taxa.

A comparison among the Phanerozoic origination rates and patterns of the three major vascular plant groups reveals that the early origination of the angiosperms is unexceptional in its tempo when compared to that of the pteridophytes or gymnosperms (fig. 6.9). The data indicate that each group radiates rapidly early in its evolutionary history and then subsequently declines in tempo. Likewise, the highest origination rates of the pteridophytes and the gymnosperms are comparable to those of the angiosperms. The only two differences revealed by this comparison that can be judged to be numerically meaningful are (1) standing diversity at the time angiosperms diversified early on was high (such that high rates of origination when normalized to standing diversity are high indeed) and (2) the angiosperms have a sustained tempo of speciation throughout their evolutionary history. For example, the limited data indicate that angiosperm origination peaks in the mid-Cretaceous (Turonian) and in the Early Tertiary (Lower Eocene).

This unique feature of the angiosperm origination pattern is open to numerous interpretations. However, the simplest and most plausible interpretation, particularly in light of what we know about angiosperm biology in general, is that the flowering plants continued to evolve morphological and reproductive innovations and "reinvent" themselves, whereas the distinguishing attributes of the pteridophytes and various gymnosperms became developmentally canalized early in their evolutionary history. It must be borne in mind that species richness is governed as much by extinction rates as it is by origination rates. Thus, an alternative interpretation is that the evolutionary innovations of the angiosperms reduce the probability of extinction as well as increase the probability of speciation. For example, significant changes in angiosperm floral trait shifts may have initiated concomitant changes in fruit-types that resulted in changes in dispersal mechanisms. One of these changes appears to have occurred in the Early Tertiary when seed and fruit size conspicuously increased to modern-day standards. Two explanations have been advanced to account for this significant increase in propagule size. One emphasizes the adaptive radiation of mammals and birds (Tiffney 1984), while the other focuses on climatic changes that permitted the spread of multistratal closed-canopy forests (Eriksson et al. 2000). These two explanations are not mutually exclusive. The adaptive radiation of frugivores capable of larger dispersal ranges would have been favored by vegetation dominated by plants producing larger edible disseminules. In turn, these species would have been well adapted to closed-canopy communities perhaps made possible by climatic changes. The reciprocity between climatic and biotic changes during the early Tertiary would in turn have resulted in runaway selection favoring an increase in the size-ranges of frugivores and edible angio-sperm disseminules, even if there were no net gain in relative fitness.

Turning to the same data set used to calculate origination rates and calculating extinction rates as the number of species not continu-ing from one geological period to the next younger per million years per species standing diversity, we see that the angiosperms tend to have significantly lower extinction rates compared to their pterido-phyte and gymnosperm temporal counterparts (fig. 6.9 b).

It must be reemphasized that the data set used to estimate origi-nation and extinction rates is not quite as robust as it could be in light of recent discoveries of new species in the fossil record, particularly of angiosperms. Nonetheless, the nature of the data set (particularly the number of morphospecies that appear and disappear over time) is not going to be radically affected by these new discoveries because, with respect to angiosperms, they are qualitatively different. Even if we limit ourselves to qualitative aspects of the patterns shown in fig. 6.9, two statements are justified: (1) the rate at which new angiosperm species enter the fossil record is *consistently* and continuously higher than that of any other group, and (2) the rate at which these species die out is *consistently* lower. Therefore, any robust hypothesis must fo-cus on those biological features that permit species to resist extinction as well as those that foster speciation.

Species Selection and the Null Hypothesis

Arguably, any hypothesis attempting to explain the rapid taxonomic and ecological diversification of any clade requires a null hypothesis that permits us to focus on the tempo of the appearance of evolution-arily novel characters in the fossil record and the extent to which this tempo correlates with changes in standing species diversity. The logic of this kind of null hypothesis rests on three propositions: (1) most species are (and have been) identified on the basis of their morpho-logical- anatomical phenotypic character states as opposed to their molecular-genomic biology or their reproductive biology (the majority of extant species and all fossil species are morphospecies; see table 6.1), (2) the ability to recognize taxonomic distinctions among morphospecies increases in proportion to the number of their phenotypic attributes, and (3), every clade can be described by some number N of characters that are adequate to distinguish the clade's component species. To illustrate this numerically, note that every clade is taxonomically defined by (and identifiable on the basis of) some number N of unique characters. If we make the simplifying assumption that each unique character has only two character states, it follows that the number of unique character state combinations (the number of potential phenotypic species) in any clade is given by the formula 2^N. Thus, in the case of one clade distinguishable from all others based on 5 unique characters, the maximum number of theoretical morphospecies is $2^5 = 32$, whereas for another clade defined by 10 unique characters, the number is $2^{10} = 1,024$. Clearly, the number of theoretical morphospecies will increase rapidly as the number of character states per unique character in each clade increases. For example, if each character has three states, the two hypothetical clades have $3^5 = 243$ and $3^{10} > 59,000$ potential morphospecies.

As noted, this approach has distractions and it must be used cautiously for a number of reasons. For example, the number of theoretically possible phenotypic species provides only an upper limit on the predicted number of real species. Some character state combinations may be physically impossible or highly improbable (bilaterally symmetrical corollas with fused spirally arranged petals are possible but unlikely for geometric and developmental reasons). All biologically possible combinations may not appear even in an ancient clade due to genomic and epigenetic constraints, whereas, in an evolutionarily young clade, there may be insufficient time for some or even many phenotypes to evolve. However, in theory, we might expect the number of phenotypic species to (1) increase rapidly early in the evolutionary history of a clade (as the numbers of characters and character states unique to a clade increases, or as unique character combinations appear), (2) subsequently plateau (as developmental and genetic potential become canalized), and (3) at some point late in the history of the clade, dwindle (as unique character permutations are eliminated as a consequence of extinction events). If this prediction holds true, the correlation between the number of unique character state combinations (theoretical phenotypic species) and the number of real phenotypic species should be stronger early in the history of the clade than at any other time.

This null hypothesis says nothing about the kind of characters or the number of their states that should be used to construct a morphospace. Obviously this will vary depending on the group being examined. It will also depend on research predilections and the kinds of questions being asked. However, when dealing with the flowering plants, we are immediately drawn to floral characters and character states, particularly in light of what appears to be a correlation between the appearance of new floral characters and the standing diversity of extant species in angiosperm families coappearing in the fossil record between the Aptian and the Upper Paleocene.

A Neo-Goldschmidtian Perspective on Macroevolution

As mentioned earlier, contra the neo-Darwinists, Richard Goldschmidt argued that speciation is a rapid process involving dramatic genomic reorganization. The occurrence of rapid changes in critical reproductive character states is well documented. By way of an example, species within the family Asteraceae are distinguished in part by whether their inflorescences contain radially symmetrical "disk" flowers, bilaterally symmetrical "ray" flowers, or both. Yet, as discussed in chapter 5, by performing artificial crosses between two species of *Haplopappus* that have rayed and rayless florets (*H. aureus* and *H. venetus* subspecies *venetus*, respectively), research reveals that the presence or absence of ray flowers is controlled by a single gene, which can mutate to effect phenotypic differences reflected by the two species.

Perhaps the best-known examples of single gene plant mutations with significant phenotypic effects are those altering homeotic genes

(genes that contain the genetic information required to direct development along a particular morphogenetic pathway), which have the ability to shift the developmental fate of cells, tissues, or entire organs. In the majority of cases, mutations of homeotic loci change the type (rather than the number) of organs produced, which suggests that the developmental patterns affected by these mutations involve genes that regulate organ identity and not those that regulate organ number. The most extensively studied floral homeotic mutations occur in the mouse-ear cress, *Arabidopsis*, and in the snapdragon, *Antirrhinum*. Like many angiosperms with "perfect flowers," these plants have four whorls of floral organs of which the outermost develop into sepals and the innermost develop into carpels. Mutations of *AP3* and *PI* genes of *Arabidopsis* or the *DEF* gene of *Antirrhinum* cause petals to be replaced by sepals and stamens by carpels, which results in "imperfect" flowers incapable of self-fertilization. Mutations in the *AG* gene of *Arabidopsis* and the *PLENI* gene of the snapdragon convert stamens into petals and carpels into sepals, respectively, either of which eliminates the capacity for self-fertilization. Because homeotic mutations such as these have the potential to establish reproductive barriers, they can serve as a genomic vehicle for character displacement, genetic divergence, and in theory the eventual appearance of new species.

The foregoing examples showing how single-gene mutations can produce major phenotypic shifts can evoke a neo-Goldschmidtian hypothesis for the rapid speciation of the angiosperms. However, enthusiasm for this hypothesis must be tempered in light of the low probability that any allele of this kind will become fixed in a population. Specifically, in general, a new mutant arising as a single copy in a diploid population of size N has a probability of fixation P given by the formula $P = (1 - e^{-(2Ne)s/N})/(1 - e^{-(4Ne)s})$, where Ne is the effective population size and s is the selective advantage of the allele. Note that this formula reduces to $P = (1 - e^{-2s})/(1 - e^{-(4N)s})$ when $N = Ne$, and becomes $P = 1/2N$ when $s = 0.0$ (a neutral mutation). Therefore, in the case of a hypothetical mutant with a selective advantage of $s = 0.01$ appearing in a population of 1,000 individuals, we see that

$P = (1 - e^{-0.002})/(1-e^{-4}) = 0.00199$ or roughly 0.2%. Likewise, if the mutant is selectively neutral, we find that $P = 0.05\%$. This hypothetical case illustrates that Goldschmidtian "hopeful monsters" created even by advantageous or neutral allelic changes (which arguably represent best-case scenarios) have exceedingly low probabilities of becoming fixed in a population. Indeed, if our hypothetical mutant is even slightly deleterious (e.g., 0.001), we see that $P = 0.004\%$ when $N = 1,000$.

Small Populations of the Annual Growth Form

Even though the fixation of single-gene mutations has a low probability, a number of theories (such as genetic drift, shifting balance, and crash-founder mathematical models) have been proposed that predict the rapid spread of novel alleles that can quickly become embedded in adaptive gene complexes. Interestingly, all of these theories require two conditions: (1) recurrent mutations of the same alleles, and (2) very small, isolated populations. Little can be said about the natural mutation rates of homeotic plant genes other than to point out that they are likely to be faster in annual rather than perennial species simply as a consequence of the disparity between the life spans, generation times, and growth rates of these two life-forms (fig. 6.10). But it is not unreasonable to suggest that annual species are more likely to exist in smaller and more isolated populations than perennial woody species. If both of these conjectures are generally true, it is reasonable to suggest that the mutation and fixation rates of alleles are probably faster among annual as opposed to perennial species such that speciation attributable to Goldschmidtian hopeful monsters might be faster among annual as opposed to perennial species.

In light of this speculation, it is intriguing to note that early studies report that (1) many flowering plant families, particularly the most species-rich, are dominated by annual species (such as the Asteraceae and Rubiaceae), (2) rates of chromosome number changes are reported to be higher among angiosperms with an annual life

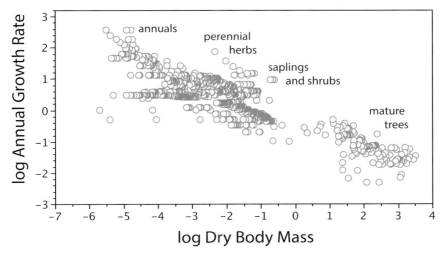

Figure 6.10. A bivariate log-log plot showing the relationship between annual growth rates (original units kg dry mass per individual plant per year) and standing dry body mass (kg dry mass per individual plant). Note that annual growth rates per mature plant are higher for mature annuals than for any of the other three categories of growth forms.

form compared to species with other growth habits, and (3) annuals also have the highest level of incompatibility and hybrid sterility among all angiosperm life forms. For example, although it is biased in terms of the temperate radiation of herbaceous species from what were probably tropical woody ancestors, a size frequency histogram of 1,133 genera (excluding those dominated by vines, lianas, or aquatics) listed in the Britten and Brown illustrated flora of the northeastern United States and Canada shows substantially many more genera of herbaceous annuals than woody perennial species (fig. 6.11). When considering variability in chromosome number within genera whose first appearances are known approximately from the fossil record, chromosome number diversity per lineage per unit time conforms to the series herbaceous angiosperms > woody angiosperms > conifers > cycads, and, when rates of increase in species diversity were estimated in a similar manner, they were strongly correlated with karyotypic rates (Levin and Wilson 1976). Some studies also indicate that arborescent angiosperm species have slower rates of molecular evolution

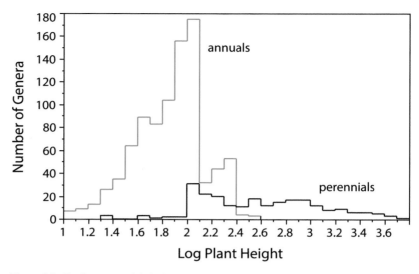

Figure 6.11. The frequency distributions of North American genera with an annual growth habit and those with a perennial growth habit plotted as a function of average plant height. The preponderance of genera is small annuals.

than herbaceous species (for example, Smith and Donoghue 2008). These interpretations of trends must be tempered by recognizing that although lineages such as the cycads (and mosses) are ancient, there is ample molecular evidence to suggest that many of the species in these lineages have evolved fairly recently.

Although a number of biological factors undoubtedly come into play when considering these trends, evolutionary rate differences between major groups of seed plants are explicable to a large extent in terms of the breeding structures of populations. In general, herbs tend to have small to moderate effective population sizes and relatively high dispersability, whereas woody angiosperms and gymnosperms are usually obligate outbreeders with large effective population sizes and low dispersability. Exceptions to each of these generalizations exist as prior work also suggests that annuals, on average, have low hybrid potential and high hybrid sterility compared to other angiosperm life forms. Clearly, more studies using modern genomic techniques are required to explore whether these generalizations are valid.

However, on the basis of the available data, the aforementioned trends are consistent with the supposition that the probability of dispersing, fixing, and keeping new karyotypes or novel character combinations in populations as a result of reproductive isolation is higher in herbs than in other seed plant life forms.

Polyploidy and Agamospermy

The role of polyploidy in plant evolution and speciation has been reviewed and discussed many times, particularly in terms of the distribution of species possessing at least one diploid set of chromosomes from each of the parent species (amphiploidy) among genera containing species with different life forms. Three very broad generalizations emerge from these surveys: (1) polyploidy is far more common among plants, particularly pteridophytes, mosses, and angiosperms, than animals, (2) among the angiosperms, it typically takes the form of amphiploidy (having at least one complete diploid set of chromosomes derived from each parent species), and (3) this provides a genetic system for perpetuating adaptive hybrid genotypes by means of sexual reproduction. In addition, there is ample evidence to suggest that polyploidy is more frequent among species with long-lived herbaceous life forms coupled with some means of vegetative propagation than among species with other life forms. It is noteworthy that polyploidy is absent in some ancient plant lineages, such as the cycads, which argues against the proposition that the frequency of polyploidy in a lineage is merely a symptom of lineage age (fig. 6.12).

It is important to draw attention to two aspects of polyploidy in general and amphiploidy in particular. First, polyploidy, particularly amphiploidy, and the genetic systems that can perpetuate adaptive hybrids are far more commonly encountered among angiosperms than any other embryophyte (or any other) lineage, although as a syndrome they are not unknown among other seed plant lineages. For example, *Sequoia sempervirens* is one of the very few known naturally occurring polyploid conifers, and, second, amphiploidy permits spe-

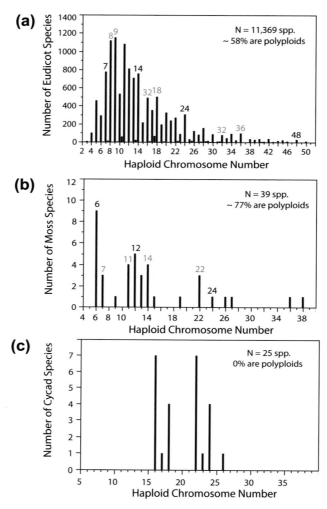

Figure 6.12. Frequency distributions of haploid chromosome numbers for eudicot (a), moss (b), and cycad (c) species. Supernumerary or B chromosomes not counted; odd numbers are arithmetic means (for example, 2N = 7 is N = 3.5). Multiples of the lower numbers are assumed to reflect autopolyploidy.

cies to bypass numerous sterility barriers and thus breed true, which, in turn, enables the exploitation of the advantages conferred by highly heterozygous genotypes. It is perhaps also informative to note that, among angiosperms, polyploidy tends to increase toward higher elevations and latitudes. Thus, it is possible that polyploid plants have an ecological advantage in cold environments. If true, this feature might also isolate them from conspecifics in subpopulations prone to speciation.

Previously, we spoke about the trade-offs between asexual and sexual reproduction and saw that asexual reproduction can benefit a species capable of sexual reproduction. Embryo formation without benefit of fertilization (agamospermy) is prevalent among angiosperms, ferns, and mosses, which are the three largest species-rich embryophyte groups. This phenomenon also permits the proliferation of hybrid genotypes asexually and provides a strategy allowing genotypes to "escape from sterility," as for example among hybrid triploids or pentaploids. However, agamospermy also occurs among species that are sexually fertile (such as *Citrus* hybrids). Thus, sexually fertile agamospermous organisms can produce genetically different progeny, which favors the colonization of new habitats and adaptive evolution by means of genetic recombination, but they can also reproduce asexually to produce disseminules that are as adapted to local environmental conditions as they are. These disseminules have all the advantages of being encapsulated in seeds and fruits, which may have specialized biotic dispersal agents. To the best of my knowledge, this "agamospermous syndrome" occurs nowhere in the plant kingdom other than among the flowering plants.

Reinventing Themselves

The pattern of angiosperm diversification preserved in the fossil record is not diagnostic of a monotonic process, one in which species numbers increase steadily at the same pace. Rather, it is characterized by episodes of comparatively rapid species diversification followed by periods of quiescence (see fig. 6.9 c). Although the intrinsic limitations of the fossil record should never be ignored, particularly the poor fidelity of its signal in understudied or poorly documented geological periods, this pattern can be interpreted as evidence for saltational evolution or punctuated equilibrium resulting from the appearance of new functional traits, which opened the door to new niches or different modes of more rapid speciation. The acquisition of the flower and the subsequent elaboration of phenotypic innovations that facilitated animal-assisted pollination have been asserted historically as the causative agents underwriting the early burst in angiosperm diversification during the Cretaceous. Likewise, the evolutionary acquisition, exploitation, and "rediscovery" of the annual growth form, which is entirely absent among living gymnosperms, in tandem with continued coevolution with animal pollinators and dispersal vectors may have been important ingredients in the subsequent evolutionary success of the flowering plants during the Cenozoic (Crepet 2008).

Seen in this light, the hallmarks of the angiosperms include a capacity to "reinvent themselves" by mixing old vegetative and reproductive traits into different combinations in addition to their ability to innovate new traits throughout their history. These characteristics surface in neobotanical studies that have drawn particular attention to the phenomenon called phenotypic plasticity, which is nowhere better revealed than through studies of angiosperms. Because phenotypic plasticity confers potential adaptive diversity to individual genotypes, it is likely to increase both the ecological distribution of a taxon and its pattern of diversification. Provided that each genotype within an individual species is sufficiently plastic to be broadly tolerant of environmental diversity, the ecological range of the species is expected in increase in proportion to the response breadth of its individual genotypes. As the geographic range of a species increases, the potential for ecotypic divergence and genetic isolation of subpopulations is also expected to increase, which can, in theory, foster the appearance of protospecies, particularly among plants with an annual life form. This hypothesis requires extensive research. However, a few studies indicate that more plastic traits can evolve more rapidly than less plastic ones.

Curiously, phenotypic plasticity may contribute to survival as well as rapid speciation. Because many aspects of adaptive differentiation may be obviated in taxa manifesting functionally appropriate phenotypes in response to key environmental pressures, phenotypic plasticity can shield genetic diversity from the effects of natural selection and it can enhance the long-term survival of taxa by means of species selection.

Darwin's "abominable mystery" is likely to remain unresolved as long as a single explanation is sought for the rapid diversification and subsequent ecological success of the angiosperms. Comparisons among the diversification patterns of angiosperms, gymnosperms, and pteridophytes reveal that all three groups experienced comparably high diversification rates early in their respective evolutionary histories. The only unique feature of the angiosperm pattern revealed by these comparisons is that the group exhibits repeated instances of high diversification rates subsequent to its origin. In contrast, the diversification patterns of the pteridophytes and gymnosperms are characterized by high initial rates followed by a more or less steady decline in the appearance of new species. The angiosperm pattern is all the more remarkable considering that the "pteridophytes" and the "gymnosperms" each designate a grade of reproductive organization (rather than a clade) that reappears at different times in the early history of the embryophytes. If each of the lineages within these two grades of reproductive organization manifested an initial diversification "burst," we would expect the overall pattern of each grade to look much more like the diversification pattern of the angiosperms. This dissimilarity tells us that the evolutionary history of the angiosperms is truly unique not because of the features angiosperms possessed when they "burst onto the evolutionary scene" but because they sustained their initial tempo of speciation and, at times, exceeded it considerably.

This impression is reinforced by insights gained from population and evolutionary theory and from neobotanical investigations, which in tandem reveal a number of factors that can sustain and increase the tempo and mode of speciation. These insights indicate that the success of the angiosperms is *Vielzeitigkeit* (multifaceted) and not the product of any one functional trait or syndrome of traits. Indeed, the available evidence suggests that the angiosperms have an unparalleled capacity for evolutionarily "reinventing themselves" and that each re-invention has allowed them to reiterate their pattern of species diversification and ecological success.

It is clear that plant-animal interactions were critical to the success of the earliest flowering plants in light of a reciprocal driving mechanism for angiosperm and animal diversifications. However, if this were the single evolutionary innovation of the flowering plants, we would expect to see a monotonically decreasing pattern in the rate of angiosperm diversification throughout the Cretaceous-Tertiary. In contrast, the available (albeit limited and therefore suspect) paleobotanical data indicate a pattern of saltational evolution and species diversification, which is more in keeping with repeated bursts of phenotypic and behavioral innovations resulting from the evolutionary introduction of novel functional traits throughout the history of the flowering plants. Whether these bursts indicate adaptive evolution sensu stricto remains problematic in light of the suggestion that multicellular organisms are generally little affected by intense natural selection. We have discussed only a few among the many traits for special consideration only because neobotanical studies have provided sufficient evidence that each can foster adaptive phenotypic divergence within populations. Some among these traits provide the raw materials for phenotypic innovation (as for example single-gene mutations capable of producing striking morphological changes in a single generation); others provide an avenue for fixing these traits in plant populations or escaping extinction (for example, the annual growth habit, substantial phenotypic plasticity, and asexual/sexual reproductive syndromes).

Suggested Readings

Arnold, M. L. 1997. *Natural hybridization and evolution*. Oxford: Oxford University Press.

Crepet, W. L. 2008. The fossil record of angiosperms: Requiem or renaissance? *Ann. Missouri Bot. Gar.* 95: 3–33.

Crepet, W. L., K. C. Nixon, and M. A. Gandolfo. 2004. Fossil evidence and phylogeny: The age of major angiosperm clades based on mesofossil and macrofossil evidence from Cretaceous deposits. *Amer. J. Bot.* 91: 1666–82.

Dilcher, D. L. 2000. Toward a new synthesis: Major evolutionary trends in the angiosperm fossil record. *Proc. Natl. Acad. USA* 97: 7030–36.

Eriksson, O., E. M. Friis, and P. Löfgren. 2000. Seed size, fruit size, and dispersal systems in angiosperms from the early Cretaceous to the late Tertiary. *Amer. Nat.* 156: 47–58.

Faegri, K., and L. van der Pijl. 1979. *The principles of pollination ecology.* Oxford: Pergamon.

Grant, V. 1949. Pollination systems as isolating mechanisms in flowering plants. *Evolution* 3: 82–97.

Hilu, K. W. 1983. The role of single-gene mutations in the evolution of flowering plants. In *Evolutionary Biology*, vol. 16, ed. M. K. Hecht, B. Wallace, and G. T. Prance, 97–128. New York: Plenum.

LaBandeira, C. C., D. L. Dilcher, D. R. Davis, and D. L. Wagner. 1994. Ninety seven million years of angiosperm insect association: Paleobiological insights into the meaning of co-evolution. *Proc. Natl. Acad. Sci. USA* 91: 12278–82.

Levin, D. A., and A. C. Wilson. 1976. Rates of evolution in seed plants: Net increase in diversity of chromosome numbers and species numbers though time. *Proc. Natl. Acad. Sci. USA* 6: 2086–90.

Maynard Smith, J. 1978. *The evolution of sex.* Cambridge: Cambridge University Press.

Niklas, K. J., B. H. Tiffney, and A. H. Knoll. 1980. Apparent changes in the diversity of fossil plants: A preliminary assessment. *Evol. Biol.* 12: 1–89.

Raven, P. H. 1977. A suggestion concerning the Cretaceous rise to dominance of the angiosperms. *Evolution* 31: 451–52.

Silvestro D., C. D. Cascales-Minana Bacon, and A. Antonelli. 2015. Revisiting the origin and diversification of vascular plants through a comprehensive Bayesian analysis of the fossil record. *New Phytol.* 207: 425–36.

Smith, S. A, and M. J. Donoghue. 2008. Rates of molecular evolution are linked to life history in flowering plants. *Science* 322: 86–89.

Soltis, P. S., and D. E. Soltis. 2004. The origin and diversification of angiosperms. *Amer. J. Bot.* 91: 1614–26.

Tiffney, B. H. 1984. Seed size, dispersal syndromes, and the rise of the angiosperms. *Ann. Missouri Bot. Gard.* 71: 551–76.

7

The Evolution of Multicellularity

All organic beings have been formed on two great laws—unity of type and the conditions of existence. By unity of type is meant that fundamental agreement in structure, which we see in organic beings of the same class and which is quite independent of their habits of life. The expression of conditions of existence . . . is fully embraced by the principle of natural selection [which] acts by either now adapting the varying parts of each being to its conditions of life; or by having adapted them in long-past periods of time.
—CHARLES DARWIN, *Origin of Species* (1859)

The concept of the "body plan" or "archetype" can be traced to the work of Georges Cuvier (1769–1832), Richard Owen (1804–1892), and other nineteenth-century comparative morphologists who showed that organisms can be classified according to their shared structural and anatomical traits, many or some of which have no obvious connection to the ecological lifestyles of the organisms sharing them. Most of the groups of organisms that we see in our daily lives, such as birds and mammals, are identifiable by virtue of their shared physical architecture, or "building plan" (*Bauplan*). The concept is easily grasped intuitively. If we see an organism with an exoskeleton, six legs, and a head-thorax-abdomen construction, we immediately recognize that it is an insect. If, on the other hand, we see an organism with an exoskeleton, but one that has eight rather than six legs, we might surmise that it is a spider. Likewise, if we see a plant bearing flowers, we know it is an angiosperm. Indeed, before the advent of molecular techniques, taxonomists grouped organisms into their various families and higher taxonomic levels primarily using what Darwin referred to as their *unity of type*. This is not to say that considerable variation does not exist within each type. Among the flowering plants, some families contain species with their floral parts arranged in threes or in multiples of threes, whereas other families contain species with their floral parts arranged in fours or fives (fig. 7.1). This variation reflects the long evolutionary history of flowers going back at least to the Cretaceous (145 to 66 Mya) during which genetic innovations and changes in floral development attended the diversification of flowering plants. Yet, the variation in floral structure seen among extant flowering plants does not cloud the fact that all angiosperms share homologous reproductive structures.

The strong stamp of a unity of type is not generally lost even when organisms adapt to dramatically different environmental conditions. Consider for example aquatic plants or animals. Vascular plants that take on such a lifestyle change typically manifest a reduction in xylem tissue since it is unnecessary when roots, stems, and leaves are submerged in water and when the force of gravity is negated by the aqueous medium. Likewise, vertebrates as divergent as mammals, birds, reptiles, and fish adapted to living in water share a number of morphological traits. One of the most obvious shared adaptations to their *conditions of existence* is the streamlining of their body plans, and, in the case of some animals, increasing the dorsal surface with respect to the ventral surface, which creates lift when these animals swim (fig. 7.2). Both of these evolutionary modifications reflect the fact that water is roughly one thousand times denser than air and thus provides considerable resistance to forward motion. Nevertheless,

Figure 7.1. Variation in floral structure, size, and appearance. Families are given in parentheses. a. *Helleborus* (Ranunculaceae). b. *Cucurbita pepo* (Cucurbitaceae) c. *Chelone lyonii* (Plantaginaceae). d. *Rosa rugosa* (Roasaceae). e. *Crocus tomasinianus* (Iridaceae). f. *Solanum dulcamara* (Solanaceae).

aquatic reptiles, birds, and mammals are still recognizable as sharing the same unity of type as their terrestrial counterparts.

Darwin clearly recognized the importance of the concept of the *unity of type*, since this is evidence for organisms sharing a common ancestor and thus a shared heredity. He also saw the importance of the concept of the *conditions of existence*, since organisms sharing the same conditions converge on the same or similar functional modifications,

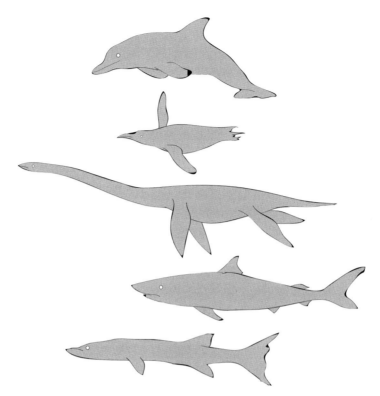

Figure 7.2. Examples of adaptive convergence to swimming in water. Note that all examples show streamlining (a smoothing of the surface and a reduction in the area projected in the direction of forward motion). From top to bottom: dolphin, penguin, plesiosaur (an ichthyosaur would be more exemplary would it not be for the author's inabilities to draw one), shark, and bony fish. Some examples also show dorsal surfaces that exceed ventral surfaces. This "blowing" accelerates water passing over the dorsal surface with respect to the ventral surface, thereby causing lift (dolphin, penguin, and shark).

which provides evidence for natural selection. Yet, even though it is intuitively easy to grasp, body plans are not easily defined for each of the major groups of organisms. One definition for body plan is *the constellation of morphological traits shared by the majority of organisms assigned to a specified taxonomic level* (such as the Class or Phylum). This definition may not appeal to everyone because it ignores two important interrelated features. First, body plans evolve such that the morphological traits shared by closely related species today need not reflect ancestral traits, and, second, morphology does not necessar-

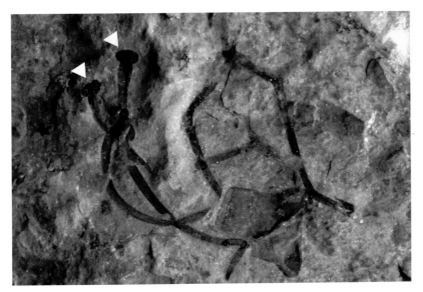

Figure 7.3. A compression fossil assigned to the vascular plant *Cooksonia caledonica* (a sporophyte) This genus is characterized by its small size, more or less equal branching of leafless axes, and presence of terminal spheroidal sporangia (two are indicated by arrows in the upper left). Nothing conclusive is known about the base of *Cooksonia*. Close inspection of this specimen, however, reveals no evidence for branched axes. What appear to be branched axes are overlying unbranched axes. Cornell University fossil plant collections.

ily shed direct light on the developmental biology underlying shared traits. A botanist might argue that the body plan of all vascular land plants (those possessing the vascular tissues xylem and phloem) involves a leaf-stem-root organization, or, in the case of parasitic plants, the reduction or modification of one or more of these three organ types. Yet, the body plans of vascular plants have evolved, sometimes dramatically. A paleobotanist would have to point out that the most ancient fossils of vascular plants, such as *Cooksonia* and *Rhynia*, lacked the organographic distinctions among leaves, stems, and roots. The sporophytes of these plants had naked axes that functioned as mechanical supports and photosynthetic organs (fig. 7.3). Along a related line of reasoning, how do we know that morphological traits shared among extant organisms are homologous? Even if we are interested only in the body plans of modern-day plants, how do we know that all leaves, stems, and roots are the same things? Homologies are very hard to prove in the absence of careful developmental and paleontological study. For example, the fossil record provides evidence that the structures botanists call leaves have evolved at least four times (independently in the lycopods, horsetails, ferns, and seed plants). Likewise, roots have evolved several times (at least once in the lycopods and perhaps as many as three times independently in other vascular plant lineages). Without a clear understanding of the developmental mechanisms underlying the ontogeny of leaves and roots, we cannot legitimately say whether organs sharing similar phenotypes are truly "shared" among the organisms bearing them, particularly if morphology is used exclusively to classify organisms.

We will be exploring these and other issues in this chapter. The goal is not to define the various plant body plans, even though this will be done. Rather, our goal is to explore the diversity of body plans and to understand why this diversity exists. It must be noted that for the purposes of this chapter the word "plant" is extended to include all photosynthetic eukaryotes. This definition allows us to include a large group of polyphyletic organisms collectively called algae as well as the monophyletic land plants sensu stricto (embryophytes). This broadening of what is typically meant by the word plant is required to fully understand how body plans in photosynthetic organisms evolved and perhaps why they evolved in the ways that they did. The reasoning behind this statement is laid out in the following section.

Why Include the Algae?

In contrast to the large number of papers and book chapters devoted to the body plans of animals, comparatively little has been said or written about the body plans of plants. Further, with the exception of the seminal papers of Donald R. Kaplan and Wolfgang Hagemann, the literature dealing with this topic focuses primarily on the flowering plants, despite the fact that the land plants evolved from a green algal ancestor that was similar in many ways to modern-day charophycean

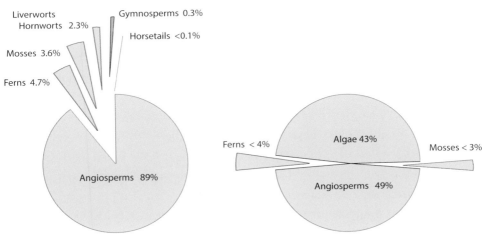

Figure 7.4. Pie diagrams showing the number of species (expressed as percentages of the total number of all known extant species) currently assigned to the different land plant lineages (left) and to the different plant groups including the polyphyletic algae (right). The most species-rich group of the monophyletic land plants is the angiosperms. The ferns and mosses follow in their percentage contributions. However, when the algae are included in the collective of all species numbers, the angiosperms comprise slightly less than half of all species and the number of algal species is comparable to that of the angiosperms. It should be noted that the land surface has been more intensely explored and catalogued for plant life compared to surveying aquatic habitats. Therefore, the percentage contribution of the algae to the total number of species is likely an underestimate.

algae and that the flowering plants are one among many land plant lineages. This angiocentric focus is explained (but not condoned) when we examine the relative abundances of modern-day land plants. Nearly 90% of all extant land plant species are angiosperms (fig. 7.4). The next two most species-rich groups are the ferns (4.7%) and the mosses (3.6%). Perhaps more surprising, the gymnosperms comprise less than 0.5% of all extant terrestrial plant species despite the fact that some groups, such as the conifers, are the dominant vegetation in certain high elevation or latitude habitats, which shows that species-richness and ecological importance are not invariably correlated.

The numerical supremacy of angiosperm species is significantly diminished when the number of algal species is added to the total of all known extant photosynthetic eukaryotic species. When all such species are considered, the angiosperms are seen to encompass less than 50% of all known extant species. This percentage does not differ significantly from that of all algal species (fig. 7.4). It must be emphasized that "all algal species" do not belong to the same lineage. The flowering plants are undoubtedly the most species-rich of any of the monophyletic plant lineages. However, the rationale for comparing these different percentages of species is simple—it is neither advisable nor admissible to classify or discuss plant body plans without including the algae since they are collectively as species-rich as the angiosperms.

There is another reason for taking the algae into consideration (other than that the land plants evolved from an ancient plexus of green algae similar to modern-day charophycean algae). The algae are as morphologically, developmentally, and ecologically diverse as the angiosperms (indeed, more so in some cases). Also, as noted, they are polyphyletic. The importance of this last statement should not be overlooked. It is critical. Consider that in his book *Wonderful Life*, Stephen J. Gould famously said, "Any replay of the tape [of life] would lead evolution down a pathway radically different from the road actually taken." It is certainly true that the history of life is unique. It is also true that much of evolutionary history reflects seemingly random events, some of which were catastrophic. However, it is debatable as to whether a replay of life's history would be *radically different*. It is difficult to know exactly what Gould meant by radically different. However, as we will see in chapter 8, physical laws and processes cannot be obviated by any organism. As a result, they have exerted predictable and persistent selection on all forms of life as attested by numerous examples of convergent evolution (see fig. 7.2). It is reasonable to suppose therefore that unrelated organisms will respond to physical forces in similar ways. This is certainly true for the algae. Recall that, when discussing primary, secondary, and tertiary endosymbiosis (see chapter 3), we saw that the major algal groups are separate "experiments" down the eukaryotic evolutionary pathway

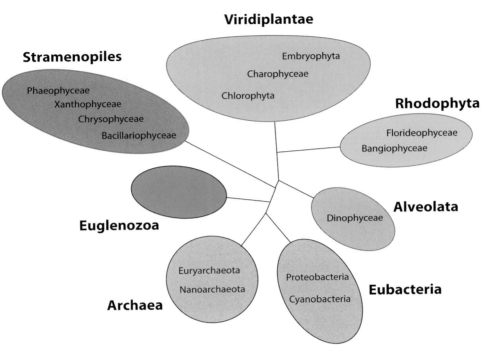

Figure 7.5. Schematic of the broad phylogenetic relationships among the prokaryotes (Archaea and Eubacteria) and the five major groups containing eukaryotic photosynthetic organisms (Euglenozoa, Stramenopiles, Viridiplantae, Rhodophyta, and Alveolata). The Stramenopiles include the Phaeophyceae (brown algae), Xanthophyceae (yellow-green algae), Chrysophyceae (golden-brown algae), and the Bacillariophyceae (the diatoms). The Viridiplantae includes all green plants. The chloroplasts of the Stramenopiles, Viridiplantae, and the Rhodophyta (the red algae) evolved as the result of a primary endosymbiotic event from a cyanobacteria-like ancestor. The chloroplasts of the Euglenozoa (euglenoids) and Dinophyceae (the dinoflagellates) evolved as a result of secondary or tertiary endosymbiotic events (see chapter 2).

(fig. 7.5). In addition, some of these experiments began billions of years ago (fig. 7.6). These two facts allow us to determine whether Gould's prediction is correct by examining whether the very different algal lineages followed the same pathways. We can also explore why perhaps some evolutionary pathways were common while others are rare by replaying the *tapes of plant life*.

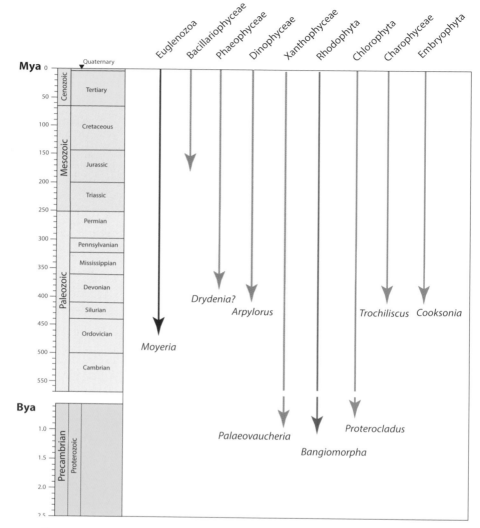

Figure 7.6. Approximate first occurrences of some of the major algal lineages (see fig. 7.5 for phyletic relationships) and the land plants (Embryophyta) plotted against the geological column. Note the different age units: billions of years for the Precambrian and millions of years for the Phanerozoic (the Paleozoic, Mesozoic, and Cenozoic). Based on fossil evidence, the oldest algal lineages are Xanthophyceae (yellow-brown algae), the Rhodophyta (red algae), and the Chlorophyta (green algae). The oldest fossil yellow-brown alga is *Paleovaucheria*, which dates to ≈1,000 Mya. The oldest fossil red alga is *Bangiomorpha*, which dates to 1,100–1,200 Mya. The oldest fossil green alga is *Proterocladus*, which is found in sediments dated to 800 Mya. The oldest charophycean algae and vascular land plants (*Trochiliscus* and *Cooksonia*, respectively) are found in the Upper Silurian, ≈420 Mya. All of these dates reflect minimum ages of first occurrences, since older fossils may be found in the future.

Classifying Body Plans

The phycologist Adolf Pascher (1881–1945) was one of the first to suggest using pigments, photosynthates, cell wall composition, and cell motility (if any) to identify and classify the major algal lineages. Pascher's four criteria (which show that the green algae and the land plants are a monophyletic group of plants) eschew using size, morphology, tissue-type, or any easily observed morphological or anatomical trait with which to classify algae. His use of chemistry and mobility was extremely astute. Even a brief survey of the physiological, morphological, and ecological diversity of photosynthetic eukaryotes quickly reveals why this is so and why plant body plans cannot be classified easily or reliably on the basis of body size, shape, anatomy, or physiology. For example, the algae range in size from unicellular picoplanktonic forms (< 2 μm in diameter) such as *Nannochloris* to some of the largest organisms in the sea such as the giant kelp *Macrocystis* (< 45 meters in length). Additionally, this size range exists within many lineages. The extent to which the different lineages have converged in appearance and lifestyle is also extensive. For example, certain species of green and red algae are capable of precipitating calcium carbonate dissolved in seawater to construct skeletal-like structural elements that provide support for photosynthetic tissue. Many of these multicellular organisms have converged in their general appearance, yet evolved independently from unicellular ancestors roughly one billion years ago (fig. 7.7).

Give the extensive convergence in so many traits, how can the plant body plans be classified? The tactic taken here is to identify the developmental mechanisms that allow plants to organize and construct themselves. The goal is to identify the "motifs" that run throughout all of the various lineages. There is no reason to believe a priori that each motif shares the same underlying developmental basis across all the lineages, although some homologies may exist among closely related groups of plants.

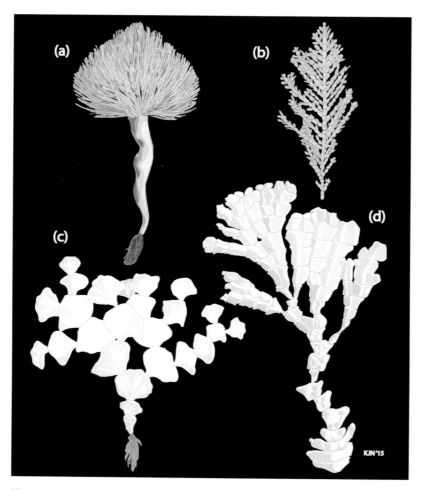

Figure 7.7. Four "coralline" algae drawn in black and white so as not to immediately reveal which are green algae and which are red algae. Each of these plants grows in a wave-swept habitat. Each produces a holdfast anchoring it to a solid substrate and each possesses a jointed skeletal-like architecture that permits it to flex and bend in moving water, thereby reducing drag forces. a. *Penicillus capitatus* (a green alga). b. *Corallina mediterranea* (a red alga). c. *Halimeda opuntia* (a green alga). d. *Bossiella compressa* (a red alga).

This approach identifies four basic body plans: the unicellular, colonial, coenocytic, and multicellular body plans that can be distinguished by four basic developmental motifs (fig. 7.8): (1) the presence or absence of synchronous nuclear and cytoplasmic division, which determines whether cells are uni- or multinucleate (*Chlamydomonas* and

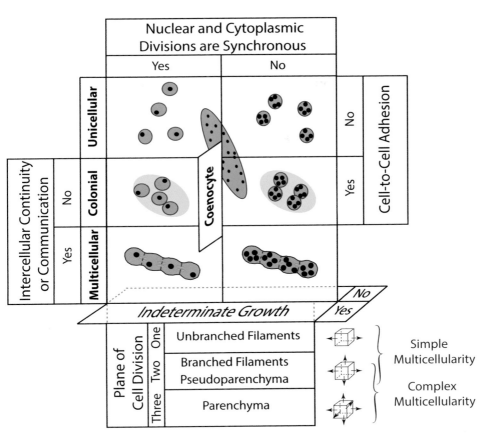

Figure 7.8. Classification scheme for plant body plans based on a binary ("yes" or "no") grid of four developmental motifs that move from top to bottom and from left to right: (1) (top) nuclear and cytoplasmic divisions are or are not synchronous, (2) (right) cells remain or do not remain attached when cells divide, (3) (left) intercellular continuity or communication is or is not maintained after cells divide and adhere, and (4) (below) cells continue or do not continue to increase in size. The various permutations of these four motifs identify four basic body plans (shown in bold type): unicellular, colonial, coenocyte, and multicellular. Three variants of the multicellular body plan (shown below the binary matrix) emerge as a result of whether cell division is confined to one, two, or three planes with respect to the body axis (shown below the grid display of developmental motifs). Cell division confined to one plane of reference with respect to the body axis results in unbranched filaments. Cell division confined to two planes results in branched filaments (or the interweaving of filaments to produce a pseudoparenchymatous tissue). Cell division confined to three planes results in a parenchymatous tissue. Simple multicellularity refers to organisms whose every cell is in contact with the surrounding medium (air or water); complex multicellularity refers to organisms with some cells lacking direct contact with the surrounding medium.

Bryopsis, respectively); (2) the presence or absence of cellular aggregation after cell divisions, which determines whether a plant has a unicellular or colonial body plan (*Calcidiscus* or *Phaeocystis*, respectively);
(3) indeterminate growth of the multinucleate cell, which results in
the coenocytic body plan (*Caulerpa*); and (4) intercellular continuity
or interdependent communication among cells during and after cell
divisions, which establishes the multicellular body plan (many algae
and all land plants).

The multicellular body plan has three basic tissue constructions
that can be described in terms of the extent to which cell divisions are
confined to one, two, or three planes of reference with respect to the
main axis of the body plan (fig. 7.8): (1) unbranched filaments result
when cell divisions are confined to one plane of reference (*Spirogyra*);
(2) branched filaments and a pseudoparenchymatous tissue construction are possible when cell divisions are confined to two orientations
(*Stigeoclonium* and the fruiting bodies of many fungi); and (3) a parenchymatous tissue construction becomes possible when cell division
occurs in all three directions (*Fritschiella*).

A review of the literature treating the algae reveals that all but
two of the 22 theoretically possible combinations of body plans, tissue constructions, and determinant versus indeterminate growth
patterns are represented by one or more extant species. The two
permutations that are not represented by any extant or extinct species are the uninucleate unicellular body plan with indeterminate
growth and the parenchymatous body plan composed entirely of
multinucleated cells. The absence of the former may be the result
of physiological constraints imposed by the volume of cytoplasm
that a single nucleus can sustain, a hypothesis proposed by Julius
von Sachs (1832–1897). A convincing explanation for the absence
of the second of these two possible body plans remains problematic,
although reasoning along the same lines as von Sachs may illuminate
the way to an explanation.

Table 7.1. Distribution of plant body plans and tissue constructions in representative lineages of photosynthetic eukaryotes

	Coenocyte	Unicellular	Colonial	Multicellular		
				Fil.	Pseudo.	Paren.
Embryophyta	+[a]	−[b]	−	+	+[c]	+
Charophyceae	−	+	+	+	+	+[d]
Chlorophyta	+	+	+	+	+	+
Rhodophyta	−	+	+	+	+	+
Phaeophyceae	−	+	+	+	+	+
Xanthophyceae	+	+	+	+	+	+
Chrysophyceae	−	+	+	+	+	+
Bacillariophyceae	−	+	+	−	−	−
Dinophyceae	−	+	+	−	−	−
Euglenozoa	−	+	+	−	−	−

Note. Fil. = filamentous. Pseudo. = pseudoparenchymmatous. Paren. = parenchymatous.
[a] Coenocytes occur in the early stages of endosperm and megagametophyte development.
[b] Gametes are legitimate examples of a unicellular body plan, albeit a short-lived one.
[c] As for example, within the locules of citrus fruits.
[d] Reported by Linda E. Graham for *Coleochaete scutata* and *C. orbicularis*.

Phyletic Distributions of the Body Plans

As noted, with only two exceptions, all of the hypothetical body plan variants enumerated in fig. 7.8 are known to exist. However, the question is, How are these variants distributed among the various plant lineages?

A survey of the literature shows that the unicellular and colonial body plans occur in all plant lineages (table 7.1). This is not surprising since the unicellular body plan is considered to be the ancestral condition in all of the major plant, animal, and fungal lineages, and the ability to form colonies requires the evolution of cell-to-cell adhesives, which are prefigured in many cases by the adhesives required for the union of gametes or the adoption of a sessile lifestyle. Simple multicellularity occurs in the majority of lineages that evolved as a consequence of primary or secondary endosymbiotic events (the green, red, and brown algal lineages). It is absent, however, in the diatoms (Bacillariophyceae), perhaps because the nature of the silica-rich walls

characterizing this lineage precludes intercellular cytoplasmic connections. Multicellularity is also absent in lineages resulting from tertiary
endosymbiosis (dinoflagellates). The reasons for this are unclear since
the lineages failing to achieve multicellularity are unique in many
other respects and thus no one factor can be posited as causative. For
example, most dinoflagellates have an unusual form of nucleus, called
a dinokaryon, in which the chromosomes are attached to the nuclear
membrane, lack histones, and remain condensed throughout the cell
cycle. The coenocyte is the rarest of the body plans. It occurs only in
the Chlorophyta (specifically in the Ulvophyceae) and in the Xanthophyceae, and only within taxonomically confined branches within each
lineage. One possible explanation is that this body plan is susceptible
to microbial or viral attack. Once an aggressive pathogen gains entry
into the cytoplasm, it can sweep through the organism rapidly because
there are no cell walls to stop or hinder its advance.

The distribution shown in table 7.1 does not shed light on evolutionary trends within individual lineages, nor does it reveal anything
about why this distribution exists. Deeper insights into the evolution
of body plans requires greater taxonomic resolution within individual
lineages and the application of a posteriori reasoning. It also requires
an unambiguously well-defined ancestral condition.

Consider, for example, the DNA-based phylogenies of the two
algal lineages in which the coenocytic body plan exists, the Xanthophyceae and Ulvophyceae. Inspection of the cladogram for the
Xanthophyceae reveals that the unicellular and colonial body plans
occur in the early divergent portions of the lineage represented by
present-day *Chlorellidium* and *Heterococcus* (fig. 7.9). This is consistent
with the proposition that the unicellular body plan is the ancestral
condition in all lineages. However, later divergent branches reveal
the coenocytic body plan represented by *Asterosiphon* and *Vaucheria*.
Later divergent branches in this cladogram indicate that the siphonous body plan is absent and replaced in some instances by a multicellular body plan, although the ancestral unicellular and colonial body
plans reemerge even in the near-last divergent branch as represented

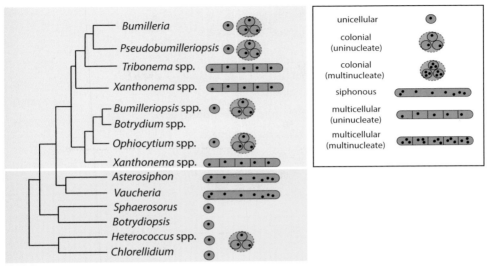

Figure 7.9. Plant body plans (see insert) mapped onto a redacted molecular-based phylogeny for the Xanthophyceae based on the analyses of Maistro et al. (2009). The unicellular and colonial body plans occur on the early most divergent branches (represented by *Chlorellidium* and *Heterococcus*), whereas the multicellular body plan occurs in the last divergent branch (represented by *Xanthonema* and *Bumilleria*). The coenocytic body plan, represented by *Asterosiphon* and *Vaucheria* (shown in the lower shaded rectangle) occurs earlier than the first appearance of multicellularity, represented by *Xanthonema* (shown in the upper shaded rectangle). This topology permits two interpretations. The coenocytic body plan is either ancestral to the multicellular body plan, or the taxa representing the coenocytic body plan are morphologically derived as a consequence of ecological specialization. For details concerning the molecular-based phylogeny, see Maistro et al. (2009).

by *Pseudobumilleriopsis* and *Bumilleria*. Bearing in mind that every cladistic analysis is a hypothesis that can change as additional taxa and data are added to the analysis, this cladogram can be nevertheless interpreted in two ways—the coenocytic body plan either prefigures the multicellular body plan, or it is a derived condition reflecting the ecological specialization of the taxa selected to represent early divergent branches in this phylogeny. Recall from previous discussions that the morphological traits of extant taxa, such as *Asterosiphon* and *Vaucheria*, need not invariably reflect the ancestral condition if the taxa in question have undergone intense or prolonged selection in ecologically specialized habitats. Certainly, the presumption of any

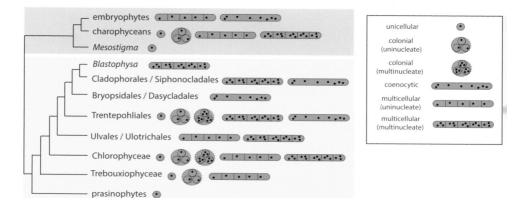

Figure 7.10. Plant body plans (see insert) mapped onto a simplified molecular-based phylogeny for the green plant lineages (the charophycean algae and land plant lineage, shown in blue; the Ulvophyceae shown in yellow) based on the analyses of Cocquyt et al. (2010). In each branch of this phylogeny, the unicellular body plan occurs in the first divergent branch and multicellularity precedes the appearance of the coenocytic body plan. This patterns lends credence to the inference that the coenocytic body plan is a derived condition. For details concerning the molecular-based phylogeny, see Cocquyt et al. (2010).

cladistic analysis is that the taxa selected are representative of their lineage's phylogeny. But this need not always be true.

This caveat holds as well for interpreting the molecular-based phylogeny of the Ulvophyceae, which is arguably less ambiguous in revealing trends in body plan evolution even when linked to a highly simplified phylogeny for the charophycean algae and the embryophytes (fig. 7.10). Inspection of this cladogram shows that the unicellular body plan occurs in representatives of earliest divergent branches (the prasinophyte algae and the genus *Mesostigma*, which thus far has been consistently identified as a sister taxon to the charophytes and embryophytes). The unicellular, colonial, and multicellular body plans as well as the multinucleate condition are represented on subsequently diverging branches leading to the Ulvales and Trentepohliales, whereas the coenocytic body plan appears only in the Bryopsidales and Dasycladales. A similar trend is seen in the *Mesostigma*—charophycean algae—embryophyte lineage wherein the coenocytic body plan is only seen during the early "free nuclear" stage of endosperm and megagametophyte development and the early stages of conifer embryogenesis. Once again, bearing in mind that every cladogram is a hypothesis awaiting further confirmation, the phylogeny of the green plant lineages shown in fig. 7.10 is consistent with the supposition that the coenocytic body plan is a derived rather than an ancestral condition.

The Motifs for Multicellularity

Multicellularity has arisen at least once in every major eukaryotic clade (plants, fungi, and animals) and is found in organisms differing in their ploidy levels. It has also evolved numerous times among the prokaryotes. In the majority of cases, the available evidence indicates that each lineage in which multicellularity appears has undergone a "unicellular → colonial → multicellular" body plan transformation series (for example, see figs. 7.9 and 7.10). The available evidence also indicates that the mechanisms whereby this transformation series was achieved differ among the various multicellular lineages. This evidence is consistent with the view that selection acts on functional traits rather than on the mechanisms that generate them ("many roads lead to Rome"): the developmental mobilization of very dissimilar genomic systems or processes can result in similar phenotypes.

   In the context of the evolution of development, this dictum has been formalized by Stuart A. Newman and coworkers, who propose a framework for conceptualizing the development and evolution of multicellular animals based on what they call dynamical patterning modules (DPMs). Each DPM involves one or more sets of shared gene networks, their products, and physical processes common to all living things (such as cohesion, viscoelasticity, diffusion, tensile and compressive mechanical forces, and activator-inhibitor Turing dynamics). Within the DPM framework, generic as opposed to genetic processes operating in tandem with gene networks and their products can act

Table 7.2. Examples of representative cell-to-cell adhesives isolated from the cyanobacteria and the major eukaryotic lineages

Lineage or Clade	Representative Intercellular Adhesive
Cyanobacteria	glycoproteins, lipopolysaccarides
Chlorophyta	hydroxyproline-rich glycoproteins
Embryophyta	Ca^{+2}-rhamnogalacturonic-dominated pectins
Rhodophyta	sulfated galactan polymers
Phaeophyceae	β-D-mannuronic and α-L-guluronic acid polymers
Fungi	glycoprotein-based "glues"
Metazoa	type-1 transmembrane cadherin proteins

in isolation or in combination to give rise to the basic body plans of multicellular animals. In this context, *generic* refers to mechanisms and processes that affect nonliving as well as living things.

In theory, these DPMs can operate in plants and fungi as well as animals because of fundamental similarities among all eukaryotic cells. Consider, for example, cell-to-cell adhesives. All eukaryotic cells have the capacity to secrete polysaccharides and structural glycoproteins that self-assemble to form extracellular matrices around plant as well as animal cells (table 7.2). Both types of matrices contain interpenetrating polymeric networks that employ hydroxyproline-rich glycoproteins as major scaffolding components (the HRGP extensin superfamily in various algae and in the embryophytes). These proteins generally form elongated, flexible, rod-like molecules with marked peptide periodicity with repeat motifs dominated by hydroxyproline in a polyproline II helical conformation extensively modified by arabinosyl/galactosyl side chains. It is possible therefore that this "superfamily" of cell-to-cell adhesives evolved by the co-option of an ancestral gamete-gamete self-recognition or cell-adhesion-to-substratum toolkit that nevertheless produces different adhesives in different organisms. Likewise, the wide array of microtubule-associated proteins in algae, embryophytes, fungi, and metazoans reflects the co-option of genomic toolkits from a shared last ancestor.

Nevertheless, the DPMs identified by Newman and coworkers cannot be applied directly to plant development because of substantive differences among plants, animals, and fungi. For example, during animal development, cells are typically free to migrate and slide past one another in ways that permit differential adhesion, cortical tension, and other processes that can facilitate the sorting and assembly of some tissues. In contrast, the cell walls of land plants and many different kinds of multicellular algae are firmly fixed to one another. Likewise, signaling molecules in plants can also act intercellularly as well as intracellularly as transcriptional modulators and determinants of tissue as well as cell fate, thereby blurring the functional separation of gene regulatory networks affecting multi- as opposed to single-cell differentiation. Although the transport of developmental transcription factors from one cell to another is not unknown in animal systems (as for example in *Drosophila* nurse cells), it is very rare. Further, cell polarity in land plants involves PIN and PAN1 proteins, whereas animal cell polarity involves integrin, cadherin, and PAR or CDC42 proteins. Finally, cell division mechanics and the deposition of cell walls differ even among closely related species in the same lineage, as for example among the different desmids or in different filamentous ascomycetes.

In light of these and other issues, Valeria Hernández-Hernández, Mariana Benítez, and coworkers proposed a set of six DPMs associated with plant developmental processes. Only four of these are relevant to the evolution of multicellularity: (1) the formation and orientation of a future cell wall (FCW), (2) the production of cell-to-cell adhesives (ADH), (3) the formation of intercellular lines of communication that engender spatial-dependent patterns of differentiation (DIFF), and (4) the establishment of axial and lateral polarity (POL).

These four modules operate in tandem. For example, among embryophytes, the presence of adhesive pectin polysaccharides in the middle lamella, which bonds neighboring cells together, is associated with the deposition of the future primary cell walls of adjoining cells. The cell wall begins to be formed from cell plates during the division of the cytoplasm (cytokinesis), such that cell adhesion is the default state. Additionally, the proportion and chemical state (for example, the level of esterification) of each of the cell wall components is regulated over the course of development, locally as well as globally, adjusting the mechanical properties of cells and tissues, and contributing to the regulation of cell and organ growth in size. A somewhat analogous system operates during the tip-growth of fungal hyphae. The DIFF and POL modules also function together because both are required for cell-type specification and intercellular communication. For example, among embryophytes, DIFF and POL involve the transport of metabolites, transcription factors, and phytohormones through intercellular bridges, called plasmodesmata (fig. 7.11).

Experimental evidence from studies of the mouse-ear cress, *Arabidopsis thaliana*, and other model systems shows that the flow of the phytohormone called auxin and cell wall mechanical forces reciprocally interact during the emergence of polarity. Auxin promotes cell expansion by cell wall loosening, involving the acidification of the cell wall and the concomitant disruption of noncovalent bonds among cell wall polysaccharides. The preferential localization of auxin transport proteins, called PINs (or their transporting vesicles), determines auxin fluxes and targets the locations for future cell wall loosening.

Each of the FCW, ADH, DIFF, POL modules involves the participation of generic physical mechanisms such as mechanical forces. Consider, for example, how the FCW module operates in land plants in response to mechanical stresses. It has long been known that as a cell enters the cell division cycle (the preprophase stage), the orientation of the future cell wall's position is prefigured by the orientation of a structure composed of microtubules, called the preprophase band. What remained unclear was the mechanism that prefigures the preprophase band's orientation. However, centrifugation experiments of both haploid and diploid land plant cells showed that the position of the interphase nucleus establishes the location of the future division plane (fig. 7.12). Building on these and other observations, it has been proposed that the microtubules tethering the nucleus to the various inner surfaces of the cell membrane center the nucleus on the basis of

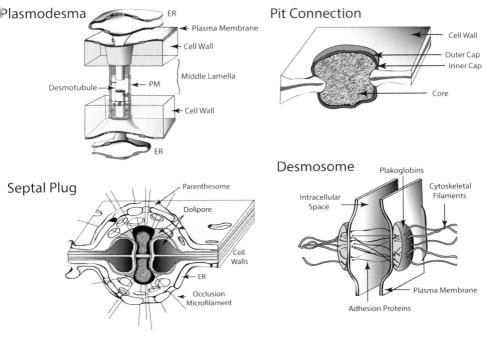

Figure 7.11. Schematics of four ways in which neighboring eukaryotic cells adhere and communicate with one another. Plasmodesmata (singular: plasmodesma), found in land plants and some charophycean algae, are comparatively small (≈50 to 60 μm in diameter) and consist of the plasma membrane (PM), a cytoplasmic "sleeve" composed of the endoplasmic reticulum (ER), and the desmotubule. Pit connections, found in the red algae, consist of an outer and upper cap surrounding a proteinaceous core. It is unclear whether this structure permits the transport of metabolites from one cell to another, although cell-to-cell signaling may occur. Septal plugs or "pores," which occur in the basidiomycetes, are characterized by a swelling around a central pore (dolipore) and a perforated parenthosome that permits cytoplasmic continuity between adjacent cells (but restricts intercellular organelle transport). Desmosomes, which are one of many different structures that cause animal cell adhesion, are found in simple and stratified squamous epithelia. They consist of cadherin protein complexes that attach cell surface adhesion proteins to intracellular cytoskeletal filaments. The intercellular spaces at desmosomes are ≈30 nm wide.

differences in the tensile forces generated within them. Collectively, shorter as opposed to longer microtubules are favored to achieve an equilibrium configuration that would axiomatically coincide with the minimal surface area of the future cell wall. Cells that are too large would have microtubules that would be unable to efficiently tether the nucleus to some cell membrane sites; cells that are too small

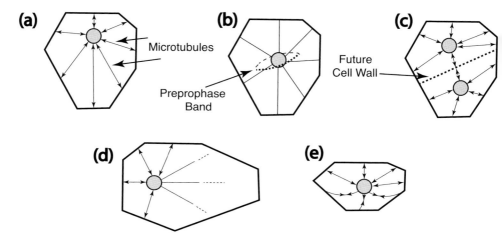

Figure 7.12. Schematic of a hypothesis accounting for symmetrical embryophyte cell division. a–c. It is hypothesized that the location of the nucleus (and thus the preprophase band, which prefigures the future cell wall) is determined by microtubules achieving equilibrium lengths minimizing the distance between their attachments to the cell wall and the nucleus. The preprophase band is established around the nucleus before mitosis and prefigures the location of the phragmoplast, which is involved in the deposition the future cell wall. d–e. If the microtubules tethering the nucleus to the walls of cells are too large or too small, they are unable to achieve a stable equilibrium configuration unless genomic mechanisms are brought into play during the cell division cycle.

would have microtubules experiencing insufficient tensile forces. An important deficiency of this model is its current inability to explain how asymmetric cell divisions are achieved.

Although genomic components are required for the operation of the FCW module as revealed by the ubiquitous participation of subfamily III leucine-rich repeat-receptor-like kinases in symmetric and asymmetric cell divisions, it is reasonable to surmise that the developmental choreography of ancient unicellular and colonial organisms might have relied on the mobilization of physical generic forces to establish simple default developmental conditions. Examples of default states are a spheroidal cell shape, symmetric cell division, and native cellulose crystallinity. However, over evolutionary time, selection to survive and adapt to different conditions required the evolution of more elaborate developmental responses that in turn required the innovation of mechanisms that modified the operation of physical forces and processes to achieve alternative developmental outcomes, such as nonspherical cell shapes and asymmetric cell divisions. This speculation holds true for fungal and animal evolution. For example, the formation of the structures prefiguring the appearance of villi in the gut of the chicken embryo relies on compressive mechanical forces generated by the differentiation of nearby smooth muscle tissue that causes the buckling of endoderm and mesenchyme.

Simple versus Complex Multicellularity

It is sometimes useful to distinguish between "simple" and "complex" multicellularity, particularly when considering the consequences of evolving a multicellular body plan, because the distinction between the two focuses on whether all cells make contact with their external environment, or whether some are internalized. According to this definition, unbranched filaments and sheets of cells one cell thick are examples of simple multicellularity, whereas solid spherical or cylindrical masses of cells are examples of complex multicellularity (fig. 7.13). This difference is important because (1) it is correlated with differences in cell specialization, energy consumption per gene expressed, and increases in non-protein-coding DNA, (2) it helps to assess the likelihood that a multicellular organism can evolutionarily revert to a unicellular state (a complex multicellular to unicellular transition is far less likely than a reversion to a unicellular state from a simple multicellular state), and (3) it helps to identify when a simple multicellular organism evolutionarily reaches a size or morphology that necessitates tissue specialization for the bulk transport of nutrients.

Consider the consequence of size on passive diffusion as shown by a variant of Fick's law of diffusivity, which estimates the time required for the concentration of a neutral molecule i initially absent within a cell to reach 50% of the external concentration (see table 8.1 and attending discussion). This formula shows that the time it takes i to

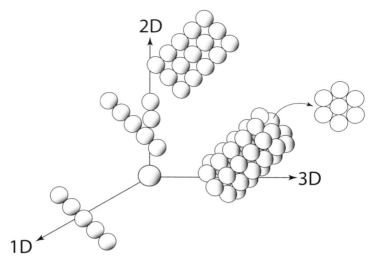

Figure 7.13. Schematic of simple and complex multicellularity plotted on a coordinate system identifying the ability of cells to divide in one, two, or three directions with respect to the principal body axis (1D, 2D, and 3D, respectively; also see fig. 7.8). Simple multicellularity is defined as a condition where every cell is in direct contact with the external medium. Examples are unbranched and branched filaments, and sheets of cells one cell thick (which are achieved by 1D and 2D). Complex multicellularity is defined as the condition where some cells lack direct contact with the external medium because they are surrounded by neighboring cells. Examples are solid spherical or cylindrical aggregates of cells (which are achieved by 3D).

diffuse into a spherical cell from all directions and reach half the external concentration is linearly proportional to cell radius, increases with increasing volume, and decreases with increasing surface area. Consequently, passive diffusion can be sufficient for the metabolic demands of small or attenuated cells, filaments, or thin sheets of cells, but it becomes increasingly insufficient as cells or cell aggregates increase in size.

Diffusion can be an important mode of communication that can drive cell specialization because increasingly steep gradients in nutrient-availability resulting from increasing size or distance can result in a reaction–diffusion (R–D) morphogenic system. For example, the filaments of the cyanobacteria *Anabena* or *Nostoc* are

1-dimensional R–D systems. Heterocysts, which are specialized cells capable of fixing atmospheric nitrogen and which can exist in a dormant state of suspended animation for decades, manufacture a diffusible inhibitor that prevents heterocyst formation unless its concentration falls below a specific threshold. As the distance between two heterocysts increases (due to the division of intervening cells), the first undifferentiated cell to be triggered midway develops into a heterocyst, releases the inhibitor, and reiterates the process. Analogous R–D systems operating in 2- or 3-dimensions are posited for the development of stomata and root hairs. Even more complex examples can be drawn from the evolution of land plants or extant macroscopic algae. As early embryophytes evolved greater height, passive diffusion eventually failed to provide water to aerial tissues at sufficient rates, thereby requiring bulk water flow via xylem. Likewise, the inner cortex of the stipes of the great kelp *Macrocystis* contain specialized "trumpet" cells that transport photosynthates much like the phloem of vascular plants.

Nevertheless, the organization of differentiated cells into 2- and 3-dimensional patterns, shapes, and forms need not reflect the immediate consequences of selection. Although the adaptive advantage of a morphological novelty is clear (such as in the development of plant and animal vascular systems), morphological or behavioral novelties can also arise via phenotypic plasticity, or as a consequence of the inherent physical properties of embryonic tissues and condition-dependent developmental systems, only to become assimilated subsequently into the organism's developmental program. The former may help to explain why colonial life-forms often presage the appearance of multicellularity in some lineages, since phenotypic plasticity among genetically identical cells contributes to fitness. Likewise, the colonial body plan has distinct advantages over the unicellular life form (for example, nondefective cells can compensate for the defects of neighboring cells, or provide resources to other cells that are then free to specialize).

Multilevel Selection Theory and the Alignment-of-Fitness

As noted, the traditional explanation for the evolution of multicellularity posits that it involves a "unicellular → colonial → multicellular" body plan transformation series. This perspective is consistent with molecular phylogenies of the major eukaryotic clades, which identify unicellular organisms in early divergent lineages and colonial and multicellular taxa on subsequently diverging branches of the phylogeny (see figs. 7.9 and 7.10). A still unanswered question is how this transformation series is achieved and why multicellularity arose in some instances and not in others.

Multilevel selection theory has been used to answer the *how* but not the *why* of multicellularity. As reviewed by Henri J. Folse III and Joan Roughgarden in 2010, multilevel selection theory identifies two evolutionary stages that are prerequisites for the evolution of multicellularity—an *alignment-of-fitness* phase in which genetic similarity among adjoining cells prevents cell-cell conflict and an *export-of-fitness* phase in which cells become interdependent, collaborate in a sustained physiological and reproductive effort, and largely lose their self-identity. Phyletic analyses of lineages in which obligate multicellularity has evolved are consistent with this multilevel selection model, and some indicate that lineages characterized by species with clonal group formation (a form of colonial growth) are more likely to have undergone an evolutionary transition to obligate multicellularity than lineages characterized by species with non-clonal group formation. This finding is not surprising in light of how even accidental aggregates of individuals can initiate a process called niche construction, wherein initially random, unstructured interactions among individuals evolve into nonrandom, organized interactions (box 7.1).

The alignment-of-fitness phase in this model requires cell-to-cell adhesion that in turn gives rise to the possibility of a colonial body plan. Inspection of the life cycle of a stereotypical unicellular plant, such as *Chlamydomonas*, shows that a condition prefiguring a colonial body plan is preceded by what is referred to as a "unicellular bottleneck" (the formation of a single cell that subsequently undergoes cell division to give rise to either a colony or a multicellular organism) (fig. 7.14 a). Inspection of the life cycles of multicellular organisms reveals a similar phenomenology. Depending on the kind of multicellular organism, the unicell in the life cycle is either a haploid or diploid cell, or it can be both kinds of cells (fig. 7.14 b–d). This bottleneck establishes genetic homogeneity among subsequently formed cells even among asco- and basidiomycete heterokaryonic fungi, for which experimental data indicate that genetically different nuclei compete with one another for supremacy within the same cell. For example, nuclear ratios of heterokaryons in the ascomycetes *Penicillium cyclopodium* and *Neurospora crassa* are reported to change depending on environmental conditions in ways that reflect the underlying fitness of the constituent homokaryons grown in isolation.

It must be recognized that an absence of *conflict* does not mean an absence of *competition*—and competition can be a good thing. Many developmental processes involve lateral inhibition in which neighboring cells compete to adopt the same cell fate, as for example vulval induction in *Caenorhabditis* or during the development of root hairs in *Arabidopsis*. It is equally important to recognize that mitosis does not invariably result in genetically identical derivative cells even in the absence of mutation or chromosomal aberrations. Preferential sister chromatid segregation is observed in plants, fungi, and animals. Methylation patterns of cytosine in CpG doublets and other epigenetic changes provide additional avenues for establishing genetically different groups of cells in the same organism, each of which requires the evolution of stable interdependent cell lineages sharing the same genome but expressing different gene network patterns.

Indeed, epigenetic mechanisms may be critical to *maintaining* multicellularity. Consider that the principal limitation to achieving and maintaining cooperation among adjoining cells is the appearance of "defectors" in an evolutionary game setting—that is, participants that consume resources but fail to confer any benefit to other players. An obvious example of a cellular defector is an animal neoplasm,

---

Box 7.1. Niche Construction

In the context of development (see chapter 4), we saw that the evolution of innovations (novel characters or character states) involves the transformation of the dynamics of regulatory systems (as for example, gene regulatory networks) and the transformation of the ways in which the components within regulatory systems interact with their environments (as for example, gene regulatory network responses to different external or internal environmental signals). The same can be said for other levels of biological organization ranging from randomly associated aggregates of cells to the interactions among different species within communities. The concept of niche construction focuses on the ways in which systems actively influence and change their environment and how the process of constructing new niches shapes future possible interactions. The theory underlying the concept of niche construction concerns itself formally with a *search space* in which future dynamic interactive possibilities are generated actively by the actions and properties of preexisting systems. This perspective differs conceptually from viewing niches as prefigured and unalterable possibilities that are "out there" to be occupied by organisms at some future date.

The concept of niche construction can be illustrated by considering a scenario for the evolution of the colonial body plan. In this scenario, cell-to-cell adhesives (functioning to adhere individual cells to a substrate or for intraspecific mate recognition) result in the accidental aggregation of a random number of cells sharing the same or similar genetic identities. Initially, these cells interact with one another in disordered ways such that fitness is defined at the level of individual cells. Over time, the interactions among cells become more predictable owing to selection acting on the ability of the collective to cope with internal possibly competitive metabolic dynamics and with the ability to utilize limited resources from the external environment. At this early juncture, the cell remains the unit of selection, but the aggregate of cells has constructed a new niche with

---

which may have deep genetic roots in terms of the regulation of cell proliferation and de-differentiation. A number of mechanisms capable of maintaining cooperation and reducing or eliminating defectors have been suggested, among which the effects of group selection, direct and indirect reciprocity, network structure, and tag-based donation schemes are perhaps best known. However, each of these mechanisms requires players to remember past proceedings, or to possess some method of recognizing one another as players in the same game. Epigenetic mechanisms as well as signaling pathways that connect metabolic status with nutrient availability or other environmental factors (such as the TOR signaling pathway) provide one solution to dealing

properties and dynamics that cannot be defined at the level of individual cells. Selection among the cells produced by successive generations would eliminate "cheaters" unless by cheating a cellular variant using more than its fair share of resources possessed some behavior or property that indirectly conferred an advantage to the collective (for example, a cell that consumed more nutrients but that released a bactericidal compound). In this manner, cells could evolve specific functionalities and the aggregate evolves into a colonial organism in which fitness is progressively calibrated according to the properties of the collective and not the individual cell. The evolution of this type of colonial organism into a multicellular organism would proceed along much the same way, by extending the regulatory networks among cells to simultaneously alter intercellular interactions and the interactions between the collective and its external environment. Notice that a new niche is constructed in each case by the bootstrapping of interactions that did not exist previously.

The analogy between this scenario in which an initially random aggregate of cells evolves into an entity consisting of interdependent cells that alter the environment and thus construct a new niche and a scenario in which initially random collections of organisms ecologically interact in novel ways to form communities is not too far-fetched in light of the precepts of niche construction theory. In each case, the component of the aggregate (the type of cell or organism) is not prefigured, and, in each case, it can be replaced by a different kind of individual. We must therefore avoid the temptation to characterize a colony of cells or an assemblage of organisms as a superorganism. Likewise, the novel interactions among cells in a cellular aggregate or the organisms in a community should not be viewed as adaptations per se. In each case, novelty comes about as a result of a change in the patterning dynamics among one or more interactions among the components within the whole assemblage. Such changes may confer some future advantage, or they may not.

with defectors, while the unicellular bottleneck provides, at least initially, an homogeneous collection of cooperating cells. There are other tactics as well. Theoretical models show that resource limitations can cause the rules of a game to change in ways that foster cooperation among players with no memory and no recognition of one another. Likewise, some models show that altruistic and generous strategies can sustain cooperation and reduce negative interactions. It is worth noting further that "cheater" mutants can cooperate in ways that conform to normal developmental patterns and that do not disrupt the functionality of the collective organism. Finally, there is reasonable speculation that cheaters can theoretically function as asexual

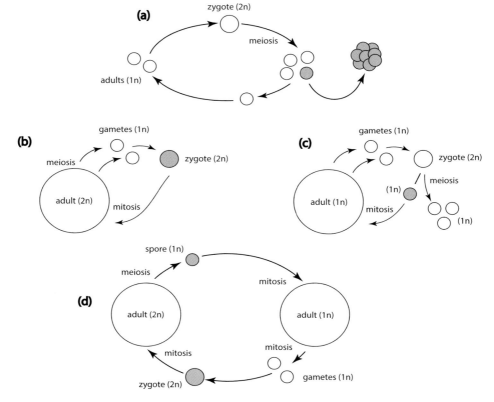

Figure 7.14. The location(s) of unicellular bottlenecks (indicated by blue circles) in each of four generic life cycles that collectively span the spectrum of life cycles found in unicellular and multicellular organisms. a. The life cycle of a unicellular plant in which a colonial body plan may appear as a result of cell-to-cell adhesion (shown to the right of the life cycle). An alternative possibility is the intercalation of a colonial body plan resulting from mitotic divisions of the diploid zygote (not shown). b. The life cycle of a typical mammal or multicellular plant, such as the brown alga *Fucus*, in which the only multicellular adult is diploid (2n). The adult produces sperm or egg cells (or both as shown in this diagram) by means of meiotic cell divisions. The gametes fuse to form the diploid zygote, which then divides mitotically to form the multicellular adult. The zygote is the unicellular bottleneck in this life cycle. c. The life cycle of a multicellular plant, such as the green alga *Chara*, in which the only multi-cellular adult is haploid (1n). Each cell that gives rise to the haploid adult is the unicellular bottleneck. The adult produces sperm or egg cells (or both as shown in this diagram) by means of mitotic cell divisions. The gametes fuse to form the diploid zygote, which then divides meiotically to form the multicellular adult. d. The life cycle of a typical land plant, such as the moss *Physcomitrella*, in which two multicellular life forms occur, a diploid organism (the sporophyte) that produces spores by means of meiotic cell divisions, and a haploid organism (the gametophyte) that produces sperm or egg cells (or both as shown in this diagram) that fuse and form the diploid zygote. This life cycle has two (as opposed to one) unicellular bottlenecks (one that precedes the mitotic cell divisions that give rise to each of the multicellular organisms in this life cycle). Note that the only real difference among all three life cycles is if and where multicellularity is expressed.

propagules in very ancient proto life cycles. In summary, cooperation among cells can evolve along a number of routes, which may explain why colonial life-forms, multinucleate cells, and multicellularity are not uncommon.

Multilevel Selection Theory and the Export-of-Fitness Phase

The emergence of a multicellular organism from a unicellular or colonial ancestor requires a shift from selection operating at the level of individual cells to the level of an organism that reproduces a functionally integrated phenotype with a heritable fitness (typically followed by some degree of cellular specialization). The key difference between the alignment-of-fitness phase and the export-of-fitness phase is that the fitness of the cell or the colony is an additive function of the fitness of individual cells, whereas the fitness of a multicellular organism is nonadditive and depends on the fitness of reproductively participating cells. Put differently, the evolution of a multicellular organism requires a means to guarantee the heritability of fitness at the emergent level of the multicellular entity.

One method to achieve this level of identity is for some components to lose their individuality. Perhaps the first example of this was the evolution of eukaryotic cells. According to the endosymbiotic theory, eukaryotic cells evolved as a result of a partnership between a prokaryotic host cell and prokaryotic endosymbionts that gradually lost their autonomy as some of their genetic information became integrated into the host's DNA. More recent examples of the loss of cell individuality can be seen in programmed cell death (apoptosis) and the loss of functional nuclei as for example the programmed enucleation of sieve tube members. In some but not all multicellular organisms, the export-of-fitness phase is accomplished by the developmental isolation of a cell germ line from a somatic cell line. A germ-soma separation may be an indirect consequence of the necessity to compensate for the increasing costs of evolving a progressively larger body size. For example, sequestering germ line cells can reduce the number of cell divisions leading up to gamete formation thereby reducing the probability of mutation arising in reproductive cells. Body size matters in this context because the probability of compounding a genetic error or mutation increases as a function of the number of cell divisions required to achieve the size of a mature organism. Small multicellular organisms have a lower probability of introducing errors into cells that will later take on a reproductive function because the construction of a small organism requires fewer cell divisions than a larger organism. Thus, progressively larger organisms must escape Muller's ratchet (the inevitable accumulation of deleterious mutations) by ultimately isolating reproductive cells from nonreproductive cells. In this context, *isolation* simply means that reproductive cells are restrained from dividing as often as their nonreproductive counterparts. This may involve geographically isolating them, as for example in a quiescent meristematic location, although this need not be the case. It is worth noting in passing that cellular specialization may be easier to evolve in larger as opposed to smaller organisms because unsuccessful attempts at specialization are more easily tolerated in larger organisms (including fungi, animals, and plants).

Nevertheless, obligate sexual reproduction is not required to override the conflict between a multicellular individual and its constituent cells. In the absence of somatic mutations, the presence of an asexually produced propagule, such as a cell or similar reproductive unit, assures a unicellular bottleneck regardless of the type of life cycle. Even with the effects of Muller's ratchet on fitness, an alignment-of-fitness occurs by means of such propagules, which can purge deleterious genomic changes as a consequence of the death of individual propagules. Likewise, multicellularity is not required for cellular specialization. Unicellular bacteria, algae, yeast, and amoebae exhibit alternative stable patterns of gene expression and thus alternative cell morphologies during their life cycles, often as a result of competing processes, such as motility versus mitosis. This feature is particularly intriguing in light of the studies showing that seemly random fluctuations in cellular dynamics may provide a simple switch for changing cell fate. For example, the bacterium *Bacillus subtilis* can exist in two stable forms, called vegetative and competent, under conditions of nutrient deficiency. Mathematical models using a stochastic algorithm can predict how and when these two cellular conditions are decided on the basis of the level of biological noise in the system. In addition, mathematical models indicate that cellular differentiation can emerge among genetically identical cells in response to the incompatibilities among competing physiological processes, or simply because of the metabolic costs of switching the tasks a cell must perform to stay alive or complete its life cycle. In more derived lineages, an alignment-of-fitness can compensate for conflicts of interest among cellular components such that a division of cellular labor becomes possible and even necessary. Even a loose "colony" of cells can have emergent biological properties that give it a collective edge in which every cell benefits. The origin of a cellular differentiation module therefore may reside in the inherent multi-stability of complex gene regulatory networks with somatic or reproductive functional roles for different cell-types possibly established by natural selection ad hoc.

Advantages and Disadvantages of Multicellularity

Table 7.1 shows that the four basic body plans appear in many, albeit not all plant lineages. This distribution raises the question, Why? More precisely, it raises this question: What are the advantages and disadvantages of being unicellular, colonial, coenocytic, or multicellular? Although every answer to this question is problematic, it can be answered deductively using logic and the first principles of physics and chemistry.

Consider the unicellular body plan. One explanation for its occurrence in every lineage is that it is the ancestral condition in every eukaryotic lineage. It is reasonable therefore to speculate that the ancestral condition will be retained by at least a few species. However, this explanation is trivial at best and very probably incorrect, because the retention of the ancestral condition in extremely ancient lineages (see fig. 7.6) and by the majority of species in these lineages presents evidence that the unicellular body plans confers an advantage. Indeed, well over 50% of aquatic plant species manifest the unicellular body plan. There are a number of reasons why this may be so. For example, small individual cells are more efficient at harvesting light and exchanging mass with their aqueous surroundings than larger cells or colonies of cells. This claim is justified by comparing the absorption cross-section of spherical cells, denoted here by $\log_{10}(I/I_b)$, differing in diameter. This parameter is a recasting of Beer's law, and allows us to predict the attenuation of light per unit horizontal area as it passes through an aqueous suspension of cells. For the simple case where light comes directly from above, the absorption cross-section is given by the formula

$$\log_{10}(I/I_b) = (\varepsilon_\lambda + nAa)b,$$

where I is the intensity of light at the water's surface, I_b is the intensity of light at distance b from the water's surface, ε_λ is the extinction coefficient of water and any dissolved materials, n is the number of cells in suspension, A is their average projected area, and a is the average fraction of incident light intercepted per cell. When $\log_{10}(I/I_b)$ is calculated for spherical cells differing in diameter and plotted as a function of photosynthetically active radiation, we see that smaller cells gather more light than their larger counterparts (fig. 7.15). They are also far more efficient than clusters of cells regardless of their diameter.

Another advantage to being small and unicellular involves the relationship between surface area and volume. Because surface area S scales as the square of an object's linear dimension and because volume V scales as the cube of the same dimension, S scales as the two-thirds of V provided that geometry and shape remain unaltered. Thus, smaller cells have disproportionally larger surface areas with respect to their volumes. We shall examine this claim in greater detail in chapter 8. However, for now, it is supported empirically by plotting the log-transformed surface area of cells differing in size against their volume and by measuring the slope of the log-log linear regression

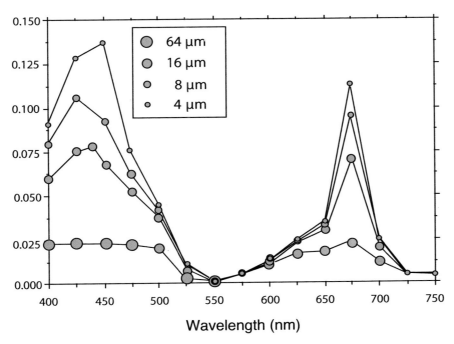

Figure 7.15. The absorption cross-section of four suspensions of spherical cells differing in diameter (units in μm, see insert) plotted as a function of the wavelength of light (units in nm). It is assumed that the direction of the light is at a right angle to the water's surface. The total volume of all cells in each suspension is modeled as equivalent. The peaks in each of the four plots reflect the two absorption maxima of chlorophyll *a* and *b* (in diethyl ether, the absorption peaks for chlorophyll *a* are at 430 nm and 662 nm, whereas the absorption peaks for chlorophyll *b* are at 453 and 642 nm). As cell diameter increases, the area under each of the four plots decreases, which indicates that cell suspensions are gathering less irradiant energy. Also, at the level of individual cells, the gathering of irradiant energy decreases as cell size increases because of the self-shading of chlorophyll molecules within each cell. Thus, smaller cells are more efficient at light interception than larger cells.

curve (fig. 7.16). Across a spectrum of cells differing over eight orders of magnitude in volume, the slope of the linear curve for log S versus log V is less than one, which indicates that surface area decreases with increases in cell volume. For perfectly spherical cells, the slope is expected to be ≈ 0.67 or 2/3. This slope is observed for most animal cells and for plant protoplasts, both of which lack rigid cell walls and thus over time tend to assume a spherical geometry, which minimizes the surface tension of their cell membranes. The slope observed for

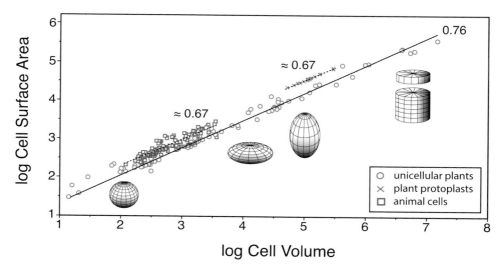

Figure 7.16. Log$_{10}$-transformed data for cell surface area plotted against log$_{10}$-transformed data for cell volume (data taken from the primary literature reporting these dimensions for plant and animal cells grown under optimal conditions; see insert for symbols). The slopes of the log-log regression curves for plant protoplasts and animal cells (dashed lines) are approximately 0.67 or 2/3, which is what would be expected for a series of spheres differing in size, because surface area increases as the square of diameter and volume increases as the cube of volume. In contrast, the slope of the log-log regression curve for unicellular plants (solid line) is approximately 0.76 or 3/4, which reflects the fact that cell geometry or shape or both are changing across cells differing in size. In this data set, the smallest unicellular plants are nearly spherical, intermediate-sized cells tend to look like oblate or prolate spheroids (shaped like doorknobs or cigars, respectively), and the largest cells tend to be disk or cylindrical in geometry (see drawings under the solid regression curve).

unicellular plants with rigid walls is 0.76 or 3/4, which reflects the fact that larger unicellular aquatic plants manifest geometries and shapes that depart from being spherical and improve on the increase in surface area. Nevertheless, the scaling of surface area with respect to cell volume does not achieve a one-to-one relationship, which indicates that there is always a disproportional loss of area with increasing cell volume. The only geometry that can maintain a one-to-one scaling relationship is a cylinder. Provided that its diameter remains constant, a cylinder can increase in length (and thus volume) without experiencing a disproportionate decrease in area.

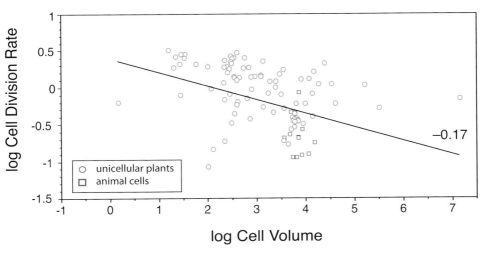

Figure 7.17. Log_{10}-transformed data for cell division rate plotted against log_{10}-transformed data for cell volume (data taken from the primary literature reporting the growth rates and cell volumes for unicellular plants and animal cells grown under optimal conditions; see insert for symbols). The slope of the log-log regression curve for cell division rate versus cell volume is (solid line) −0.17, which indicates that larger cells divide less rapidly than smaller cells.

A third advantage to being a small unicellular plant involves the relationship between growth and body size. Across otherwise very diverse plants, animals, fungi, and bacteria, cell division rates tend to decrease with increasing cell volume (fig. 7.17). Explanations for this phenomenon will be discussed in chapter 8. However, for the time, it is sufficient to take this claim as an empirical fact that has important implications for survival and reproductive success. Small organisms living in mercurial and unpredictable environments tend to be more successful than larger organisms because they grow and multiply faster than their larger counterparts. Smaller faster-growing organisms can acquire nutrients rapidly and they can take advantage of short-lived windows of opportunity for favorable reproduction.

Given the aforementioned advantages of having a unicellular body plan, it is reasonable to ask: Why should there be so many colonial and multicellular organisms? One explanation has to do with convert-

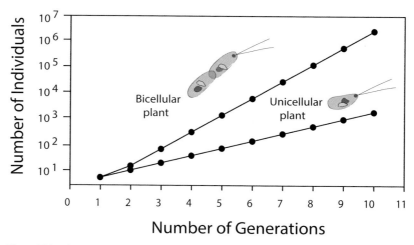

Figure 7.18. The number of individuals produced by a unicellular species and a simple multicellular species with individuals composed of only two cells, one vegetative and the other that can function as a gamete with flagella. Assuming that all of the plants of both species survive each round of reproduction and that both have comparable cell division rates, the population of the bicellular species rapidly overtakes the unicellular species in just a few generations even when both populations begin with just two individuals. After ten generations, the population of the bicellular species exceeds one million individuals, compared to the population of the unicellular species that consists of little over one thousand plants.

ing an adult into a gamete (a sperm or egg cell). Another explanation is predation. Consider the life cycle of a unicellular organism (see fig. 7.14 a). In the absence of being able to form a colony or evolving multicellularity, each "adult" organism must become a "gamete" and join with another adult cell to form a diploid zygote, which subsequently undergoes meiosis to produce four adults. As a consequence, two adults are lost in the process of reproducing sexually. Over many generations, the population size of such an organism can be overrun rapidly by even a simple multicellular organism (fig. 7.18).

Assuming that both species have the same or very similar physiology, a multicellular species can outcompete a unicellular species by acquiring nutrients and by occupying habitats suitable for both species. Unicellular organisms have adapted to this potential constraint by intercalating cell divisions into different phases of their life cycle. This form of vegetative asexual reproduction, which approximates the

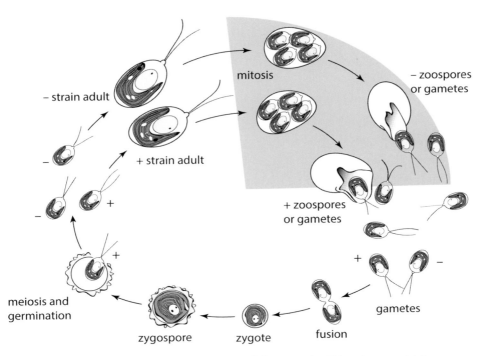

Figure 7.19. Diagram of the life cycle of the unicellular green alga *Chlamydomonas*. Adult plants have two mating types, referred to as plus and minus strains (upper left). Under some conditions, + strain and − strain adult plants encapsulate and undergo mitotic cell divisions (shown in green). After a period of dormancy, these cells emerge and function either as new adults (called zoospores), or as gametes that fuse to form zygotes, which can also undergo a period of dormancy (lower middle and left). The formation of new adult plants by means of mitotic cell divisions can compensate for the loss of two adults in every round of sexual reproduction (see fig. 7.18).

colonial body plan in some cases, increases the size of their populations beyond what would be expected had asexual reproduction not occurred (fig. 7.19).

As noted, predation is another factor that may potentially favor the evolution of multicellularity because it is more difficult to capture and eat a larger than a smaller organism. By virtue of their larger size, the colonial body plan (and the siphonous body plan) can reduce the risk of predation (that can ratchet body size upward as a result of an "arms war"). Eight-celled colonies of the green alga *Chlorella vulgaris* are established and are sustained when steady-state cultures of this normally unicellular green alga are inoculated with the phagotrophic flagellate *Ochromonas vallescia*. Likewise, the green alga *Desmodesmus* forms large colonies composed of tough-walled cells with more spiny projections when exposed to *Daphnia*. The induction of colony formation in response to predation appears to be a widespread phenomenon in some plant lineages as for example the Scenedesmaceae. Indeed, one explanation for the waves of multicellular experimentation occurring at ≈1.5 Bya in the fossil record is the expansion of predation resulting from the evolution of phagocytosis. A contributing factor is that aggregating cells into a collective can increase the acquisition and utilization of extracellular resources. For example, undifferentiated colonies resulting from incomplete cell separation of *Saccharomyces cerevisiae* appear spontaneously in cultures with low sucrose concentrations. The cells in these colonies cooperate in ways that reduce starvation and provide protection. Further, growth experiments of organisms such as the unicellular coccolithophyte *Phaeocystis* show that the colonial phenotype has faster rates of cell division compared to the unicellular phenotype.

These and other experiments indicate that a number of factors, either individually or collectively, might favor the evolution of the colonial body plan, which, according to theory, prefigures the evolution of the multicellular body plan. If true, there is no single canonical explanation for why multicellularity evolved in some lineages (or why it has been lost or never achieved in other lineages). Each group of multicellular organisms must be evaluated individually in the context of its fossil record, ecological setting, genomic capabilities, etc., when trying to answer the question, Why did multicellularity evolve? However, it is fair to say that body plan size, geometry, and shape contribute to our ability to answer this question in each case because these attributes are subject to constant selection under the auspices of physical phenomena such as passive diffusion, surface tension, and tensile and compressive forces. This theme will be explored in chapter 8.

Complexity, Cell Types, and Information Content

In chapter 4, we explored how gene regulatory networks operate synergistically with the rest of the cell and how the information stored in an organism's genome is interpreted by means of alternative splicing, proteins with intrinsically disordered domains, post-translational modifications, riboswitches, and a host of other processes and mechanisms that respond to intracellular and intercellular signals to make development fantastically complex. The question before us now is, Do these processes and mechanisms affect macroevolution and the evolution of multicellularity in their own right? Is there any evidence that organismic complexity can increase and evolve as a consequence of these developmental phenomena?

In this context, it is useful to note that many researchers have quantified organismic complexity on the basis of the number of different cell types an organism produces. This approach has three obvious merits: (1) it is indifferent to whether an organism is a fungus, plant, or animal, (2) it is insensitive to most methods of categorizing grade or clade levels of organization, and (3) it can be used to quantify the complexity of unicellular organisms because most unicellular eukaryotes manifest different cell functionalities and morphologies at different stages in their life cycles (resting cysts versus actively motile cells). Nevertheless, the use of the number of cell types as a measure of complexity has some drawbacks. For example, researchers may disagree about whether two cells differ sufficiently to be called different cell types. Gauging complexity by counting the number of different cell types an organism can produce can also become problematic when dealing with coenocytic organisms that can achieve phenotypic complexity without benefit of cellularization (the green alga *Caulerpa*), or when dealing with comparatively simple morphologies that nevertheless have highly organized patterns of nuclear or cellular placement (angiosperm megagametophytes).

There are two other important concerns. The first is that using the number of cell types to quantify complexity results in an untested tautology—"complexity increases as the number of different types of cells increases, and this number increases as an organism's complexity increases"—which requires collateral evidence to show that this reasoning is more than a self-sustaining description of a phenotypic property. The second concern is that cellular simplicity need not reflect an absence of complexity. There is no reason a priori to argue that an organism capable of producing two different types of cells is more complex than an organism capable of producing only one. And, if there is justification, is an organism composed of two cell types 50% more complex than the organism composed of one cell type? Further, the supposition that cellular simplicity denotes low levels of organismic complexity hinges on the notion that "complexity" must be defined at the level of the diversity of cell phenotypes, a notion that is challenged when considering the physiological (as opposed to the structural) complexity of parasitic organisms that often undergo morphological and anatomical reduction.

Nevertheless, intuition encourages the notion that "information" of some sort is required to produce or sustain different kinds of cells as a measure of "complexity." If true, the number of different kinds of cells is not a direct measure of complexity per se but is rather an indirect reflection of information content and thus complexity. This logic fosters a research agenda requiring the quantification of "information content" and a statistical assessment of this measure of information content with its correlation with the number of cell types that an organism is capable of producing. Only in this way can the tautology be tested.

Three candidates for biological "information" present themselves in this capacity: (1) genome size as measured by the number of base pairs, (3) proteome size as measured by the number of amino acids, and, as alluded to in the previous sections of this chapter, (3) the number of intrinsically disordered protein (IDP) domains or residues in a proteome. Recall that intrinsically disordered domains lack an equilibrium 3D structure under normal physiological conditions. Consequently, each domain can assume multiple conformations (and thus multiple developmental regulatory functions) within the same cell without increasing genome or proteome size. In this way, IDP domains contain information that is not encoded directly in the genome, a feature that helps in part to explain why genome size does not correlate with phenotypic complexity (the so-called G paradox). Regardless of preference, each of the three measurements of organismic information content can be applied to unicellular as well as to multicellular organisms, and each can be used to assess whether complexity has increased over the evolutionary history of a lineage.

However, before quantifying these measures of complexity, it is informative to examine whether any trends in different cell number are evident in successively evolutionarily divergent taxa within well-documented clades, particularly if these trends flout the conventional wisdom that "plants are invariably simpler than animals." Answering this question is not without difficulty, however. For example, the level of taxonomic resolution and the availability of sufficiently resolved phylogenies can present problems. A too-finely-resolved phylogeny may reveal no pattern because very closely related taxa may have the same or a very similar number of cell types. Conversely, a too-coarsely-resolved phylogeny may fail to reveal a pattern because distantly related taxa may have adapted to very different niches, or undergone phenotypic reduction. Two other limitations exist: (1) phylogenies resolved at the optimal level may not be available, and (2) authoritative tabulations of the number of different cell types for species in sufficiently resolved phylogenies may not exist.

These concerns are illustrated by mapping estimates of the number of different cell types reported for representative taxa onto published phylogenies on the basis of molecular data. Specifically, we use the maximum number of cell types (NCT) reported in the primary literature as tabulated by Graham Bell and Arne O. Mooers (1997) and map NCT onto the different plant lineages for which there are published phylogenies using molecular data. The data set assembled by Bell and Mooers undoubtedly reflects divergent philosophies about how to distinguish and classify animal and plant cell types (that is, the data reflect the opinions of "splitters" and "lumpers"), and thus may be biased in unforeseen ways. However, this data set has the advantage of reflecting the opinions of authorities in diverse fields of research. It also consolidates a large literature scattered in diverse journals.

Using the NCT data of Bell and Mooers and a consensus phylogeny of the chlorophycean and charophycean algae, and the land plants, we see that NCT has increased, albeit not monotonically within this important clade (fig. 7.20). The early divergent chlorophytes are unicellular and are reported to have two cell types (for example, the prasinophyte *Ostreococcus*), whereas later divergent green algae have a maximum of five different types of cells (for example, *Fritschiella*). Among the streptophytes (the charophycean algae and the land plants), the early divergent charophycean alga *Mesostigma* has two cell types, whereas the more derived charophycean alga *Chara* has ten. Turning to the land plants, specifically, we see that the maximum number of cell types among the nonvascular land plants is reported for the mosses (*Polytrichum*; NCT = 26), which is approximately twice that reported for the hornworts and liverworts (*Anthoceros* and *Monoclea*; NCT = 12 and 13, respectively). Among the lycopods, the maximum NCT is 25 for *Selaginella*, whereas among the ferns and horsetails, the number of different cell types ranges between 17 and 20 (*Psilotum* and *Azolla*, respectively). Finally, *Pinus monophylla* is reported to have the maximum number of different cell types among conifers, whereas *Fuirena ciliaris* is described as having 44 different kinds of cells, which is the maximum reported for flowering plants.

It is instructive to compare these numbers with those reported for metazoans. Among these animals, the maximum number of different cell types reported for the flowering plants exceeds that reported for Placozoa (NCT = 4) and early divergent cnidarians and ctenophorans, and is comparable to those reported for the protostomes (ecdysozoans and lophotrochozoans) (fig. 7.21). Although the resolution of the

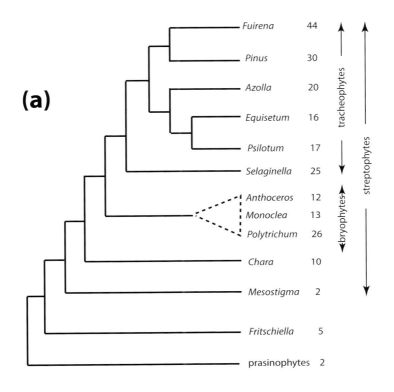

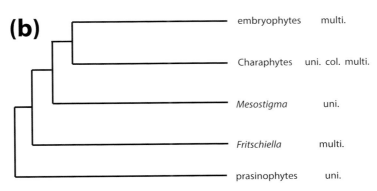

Figure 7.20. The simplified phylogenetic relationships among representative taxa within the green plants (Viridiplantae), the maximum number of cell types (NCT) reported for species within each taxon (a), and the body plans represented within each taxon (uni. = unicellular, col. = colonial, multi. = multicellular) (b). With the exception of the early divergent chlorophytes (the prasinophytes), each major taxon is represented by the genus reported to have the greatest diversity of cell types in its lineage (for example, *Chara* and *Polytrichum* are reported to have the largest number of cell types among the charophycean algae and the mosses, respectively). The uncertainty in the phylogenetic relationships among the nonvascular land plants (the bryophytes) is depicted by dashed lines. Data for NCT taken from Bell and Mooers (1997).

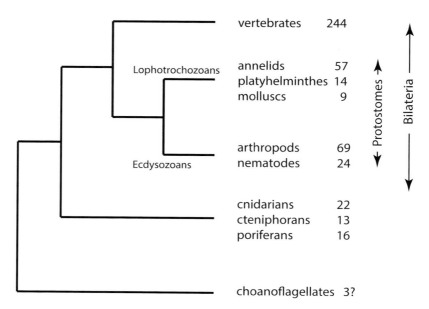

Figure 7.21. Highly simplified phylogenetic relationships among the major groups of metazoans and the maximum number of different cell types (NCT) reported for each group. Data NCT taken from Bell and Mooers (1997).

phylogeny of animals is not complete, the available data indicate that maximum cell type numbers range between 16 and 22 for the early divergent poriferans and cniderians, and between 9 and 69 for the mollusks and arthropods. The greatest number of different cell types among vertebrates is 240, which is 5.5 times that of the maximum number of angiosperm cell types.

The diversity of cell types across plants and animals is perhaps best illustrated when we compare the mean number of cell types reported for plants and metazoans (fig. 7.22). Among the plants, the red and brown algae produce a larger number of different kinds of cells compared to the green algae (chlorophytes and charophycean algae), whereas the maximum number of cell types reported for brown algae is statistically indistinguishable from that of the bryophytes. Within the Viridiplantae, mean NCT increases progressively across the bryophytes, pteridophytes, and vascular plants. The metazoans, on

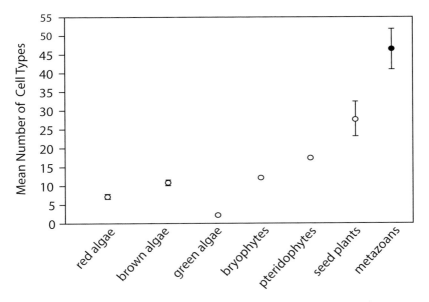

Figure 7.22. Mean number of cell types (±SE) reported for extant red, brown, and green algae, bryophytes, pteridophytes, seed plants, and metazoans. Data taken from Bell and Mooers (1997).

average, manifest the greatest diversity of cell types and the greatest range among all of the multicellular lineages.

The preceding provides only limited insight into why the numbers of cell types are distributed in the major plant lineages. It does, however, illustrate that the ability to detect patterns in NCT depends on the degree to which a lineage or clade is phylogenetically resolved. Nevertheless, it is clear that the number of different cell types does not invariably increase over the course of evolutionary history within specific clades. Consequently, if the generalization that "cell type numbers gauge complexity" is accepted as generally true, it follows that evolution has not always resulted in more complex multicellular organisms in all lineages or clades (even in the absence of the evolution of adopting a parasitic lifestyle). The data also dispel the myth that plants are invariably less complex than animals, since the angiosperms have as many different cell types as some animals (for examples, annelids or arthropods).

Cell Type Numbers and Information Content

If the maximum number of cell types does not invariably increase over the course of a lineage's evolutionary history, does it correlate across taxa with organismic information content as gauged by genome or proteome size? Importantly, the expectation that statistically significant and positive correlations exist between NCT and proteome size has thus far gone untested, even though there is evidence that alternative splicing and IDP domains (which occur widely in all eukaryotic lineages) can amplify protein functionalities without increasing genome or proteome size.

This expectation emerges from our discussion about gene regulatory networks in chapter 4 wherein we saw that alternative splicing (AS) and IDP domains produce "moonlighting" proteins (proteins that can take on a variety of functions, some of which increase the ability to diversify cell types and thus increase morphological and anatomical complexity). Alternative splicing is an adaptive and highly conserved mechanism that increases the number of proteins encoded by pre-mRNA in a context-dependent manner in both plants and animals. In turn, alternative splicing produces a disproportionate number of proteins with IDP domains. Importantly, a number of studies indicate that the majority of transcription factors, which are key players in cell fate specification, have IDP domains affected by alternative splicing. Since this enables functional and regulatory diversity while avoiding structural complications, it is reasonable to speculate that the diversity of cell types will increase as a function of the size of a proteome and (in particular) the number of IDP domains.

This speculation is consistent with preliminary data gathered from the primary literature reporting the maximum number of cell types (NCT) different plants and animals produce, their genome and proteome size, and the number of IDP domains as gauged by the number of IDP residues (IDResidues). Across this database, which consists of algae ($n = 8$), nonvascular and vascular land plants ($n = 6$), and invertebrates and vertebrates ($n = 5$), NCT correlates positively and significantly ($P < 0.0001$) with each of the three metrics of information content (fig. 7.23) as gauged by ordinary least squares protocols using log_{10}-transformed data (table 7.3). Equally informative are the numerical values of the slopes of each of the bivariate regression curves, because these slopes are scaling exponents (denoted here by α) that indicate numerically the extent to which NCT proportionally increases or decreases with respect to increasing values of each of the other variables on a log scale. Inspection of these slopes indicates that NCT fails to increase one-to-one with increasing genome size (i.e., $\alpha = 0.88$), which is a manifestation of the G paradox. In contrast, NCT scales as the 2.18 power of the number of intrinsically disordered residues, and as the 2.36 power of proteome size. Thus, the degree of cellular specialization increases dramatically as each of these two measures of protein information content increases. Collectively, these scaling relationships indicate that the number of IDResidues is a highly significant measure of information content and that the information contained in these residues helps to explain where some of the "missing" genetic information exists.

Doubts and Admonitions

A number of caveats with regard to the preceding analyses exist. The first is that correlations among variables of interest do not constitute proof for cause and effect relationships. Another concern is the nearly flat-line relationship between the number of different cell types and the information content of unicellular plants. For example, once again using log_{10}-transformed data, regression of NCT against the number of intrinsically disordered residues in the six unicellular algae yields a scaling exponent of 0.41 ($r^2 = 0.282$, $P = 0.278$), as opposed to a scaling exponent of 1.84 for the 13 multicellular plants and animals ($r^2 = 0.536$, $P = 0.004$). Taken at face value, these relationships indicate that increases in the information content of unicellular organisms have less effect on NCT compared to multicellular organisms. A third concern is that we know comparatively little about the ancestors of

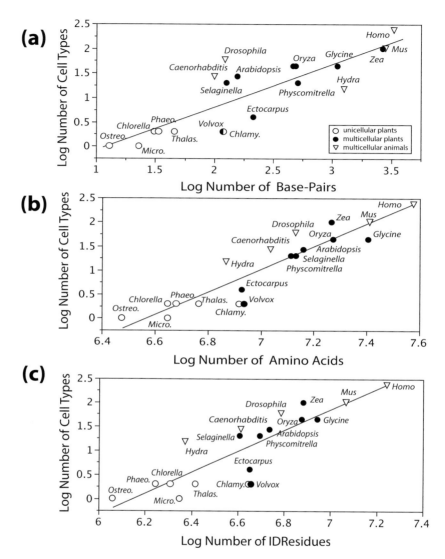

Figure 7.23. Log$_{10}$-transformed data for the maximum number of different cell types (NCT) reported for each species plotted against log$_{10}$-transformed data for genome size (number of 10^6 base-pairs) (a), proteome size (number of amino acids) (b), and protein functional diversity (as gauged by the number of intrinsically disordered protein residues, IDResidues) (c) reported for representative unicellular algae, multicellular algae and land plants, invertebrates, and vertebrates. Straight lines are ordinary least squares regression curves. Chlamy. = *Chlamydomonas reinhardtii*, Micro. = *Micromonas pusilla* NOUM 17, Ostreo. = *Ostreococcus tauri*, Phaeo. = *Phaeodactylum tricornutum*, Thalas. = *Thalassiosira pseudonana*. For statistical regression parameters, see table 7.3. Data for NCT taken from Bell and Mooers (1997).

Table 7.3. Bivariate regression parameters for $\log_{10}$-transformed data of the number of different cell types (NCT), genome size (G, in 10^6 basepairs), number of amino acids (AA), and number of intrinsically disordered residues (IDResidues) reported for 19 species of algae, land plants, and animal species

Log Y vs. Log X	Slope (α-value)	r^2	P	F
NCT vs. IDResidues	2.18	0.721	< 0.0001	44.0
NCT vs. AA	2.36	0.894	< 0.0001	110
IDResidues vs. AA	0.97	0.940	< 0.0001	46.7
NCT vs. G	0.88	0.709	< 0.0001	43.9

Note. Species composition: brown algae (n = 1 species: *Ectocarpus siliculosus*), green algae (n = 3: *Chlorella* sp., *Chlamydomonas reinhardtii*, and *Volvox carteri*), diatoms (n = 4: *Micromonas pusilla* NOUM 17, *Ostreococcus tauri*, *Phaeodactylum tricornutum*, and *Thalassiosira pseudonana*), mosses (n = 1: *Physcomitrella patens*), lycophytes (n = 1: *Selaginella moellendorfii*), flowering plants (n = 4: *Arabidopsis thaliana*, *Glycine max*, *Oryza sativa*, and *Zea mays*), and metazoans (n = 5: *Hydra attenuate*, *Caenorhabditis elegans*, *Drosophila melanogaster*, *Mus musculus*, and *Homo sapiens*).

many of the species used in these analyses. Drawing inferences about ancestral morphologies or lifestyles on the basis of the anatomy or ecology of extant species is highly problematic because we cannot discount the possibility that a seemingly simple species is the descendant of a morphologically or reproductively more complex species. Clearly, much more research is required to draw definitive conclusions. All that can be said with any certainty is that NCT increases with proteome size and with the number of intrinsically disordered protein residues for the species used in these analyses.

Suggested Readings

Bell, G., and A. O. Mooers. 1997. Size and complexity among multicellular organisms. *Biol. J. Linnean Soc.* 60: 345–63.
Bonner, J. T. 1955. *Cells and societies*. Princeton, NJ: Princeton University Press.
Bonner, J. T. 2000. *First signals: The evolution of multicellular development*. Princeton, NJ: Princeton University Press.

Buss, L. W. 1987. *The evolution of individuality*. Princeton, NJ: Princeton University Press.

Cocquyt, E., H. Verbruggen, F. Leliaert, and O. De Clerck. 2010. Evolution and cytological diversification of the green seaweeds (Ulvophyceae). *Mol. Biol. Evol.* 27: 2052–61.

Damuth, J., and I. L. Heisler. 1988. Alternative formulations of multilevel selection. *Biol. Philos.* 3: 407–30.

Folse, H. J., Jr., and J. Roughgarden. 2010. What is an individual organism? A multi-level selection perspective. *Quart. Rev. Biol.* 85: 447–72.

Graham, L. E., J. M. Graham, and Lee W. Wilcox. 2009. *Algae*. 2d ed. San Francisco: Benjamin Cummings.

Kirk, D. L. 2005. A twelve-step program for evolving multicellularity and a division of labor. *BioEssays* 27: 299–310.

Maistro, S., P. A. Briady, C. Andreoli, and E. Negrisolo. 2009. Phylogeny and taxonomy of Xanthophyceae (Stramenopiles, Chromalveolata). *Protist* 160: 412–26.

Michod, R.E. 1997. Evolution of the individual. *Amer. Nat.* 150: S5–S21.

Michod, R.E., and W. W. Anderson. 1979. Measures of genetic-relationship and the concept of inclusive fitness. *Amer. Nat.* 114: 637–47.

Michod, R.E., and A. M. Nedelcu. 2003. On the reorganization of fitness during evolutionary transitions in individuality. *ICB* 43: 64–73.

Newman, S. A., and R. Bhat. 2009. Dynamical patterning modules: a "pattern language" for development and evolution of multicellular form. *Int. J. Dev. Biol.* 53: 693–705.

Schad, E., P. Tompa, and H. Hegyi. 2011. The relationship between proteome size, structural disorder and organism complexity. *Genome Biol.* 12: R120. doi: 10.1186/gb-2011-12-12-r120.

Stanley, S. M. 1973. Ecological theory for sudden origin of multicellular life in Late Precambrian. *Proc. Natl. Acad. Sci. USA* 70: 1486–89.

Taylor, T. N., E. L. Taylor, and M. Krings. 2009. *Paleobotany: The biology and evolution of fossil plants*. 2d ed. Amsterdam: Elsevier.

Biophysics and Evolution

> Evolution may be determined—that is, completely caused in a materi-
> alistic way—and yet not rigidly predetermined from the first as to the
> course it was to follow. An equation can have multiple solutions, and
> yet each solution is determined by the equation.
>
> —GEORGE GAYLORD SIMPSON, *This View of Life* (1964)

This chapter begins with a question that was raised earlier in the Introduction. How deterministic is evolution? If we started with different
initial abiotic conditions, would carbon-based organisms identical to
the first prokaryotes that evolved on Earth follow the same pathways
as life did on Earth? Earlier, we saw that different researchers have answered this question differently, as for example Stephen J. Gould and
Christian De Duve. Recall that Gould said no, whereas De Duve said
yes. This divergence in opinion reflects the extent to which these two
researchers permitted things to differ before they perceived things
to be different, and because Gould was looking at the macroscopic
world recorded in the fossil record, whereas De Duve was looking at
the microscopic world recorded in biochemical reactions. In a very real
sense, the answer to the question posed here depends on our particular *Weltanschauung*, our worldview, which in turn depends on our

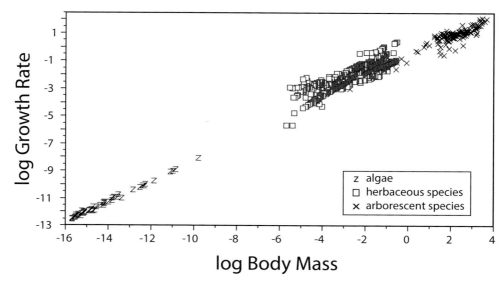

Figure 8.1. Log$_{10}$-transformed data for annual growth rates plotted against log$_{10}$-transformed data for body dry mass for a total of 1,176 unicellular algae, herbaceous (annual) species, and arborescent (woody) species (see insert for key) reported in the primary literature. Across this data set, growth rate (original units: kg dry mass per individual per year) spans over 13 orders of magnitude, whereas body mass (original units: kg dry mass per individual) spans 20 orders of magnitude. Linear regression of all of these data yields a statistically significant log-log linear curve with a slope of 0.755 ($r^2 = 0.968$, $P < 0.00005$, $F = 35,687$).

spatiotemporal perspective and what we consider to be "different" or "the same."

Nevertheless, consider three facts: (1) even the smallest bacterium is metabolically far more complex than any engineered bioreactor, (2) the largest and the smallest organisms share the same information molecules (DNA and RNA) and very much the same basic metabolic machinery to grow, and yet (3) the largest plants weigh more than 20 orders of magnitude more than a bacterium (fig. 8.1). Two conclusions can be drawn from the juxtaposition of these facts: (1) much of biodiversity is the consequence of differences in organic size, geometry, and shape, and (2) organisms must adjust their metabolism, geometry, and shape to accommodate changes in size resulting from ontogeny and growth and to accommodate differences in size resulting from evolution. George Gaylord Simpson (1902–1984), who was one of the intellectual founders of the Modern Synthesis, did not have body size in mind when he argued that evolution is both deterministic and nondeterministic. However, his reasoning pertains to the message of this chapter—that is, some of the effects of size on physical forces and processes are universal and all forms of life must comply with them. Although it is true that the manner in which a specific kind of organism deals with these forces and processes depends on its metabolic and developmental capabilities, it is also true that no organism can obviate the influence of physical forces and processes. Thus, it is reasonable to surmise that evolution travels down indeterminate and determinate pathways, but always in a context-dependent manner.

For example, consider a comparison between a hypothetical aquatic and terrestrial plant constructed out of the same tissues and of comparable size, shape, and geometry. Each is submerged in a liquid that obeys the laws of physics (the first is in water and the second is in air), and each fluid imposes a drag force when it moves relative to the plant it surrounds. Yet, water is incompressible and 1,000 times denser than is compressible air. Consequently, the laws of physics involving the phenomenon called drag (table 8.1) show that the aquatic plant will experience a drag force that is 1,000 greater than its terrestrial counterpart. Now consider the effects of size on drag forces. Physics and experimental measurements show that the rate at which a fluid flows over a surface increases to the ambient limit as we move away from the surface. Consequently, wind speed increases the higher a plant grows. Because drag increases as the square of the speed of flow (see table 8.1) a terrestrial plant will experience, on average, larger drag forces as it increases in height. These facts help to explain in part why there are few large, rigid, and anchored aquatic plants living in fast moving water (because they can experience drag forces that can dislodge or break them) and why tall trees with rigid stems can survive gale force winds (because air is so much less dense than water). The laws of physics and chemistry are universal, but the context in which they operate varies spatiotemporally.

Table 8.1. Six representative physical relationships with parameters that cannot be changed by an organism (invariant parameters) and parameters that can be modified by an organism (variant parameters) as a result of ontogenetic changes in size, geometry, or shape, or by evolutionary modifications

Physical law	Mathematical form	Invariant parameters	Variant parameters
Drag force	$D_f = 0.5\,\rho_{fl}SU^2C_D$	ρ_{fl}	S, U, C_D
Fick's law	$J_i = -D_i(dC_i/dx)$	D_i	C_i, x
Beer's law	$A_\lambda = \varepsilon_\lambda C_s x$	ε_λ, C_s	x
Hagen-Poiseuille formula	$J_w = -(\pi/8\mu)(\Delta P/l)r^4$	μ	P, l, r
Reynolds number	$Re = dU/\nu$	ν	d, U
Bending moments			
Due to gravity	$M_b = 0.5\,g\rho_b VL \sin\phi$	g	$\rho_t, V, L, \sin\phi$
Due to wind	$M_w = D_f L$	D_f	L

Note. Symbols in order of appearance: D_f = drag force; ρ_{fl} = fluid density; S = projected sail area; U = fluid speed; C_D = drag coefficient; J_i = flux of substance i; D_i = diffusivity of substance i; C_i = concentration of substance i; x = distance substance i must diffuse; A_λ = attenuation (absorbance) of light with wavelength λ; ε_λ = extinction coefficient of solute s at wavelength λ; C_s = concentration of solute s; x = depth at which light intensity is measured in the solution of solute s; J_w = volumetric flow rate through a tube; μ = dynamic viscosity; ΔP = pressure loss across both ends of the tube; l = tube length; r = tube radius; Re = Reynolds number; d = reference dimension of object blocking fluid flow; υ = kinematic viscosity of fluid; M_b = bending moment; 0.5 g = gravitational acceleration constant (9.807 m/s); ρ_b = bulk density; V = volume of tissues; L = length of lever arm; ϕ = angle of inclination from the vertical.

This chapter is devoted to viewing plant evolution through the lens of biophysics. Here, we will explore some basic biophysical formulas, construct a plant morphospace, simulate adaptive walks (sensu Sewall Wright, 1889–1988), and assess what computer scientists refer to as "opaque models." All of this requires a good deal of mathematics. Fortunately, the math is simple and can be minimized if we focus on phenotypic consequences rather than on the math itself. The specific goal is to explore adaptive evolution and to address what may seem to be an unanswerable question—Why did some evolutionary trends unfold the way they did? For example, why did body size or plant height increase in some lineages and not in others?

Asking questions such as these is subject to a well-reasoned criticism—they are examples of a posteriori reasoning. Clearly, it is very difficult to experimentally test the past directly and unequivocally. However, predictions about what happened in the past can be made to await evidence from future discoveries. Even the worse case scenario is worth pursuing—we may learn nothing about how the past unfolded they way it did, but we may gain insights into how the present works.

What Can and Cannot Be Changed

Even a cursory inspection of the mathematical expressions of some of the most fundamental biophysical relationships sheds light on why evolution manifests what appear to be trends and patterns. Inspection of each of these expressions reveals that each contains one or more "invariant" parameters (variables whose numerical values cannot be changed by modifications of phenotypic traits) and one or more "variant" parameters (variables whose numerical values can be changed as an organism grows or as a species evolves) (table 8.1). The interplay between invariant and variant parameters shows that the ability of each organism or species to cope with its environment is constrained to some degree by invariant biophysical parameters, but that variant parameters open up opportunities to develop or evolve ways to adaptively overcome these constraints.

For example, reconsider the drag forces that were mentioned in the introduction to this chapter. The formula for drag (D_f) contains one invariant parameter, fluid density (ρ_{fl}), and three variant parameters, the projected sail area (S), the drag coefficient (C_D), and fluid flow speed (U). The three variant parameters can be altered because an organism can deflect and bend, or alter its shape when subjected to an oncoming fluid and thereby reduce its projected sail area and its drag coefficient (fig. 8.2). (Note that shape and geometry are not the same things. A cylinder can be constructed by translating a circle along a linear axis. However, the shape of a cylinder depends on the diameter of the circle and the length of the translational axis.) Changing the ambient fluid flow speed is more problematic because an organism has no direct control over it, except by changing its shape, geometry, or

Figure 8.2. An example of plant morphology (shape) adjusted to minimize wind-induced drag forces (and the effects of salt spray). This fig tree (*Ficus* sp.) was growing on the South Point of the island of Hawaii where the prevailing wind moves from east to west (from the left to right in the upper panel; from the foreground to the rear in the lower panel). Wind (and possible salt) damage to the apices of branches developing on the windward side of the tree inhibited growth toward the oncoming wind. Downwind apices were shielded from the wind by the windward branches and developed into more vigorously growing branches. The combined result is a streamlined branching architecture with a reduced projected sail area (and thus, on average, reduced drag forces).

Figure 8.3. Schematics of four arborescent plants (not drawn to scale) from four very different vascular plant lineages, all of which had short, nonwoody ancestral forms. Each schematic is drawn to reflect the deflections of stems and leaves resulting from wind moving at comparable speeds from left to right (see arrow). From left to right: a lepidodendrid tree (a lycopod, Lepidodendraceae), a palm (a monocot, Arecaceae), a tree fern (Cyatheaceae), and a cycad (a gymnosperm, Cycadaceae). Lepidodendrids had long strap-shaped leaves. The other three schematics illustrate pinnately compound leaves, which can allow air to move through rather than around a leaf.

size. Size can be important because, as noted, flow speeds tend to increase away from a substrate. So, hugging a substrate by staying short reduces U and drag, whereas growing tall can increase U and drag. The fossil record shows that plant height varied within different land lineages over the course of evolution, but tended to increase among vascular plant lineages with consequences on the magnitudes of drag forces and the requirement for better anchorage and coping with leaf damage (fig. 8.3).

The interplay between invariant and variant parameters is further illustrated by considering the one-dimensional form of Fick's first law of diffusion, which describes the rate (flux) with which solutes can diffuse passively through a fluid or from one compartment into another. Fick's first law shows that the flux of a passively diffusing solute (J_i) is the product of the diffusivity constant (D_i), the concentration of the solute (C_i), and the distance the solute has to diffuse (x). The latter two parameters describe a concentration gradient ($-dC_i/dx$). (The vector calculus defines a gradient as the negative of a concentration drop.) For any particular fluid (water or air), the numerical value of the diffusivity constant is invariant for a stipulated temperature and pressure. For example, the diffusivity of carbon dioxide and oxygen in air and in water at 20°C is on the order of 10^{-5} m²/s and 10^{-9} m²/s, respectively. Thus, all other things being equal, an aquatic plant relying on passive diffusion to acquire solutes from the water around it has a "10^4 disadvantage" compared to its terrestrial counterpart. This helps to explain why aquatic plants generally employ active metabolic uptake to get CO_2. However, if we examine the mathematical expression of a concentration gradient ($-dC_i/dx$), we see immediately that solute uptake can be increased if a solute is consumed (fixed) more rapidly within the plant body, or if the distance over which the solute diffuses is reduced, each of which increases the gradient. The latter helps to explain why many aquatic plants have large surface areas relative to their volumes, as for example those constructed out of thin filaments or thin sheets of cells.

Indeed, the benefits of having large surface areas can be illustrated further because Fick's law can be manipulated easily to show the effects of surface area A and volume V on the time it takes for a solute to get into a cell. Consider the time it takes for the concentration of a nonelectrolyte solute j initially absent from the interior of a cell to reach one-half the concentration of j in the external ambient medium (air or water). Denoting this half-time as $t_{1/2}$, manipulation of Fick's law yields $t_{1/2} = (0.693/P_j)(V/A)$, where P_j is the permeability coefficient of j, which is a numerical constant for any j at a specified temperature and pressure. Thus, regardless of the permeability coefficient of j, the half time is always proportional to V/A such that any decrease in V or any increase in S decreases $t_{1/2}$, which is adaptively beneficial to growth if j is an essential nutrient.

Beer's law, which predicts the attenuation of light through dilute solutions, is also relevant to understanding evolution and ecology (table 8.1). For example, it shows why aquatic plants benefit from staying near the water's surface. The attenuation of light at any wavelength (A_λ) is the product of the extinction coefficient of a solute at any given wavelength (ε_λ), the concentration of the solute (C_s), and the depth at which light intensity is measured in the solution (x). Because naturally occurring bodies of water contain numerous different kinds of solutes (often making the water yellow, hence the term *gelbwasser*), it follows that the intensity of light is greatest near the water's surface (mathematically speaking, $A_\lambda = 0$ when $x = 0$). It is worth mentioning that there is another good reason for occupying the shallow end of the pool. Even pure water absorbs all wavelengths of light, but especially in the red, which is one of the absorption maxima of chlorophyll *a* and *b* (see fig. 7.15 and related text), which helps to explain why brown algae growing at great depths produce high concentrations of blue-absorbing pigments.

From Water to Air

The biophysical relationships listed in table 8.1 illuminate certain aspects of an important phase in plant evolution, the migration of plant life in water to plant life on land (or more properly, life in air). Consider that very thin photosynthetic structures are not at risk of dehydration when submerged in water, and Fick's law shows the tremendous benefits of having a large surface area with respect to volume if an organism relies on passive diffusion for the transport of solutes or the movement of water molecules. Likewise, it is easy to intuit that structures or organs that resist the pull of gravity are not required if organisms are neutrally or negatively buoyant, and most aquatic plants are. However, living in water comes with two disadvantages—it is hard to obtain substances dissolved in water by means of passive diffusion and light intensity is reduced even a few centimeters below the water-air interface. In contrast, gases such as carbon dioxide and oxygen diffuse 10^4 times faster in air than in water, and air is optically transparent (unless you live in a polluted environment). The only problem with living in the air is that water evaporates when the air moves, and the rate at which water is supplied to photosynthetic tissues constrains plant size, growth, and survival.

From a purely geometric perspective, living in air is made possible by doing the opposite of living in water—it requires a reduction in external surface areas with respect to volume, or some method of keeping external areas moist, as for example by growing very close to the substrate that supplies water, or by producing a coating that is impermeable to water. By and large, the nonvascular land plants adopt a "hugging" strategy and remain comparatively short in stature because, beyond a certain height, the rate at which water moves vertically from one cell to the next by means of passive diffusion fails to compensate for the rate at which water is lost from elevated photosynthetic tissues. A thin layer of water over the external surfaces of these plants can be wicked upward as water is lost by means of evaporation. However, the tensile strength and adhesion of water molecules to aerial surfaces can be exceeded during episodes of rapid evaporation. One way to circumvent these limitations is to evolutionarily transition from relying on the net movement of water down a concentration gradient to the movement of water down a pressure gradient. Human engineers use tubes to convey fluids like water, and so do vascular land plants in the form of tracheids and vessel members (fig. 8.4). Both of these cells are dead when functional. Consequently, their evolution involved the ability to control and program cell death (apoptosis), which has evolved in only a few eukaryotic lineages.

Although water flow from one tracheid to another differs in many important respects from water flow from one vessel member into another, flow through hollow tubes must abide by a number of similar rules, some of which are codified by the Hagen- Poiseuille formula and the Reynolds number (table 8.1). The former estimates the rate of fluid flow through narrow tubes, whereas the Reynolds number informs us as to whether fluid flows are dominated by viscous forces (at very low Reynolds numbers) or by inertial forces (at high Reynolds

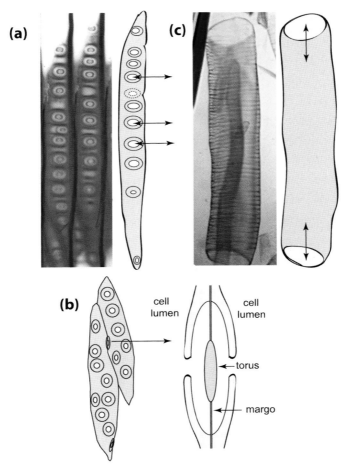

Figure 8.4. Images and side-by-side schematics of water conducting cell types in vascular plants. Two adjoining tracheids (a) with bordered pits (see b) and a single vessel member (c). Both cell types are dead at maturity and consist of a perforated cell wall. The arrows shown in (a) and (c) indicate the locations and directions of water flow from one cell into another. a. A tracheid typically is a spindle-shaped cell with numerous perforations (pits) that permit water to flow through one adjoining pit into another (called pit-pairs). b. Side view of a pit-pair, which can have valve-like structures (constructed from the middle lamella, which glues adjoining cells to one another) consisting of a disk-like torus suspended by a membranous margo. When an adjoining tracheid develops a water vapor blockage (called an embolism) in a water stressed plant, the torus-margo is appressed to the inner aperture of the neighboring cell, which confines the embolism to one cell. c. A vessel member is typically a wider cylindrical cell that, unlike a tracheid, is perforated at either end and thereby allows for the comparatively rapid bulk flow of water. Lateral walls also allow water to move into or from neighboring cells, but to a lesser extent. Vessel members are arranged end-to-end (not shown) to construct a vessel through which water can flow more rapidly than through adjoining tracheids. Although water can flow more rapidly through vessels than tracheids, tracheids are better adapted to cope with the consequences of water vapor embolisms.

numbers). The latter is important in this context because the Hagen–Poiseuille formula is accurate only when fluid flow is dominated by viscous forces (when Re < 10).

Manipulation of both formulas shows that the rate of water flow through a very narrow tube (J_w) equals $\pi \upsilon r \mathrm{Re}$, where υ is kinematic viscosity and r is the radius of the tube. For water at 20°C, υ is on the order of 10^{-6} m^2/s, whereas, for an average xylem vessel member, r is on the order of 10^{-5} m. Setting Re equal to or less than ten gives $J_w \approx 10^{-11}$ m^3/s and a flow rate roughly equal to 0.032 m/s. This value is an overestimate because it neglects the resistances incurred by water flowing over the internal wall thickenings of vessels and the resistances resulting from water flowing from one vessel into another. For example, the sapwood of oak trees (which has very wide vessels) has peak flow rates of 0.013 m/s.

Although increasing r can substantially increase J_w, this comes at a risk because the probability of a water vapor blockage (the formation of an embolism) resulting from water stress increases with increasing r. Another solution is to increase the number of water-conducting cells per cross-sectional area. Evidence from the fossil record indicates that both strategies were evolutionarily explored during the early diversification of vascular land plants (fig. 8.5). However, in the final analysis, the primary physical limitation to plant height is the rate at which water can be supplied to the uppermost reaches of a canopy of branches and leaves. At some height, the ability to supply water falls short of the rate at which water is lost through evapotranspiration, at which point stomata close. Consequently, photosynthesis slows or stops completely, and growth rates decline asymptotically such that further growth in height ceases. These relationships likely explain why vascular plants reach a maximum height—it is not mechanical stability that sets the critical limits to vascular plant height, but rather the ability to supply water to the upper reaches of a leafy canopy that defines the limits of how high photosynthetic tissues can be elevated toward the skies and still get sufficiently supplied amounts of water (per unit time).

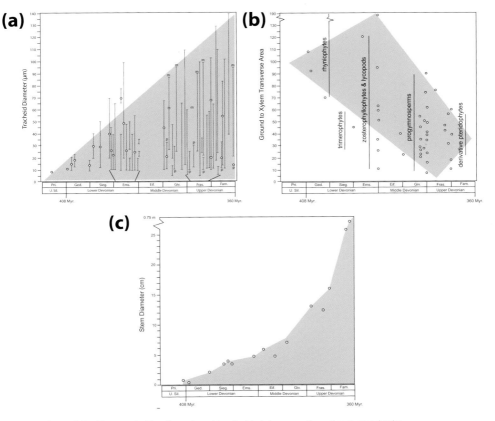

Figure 8.5. Changes in the diameter of tracheids (a), proportion of nonvascular to vascular (xylem) tissues in stems (b), and maximum stem diameter reported for plants lacking secondary growth (c) plotted as functions of geological age during the course of early vascular land plant evolution. a. The range and mean diameters of tracheids, on average, increased from the Upper Silurian (U. Sil.) through the Lower, Middle, and Upper Devonian, which indicates that the ability to transport larger amounts of water increased. p = protoxylem, m = metaxylem. b. Ground tissue (nonvascular tissues) to xylem transverse area, on average, decreased from the Upper Silurian (U. Sil.) through the Lower, Middle, and Upper Devonian, which indicates that the proportion of water-conducting tissue increased over time. c. Over the same time interval, the maximum diameter of stems reported for each geological stage increased, which indicates that plant height likely increased as the hydraulic capacity of stems increased.

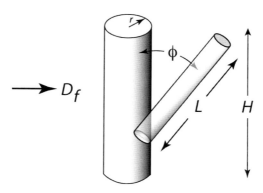

Figure 8.6. Schematic of a simple branching system illustrating the effect of orientation on the bending moments induced by gravity and by a drag force (D_f) induced by wind pressure (see arrow). The bending moment resulting from gravity on the smaller lateral branch (see table 8.1 for the complete formula) is proportional to the sine of the deflection angle from the vertical (sin ϕ) times the lever arm (L). The bending moment of the large central axis due to gravity equals zero because sin ϕ = sin 90° = 0. The bending moment due to the drag force D_f is proportional to the projected sail area of the large central axis (assuming the smaller lateral axis is completely sheltered from the oncoming wind), which is equal to $2rH$. The maximum bending moments induced by gravity and by wind drag occur at the base of each of the two branches shown in this schematic. Consequently, anchorage is required to provide counter moments.

Increasing height comes with problems other than those caused by hydraulic limitations as shown by the formula for bending moments induced by gravity (table 8.1). The formula for a bending moment (M_b) has one invariant parameter—the pull of the Earth, which is quantified by the gravitational acceleration constant, g = 9.807 m s^{-2}. The rest depends on a structure's bulk density (ρ_b), volume (V), lever arm length (L), and the orientation of the object (sin ϕ, where ϕ is the angle measured from the vertical) (fig. 8.6). Each of these four parameters is variant because each can be modified as a plant grows or as a consequence of evolutionary modification. For example, for a vertical stem, the bending moment is zero because sin 0° = 0. However, the problem with this orientation is the bending moment induced by wind rather than by gravity. In this case, the bending moment equals the force of the wind (drag) times the length of the lever arm ($M_w = D_f L$), which in this case equals height ($M_w = D_f H$). A perfectly vertical stem projects the maximum sail area toward oncoming wind parallel to the ground surface, and M_w reaches its maximum at the base of a stem. Unless this bending moment is countered by a restoring moment, the stem will become mechanically unstable. Like all sedentary organisms, anchorage is essential for land plant survival.

However, a vertical orientation is also the least efficient orientation for photosynthetic organs, such as foliage leaves. The optimum orientation for unobstructed light-gathering surfaces is horizontal. However, this orientation results in the maximum bending moment due to gravity (because sin 90° = 1.0) (fig. 8.7). The fossil record shows that the organographic distinction between what we call leaves, stems, and roots did not exist for the first vascular land plants. The body plans of the most ancient vascular land plants were constructed out of simple unadorned cylindrical "axes," which served multiple functions simultaneously, as revealed by the remarkably well preserved fossils in the Rhynie Chert flora (fig. 8.8). In retrospect, these organisms appear to have coped with the contrasting optimal orientations of mechanical and photosynthetic functionalities by producing axes oriented at angles greater than 0° but less than 90° (as for example between 45° and 60° from the vertical). However, over the long course of evolutionary history, many land plants have adapted by evolving organs that function primarily as either mechanical or photosynthetic organs (stems versus leaves).

It is worth noting that the central location of the vascular tissues that is typically observed in early vascular plant fossils is optimal for providing water to the surrounding ground tissues and for coping with the transverse distribution of tensile and compressive stresses resulting from bending. However, although the center of a beam bent under its own weight experiences neither tensile nor compressive stresses, the center experiences the maximum shearing stresses resulting from bending (fig. 8.9). Typically, plant tissues are less capable of dealing with shearing compared to their ability to resist tension or compression.

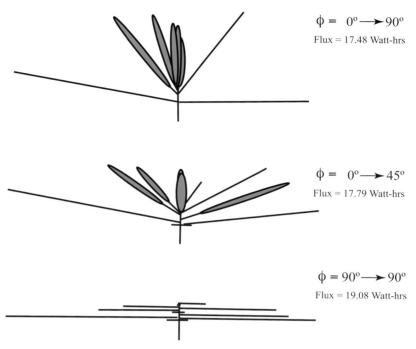

$\phi = 0° \longrightarrow 90°$

Flux = 17.48 Watt-hrs

$\phi = 0° \longrightarrow 45°$

Flux = 17.79 Watt-hrs

$\phi = 90° \longrightarrow 90°$

Flux = 19.08 Watt-hrs

Figure 8.7. Three computer simulations to assess the optimal orientation of spirally arranged leaves with different orientations with respect to the vertical. Each simulation has the same number of leaves, the same total leaf area, and the same distance between successive leaves. The light interception capability of each plant (light flux per day in units of Watt-hrs) is evaluated assuming plants grow on the equator (latitude 0°) and in full sunlight from dawn to dusk. The only difference among the three simulations is the orientation of the leaves in each rosette, which is determined by the angle of deflection from the vertical (ϕ). This angle is confined to a specified range in each simulation such that successively older leaves further from the center of a rosette deflect progressively more from the vertical, or deflect at the same angle throughout the rosette (from top to bottom: $\phi = 0°$ to 90°, $\phi = 0°$ to 45°, and $\phi = 90°$ to 90°). The simulations show that the optimal orientation of photosynthetic surfaces is $\phi = 90°$, which is the orientation that maximizes bending moments induced by gravity (see fig. 8.6).

Functions and Tradeoffs

The foregoing illustrates examples of some of the tradeoffs required when multiple functions are performed simultaneously, or when a group of organisms adapts to a different physical environment, such as an evolutionary transition from an aquatic to a terrestrial habitat. These examples show that compromises and tradeoffs are often re-

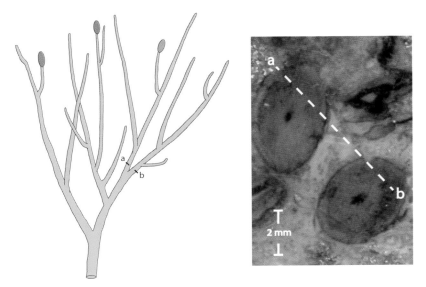

Figure 8.8. Reconstruction of the aerial portions of the Lower Devonian (≈400 Mya) vascular plant *Rhynia* (left) and a cross section through a petrification of two *Rhynia* axes (right) aligned with respect to the dashed line (a–b) in the reconstruction. The axes of early vascular plants such as *Cooksonia, Rhynia,* and other early vascular land plants were multifunctional. Vertical axes served as photosynthetic organs and held sporangia aloft (depicted on the reconstruction as ellipsoids); horizontal axes (not shown) bore clusters of rhizoids that together with symbiotic mycorrhizae absorbed water and minerals.

quired because different functions can have different requirements, as for example the optimal display of photosynthetic organs and the contrasting optimal orientation for mechanical stability, or the advantages of large surface areas to intercept sunlight and the disadvantages in terms of evapotranspiration.

The evolutionary implications of the compromises and tradeoffs that emerge because organisms perform multiple functions simultaneously to grow, survive, and reproduce can be explored with the aid of some of the biophysical relationships reviewed in table 8.1 and computer simulations of all theoretically possible morphologies (a morphospace) that can assess the extent to which each morphology performs any one function or any combination of two or more functions. This approach has five components: (1) identifying the imperative functions the group of organisms selected for study must

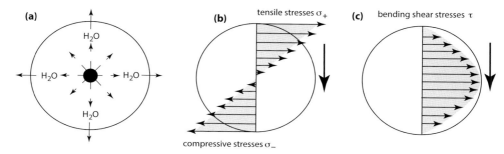

Figure 8.9. Hydraulic and mechanical tradeoffs in a cylindrical axis with a centrally located strand of water-conducting tissue (similar to that of early vascular land plants; see fig. 8.8). a. A centrally located strand of water-conducting tissue is equidistant from the surface of a cylinder from which water can be lost due to stomatal or cuticular evapotranspiration. b. This location is optimal in terms of the distribution of the tensile and compressive stresses (σ_+ and σ_-, respectively) resulting from bending (in the direction of the large downward pointing arrow) because these stresses have zero magnitude at the center of the cylinder where the strand is located (these stresses reach their maximum intensities at the surface of the cylinder as indicated by the horizontal arrows). c. However, the center of a bent beam (in the direction of the large downward pointing arrow) experiences the maximum shear stresses (τ), which have a parabolic distribution in each transverse plane.

perform, (2) constructing a morphospace of all theoretical pheno-types, (3) quantifying the ability of each morphology to perform each function and all combinations of functions, (4) exploring the adaptive landscape of the morphospace to evaluate the morphologies that successively optimize the performance of two or more functions, and (5) to use Darwin's words, exploring the synergies of the "conditions of existence" and "the unity of type."

Notice the choice of words in the phrase "*optimize* the performance of *two or more functions*." It helps to make a sharp distinction between "optimization" and "maximization," which might be otherwise over-looked. In theory, the performance of a machine that is designed to perform one function can be maximized to achieve the highest level of performance that is possible. The extent to which the ideal can be achieved is limited obviously by cost-profit considerations, the conditions of the work place, the availability of building materials, etc. But maximization is theoretically possible given specified conditions. In contrast, the performance of a machine that performs two or more

functions simultaneously can only be optimized if two or more functions have contrasting design requirements because the requirements
of each of the functions the machine performs must be reconciled.
This reconciliation precludes maximizing the performance of any one
function (box 8.1).

The analogy of a machine can be extended, albeit to a limit, when
considering the functionalities of biological entities. For example,
plants perform chemical and mechanical tasks. Regardless of whether
they are unicellular or multicellular, plants synthesize carbohydrates,
proteins, lipids, etc. from raw materials, they build cell walls, and they
cope with mechanical forces such as turgor pressure, gravity, and drag.
Each of these functions can be reduced conceptually to the operation
of chemical and physical processes, even if it is impossible to construct
a living cell using our current knowledge of how these processes work
and interact with one another.

The real concern lies in our ability to infer an organism's functional
obligations on the basis of an organism's phenotypic traits. An en-
gineer knows the design specifications of a machine a priori as well
as the conditions in which a machine must function. A piston functions as a piston. No deduction is required. However, as noted earlier
when considering adaption and adaptive evolution, the same cannot
be said when a biologist examines an organism. A leaf that develops
into a spine may function to convect heat, deter an herbivore, increase
a boundary layer, or do all or none of these things. Experimentation is
required to turn inferences about function into verifiable declarations
about the "purpose" of an organic structure or process.

With these caveats in mind, we may still ask, What are the functional obligations of a plant? It is a truism that all living things must
exchange mass and energy with their external environment to grow
and survive. However, the forms of mass and the forms of energy depend on the kind of organism. For example, purple sulfur bacteria
are photosynthetic but anoxygenic (they do not produce oxygen) and
require either hydrogen sulfide or elemental sulfur as the reducing
agent (electron donor) for photosynthesis to proceed. In the context

Box 8.1. The Optimization Process: An Analytical Illustration

The role played by optimization processes can be illustrated mathematically and graphically in a general way. A basic feature of the optimization process is that it results from a trade-off between two or more conflicting but necessary requirements or processes that are attributable to a single system. Typically, the performance of one of these requirements increases as the performance of the other decreases with respect to some shared parameter such as body mass. Under these circumstances, optimization can be conceptualized analytically by considering a hypothetical mathematical function that quantifies total performance P that is the sum of two other performance functions, P_1 and P_2, each of which is dependent on some sort of biomass investment M. For convenience, it is assumed that these investments take the form of power functions. Denoting the scaling exponents of the two performance functions as m and n and the proportionality constants as k_0 and k_1, respectively, we see that

$$P = P_1 + P_2,$$

$$P_1 = k_0/M^m,$$

$$P_2 = k_1 M^n,$$

such that

$$P = k_0/M^m + k_1 M^n.$$

In this example, M^m could represent the scaling of a bending moment (created at the base of a tree trunk by wind-induced drag) with respect to the mass of the trunk, whereas M^n could represent the ability of the canopy held aloft by the trunk to harvest light (which is some function of the mass of the trunk). A trade-off between these two performance functions exists because any reduction in the bending moment (which increases k_0/M^m) requires a reduction in either stem or leaf mass (which reduces drag but also reduces the ability to support leaves or harvest light, respectively). To find the optimal investment of biomass M^*, we take the partial derivative of P with respect to M and set the partial differential equation equal to zero:

$$\frac{\partial P}{\partial M} = -mk_0 M^{-m-1} + nk_1 M^{n-1} = 0.$$

Solving for M^* gives

$$M^* = \left(\frac{mk_0}{nk_1}\right)^{1/(m+n)}.$$

Solving for the optimal performance P^* exclusively in terms of the scaling exponents and proportionality constants gives

$$P^* = k_0 \left(\frac{mk_0}{nk_1} \right)^{-m(m+n)} + k_1 \left(\frac{mk_0}{nk_1} \right)^{n(m+n)} = k_0 \left(\frac{m+n}{n} \right) \left(\frac{nk_1}{mk_0} \right)^{m(m+n)}.$$

This derivation shows that performance functions of the type illustrated here are analogous, albeit loosely, to fitness landscapes in which P^* and M^* identify the most cost-effective biomass investments. Fig. B.8.1 illustrates this feature for the hypothetical case $k_0 = 2$, $k_1 = 3$, $m = 0.5$, and $n = 0.33$, which gives $M^* \approx 1.0$ and $P^* \approx 5$. The extent to which neighboring performance values differ numerically from (P^*, M^*) translates loosely into relative fitness. In this example, there exists an inverse relationship between the magnitudes of the scaling exponents and the relative fitness of these other strategies—that is, fitness relative to the optimal condition decreases as the magnitudes of exponents m and n increase. To visualize the analogy between the performance functions and an adaptive landscape, invert the curve for $P = P_1 + P_2$ to produce an adaptive hill. An interesting but unanswered question is whether evolution has resulted in "insensitive" performance functions—scaling relationships that yield broad, flattened performance functions like the one shown in the attending figure rather than sharply curved functions.

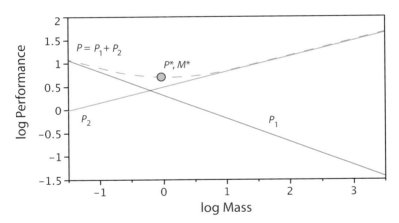

Figure B.8.1. A graphic illustration of the optimization between two performance functions P_1 and P_2 (red and green lines) measured in terms of biomass investment M plotted as a log-log transformed bivariate graph. The total performance function $P = P_1 + P_2$ (dashed blue line) reflects a tradeoff between P_1 and P_2 that identifies the optimal performance and biomass investment (P^*, M^*), which occupy an adaptive "valley." The "sensitivity" of the optimal performance function depends on the numerical values of the exponents of P_1 and P_2. In general, small exponents result in less sensitive values of P^* and M^* than large exponents.

of all eukaryotic photoautotrophs (here broadly defined as "plants"), there are four generic functions that must be performed for successful growth and survival, and for evolution to proceed: (1) they must photosynthesize and thus intercept radiant energy and exchange water, minerals, oxygen, and carbon dioxide with the fluid (water or air) surrounding them, (2) they must transport nutrients within them, even if they have a unicellular body plan, (3) they must cope with external and internal mechanical forces, and (4) they must reproduce in a manner that creates heritable diversity. Tradeoffs and compromises exist within each of these individual functions. For example, on land, photosynthesis involves the tradeoff between absorbing carbon dioxide from air and losing water. The cuticle and stomata help to reconcile this tradeoff, but neither nature nor the best engineers have invented a material that is permeable to carbon dioxide (and oxygen), which facilitate photosynthesis, and impermeable to water, which conserves water. Consequently, details such as cuticular transpiration even in the absence of photosynthesis must be considered when dealing with plant ecology and evolution.

However, if we focus on a larger, course-grained picture of functional obligations, the juxtaposition of the four functions generally defined creates a matrix of 14 functional combinatorial possibilities: 4 single-function possibilities, 6 two-function combinations, 3 three-function combinations, and 1 four-function combination (fig. 8.10). When the ability of different hypothetical morphologies to perform each of these functional tasks is numerically quantified, either in isolation or in combination, each of the 14 functional possibilities can be seen as a "landscape" over which morphological transformations can emulate maladaptive, neutral, or adaptive evolution.

In the following sections, we will attempt to visualize adaptive evolution in a manner analogous to that first proposed by Sewall Wright (1889–1988) at the 1932 International Congress of Genetics in Ithaca, New York. Wright envisioned evolution as a "walk" over a fitness "landscape." In his original paper, the walk was a sequence of genotypes leading from regions of low to high fitness, while the

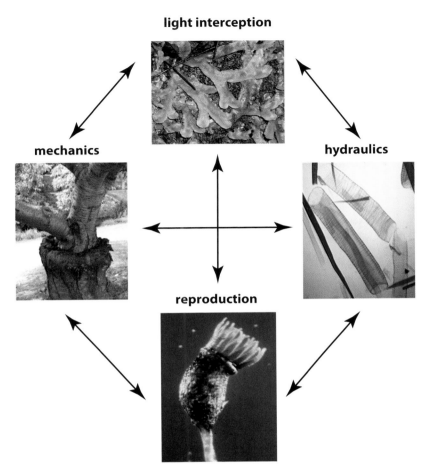

Figure 8.10. Schematic of the four broadly defined functional obligations that eukaryotic photoautotrophic organisms ("plants") must perform to successfully grow and survive, and for evolution to proceed. The performance of each of these functions involves one or more compromises and tradeoffs (not shown), as for example the loss of water vapor during the absorption of carbon dioxide for photosynthesis, and the tradeoff between conducting water rapidly in hollow tubes and the risk of forming water vapor embolisms. The juxta-position of the four main functional requirements identifies 14 functional combinatorial possibilities (or "landscapes"): 4 single-function landscapes, 6 two-function landscapes, 3 three-function landscapes, and 1 four-function landscape.

landscape was rendered as a spatial domain depicting all conceivable genotypes, each assigned a relative fitness value. As originally cast, however, this model for adaptive evolution is largely impractical. The genotypes of fossil plants are unknown, just as the genotypes of living plants are largely unknown. However, the landscape of plant evolution can be rendered in terms of phenotypes and their relative fitness, just as walks over the landscape can be seen in terms of the phenotypic alterations required to scale adaptive peaks. Regardless of how it is rendered, Wright's metaphor for adaptive evolution is a *thought experiment*, one that must be performed in full appreciation that evolution is as much a game of chance as it is structured by environmental sorting. The landscape over which plant life evolved changed unpredictably many times. Thus, plant evolution reflects the consequences of many chance events. But the landscape is also ordered and predictable—ordered in the sense that physical laws and processes influencing fitness cannot be violated, and predictable in the sense that the physical environment will dispose of less fit phenotypes and leave behind the more fit. Thus, every walk, whether cast in genotypic or in phenotypic terms, reflects the combined effects of random and nonrandom evolutionary forces.

A virtue of the phenotypic version of Wright's fitness landscape is that hypothetical adaptive walks can be compared with those preserved in the plant fossil record. Here the concern is not to rewrite evolutionary history but to discover why plants may have evolved as they did. If they evolved in accordance with physical and chemical laws and processes, then hypothetical adaptive walks based on these laws and principles should agree with the phenotypic trends seen in the fossil record. The cardinal sin threatening this test is the conceited notion that all possibilities and all effects have been anticipated. Plant life is extraordinarily complex, and manipulating a few simple shapes to construct the domain of all possible phenotypes is truly naïve. Likewise, the ways plants have dealt with the requirements for survival or reproduction are demonstrably intricate and varied. Thus, it is ingenuous to believe that we understand the complexity and intricacy of plant life well enough to simulate the precise course of evolution. But the tart response in the face of these pitfalls is "nothing ventured, nothing gained." By refusing the effort, we pass up the possibility of gaining some little insight into evolutionary history. Simulated adaptive walks may fail the test of the fossil record, but we may learn something along the way. Some preconceptions about plant evolution may be revealed as incorrect or irrelevant; the staggering diversity of plant form and structure and its relation to fitness may be quantitatively codified; and the experiment of plant life will be opened to further interpretation.

A Morphospace for Simple Sporophytes

The ability of different kinds of plants to perform in each of the 14 different functional landscapes depends on their morphology and therefore on the occupants of a morphospace. Among modern-day and ancient land plants, the basic architecture of sporophytes consists of unbranched or branched, cylindrical axes. Therefore, a sporophyte morphospace is not difficult to construct, particularly for ancient land plants lacking leaves (see fig. 8.8). Each morphological variant can be assembled from cylindrical axes with uniform girth and tissue density. The orientation of each axis with respect to the vertical and horizontal planes of reference can be specified by a bifurcation angle ϕ and a rotation angle γ, wherein each ranges between 1° and 180° (fig. 8.11). The number of axes for a particular morphology is determined by assigning a probability value for apical bifurcations p, where $p = 0$ indicates no branching and $p = 1$ indicates unlimited branching. However, for convenience, branching in each morphology is restricted to ten levels of bifurcation such that the maximum number of axial elements for each morphology is thus 2,047. Mathematically, $p = 8p_{n-(k+1)}(N + k)^{-1}$, where $p_{n-(k+1)}$ is the probability of terminating branching at the next generated (higher) level of branching and $N + k$ designates the previously generated level of branching. The length of each axis L is stipulated to be directly proportional to the probability of branching. That

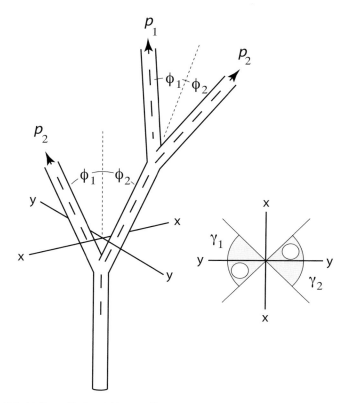

Figure 8.11. Mathematical variables used to construct a morphospace for hypothetical vascular plants. The orientation of each set of companion axes subtended by a single axis is specified by a bifurcation angle ϕ and a rotation angle γ. The probability of an axis branching subsequent to its appearance is designated as p. Morphologies in the isobifurcate subdomain of the morphospace have axes with equal branching angles $\phi_1 = \phi_2$ and rotation angles $\gamma_1 = \gamma_2$ and probabilities of branching $p_1 = p_2$ such that these morphologies can be located in a simple Cartesian coordinate system. Morphologies in the anisobifurcate subdomain have unequal angles and probabilities of branching ($\phi_1 \neq \phi_2$, $\gamma_1 \neq \gamma_2$, and $p_1 \neq p_2$).

is, axial length is a linear function of p such that the basal-most axial element is always as long as or longer than any more distal (higher level) axial element.

The morphospace has two subdomains, one containing all equally branched (isobifurcate) morphologies and another containing all unequally branched (anisobifurcate) morphologies. These two subdomains contain phenotypes that are similar to fossil plants belonging

Figure 8.12. Representative morphologies of rhyniophytes (a) and trimerophytes (b) characterized by more or less equal (isobifurcate) and unequal (anisobifurcate) branching, respectively. a. A compression fossil assigned to the genus *Cooksonia caledonica* (see also fig. 7.3) and a schematic of *Rhynia gwynne-vaughanii*, both of which produced terminal globose sporangia (scale is in millimeters). b. A compression fossil of the genus *Psilophyton crenulatum*, which produced terminal, spindle-shaped (fusiform) sporangia on lateral branching systems and a schematic of *Psilophyton dapsile*. Cornell fossil plant collections.

to the rhyniophytes and trimerophytes, respectively (fig. 8.12). This requirement emerges when we look at modern-day plants and when we consider the fossil record, which reveals that both types of branching architectures evolved at roughly the same time. Mathematically, the isobifurcate subdomain contains all variants for which $\phi_1 = \phi_2$ and $p_1 = p_2$. The anisobifurcate subdomain contains all morphologies for which $\phi_1 \neq \phi_2$ and $p_1 \neq p_2$. (See fig. 8.11). Since axial length L is a function of p, it follows that $L_1 = L_2$. Each of the two subdomains is constructed by independently varying each of the aforementioned variables. The spatial ordering of morphological variants is predetermined by assigning ascending numerical values to each of the variables. For example, bifurcation angle varies in 1° increments, and the probability of branching varies in 0.01 increments.

With some simple mathematical modifications, the biophysical relationships listed in table 8.1 can be used to assess the ability of each morphological variant in this morphospace to perform each of the four tasks enumerated in the previous section (fig. 8.13). For example, light interception can be quantified for each morphological variant by integrating the area under a curve generated by plotting the unobstructed surface area S_p of a morphology projected toward the sun divided by the total surface area S of the variant as a function of the solar angle θ for the values $1° \leq \theta \leq 180°$, using an algorithm that accounts for the self-shading. Mechanical stability is calculated by computing the maximum bending moment at the base of each morphology. The ability of each variant to conserve water can be gauged as a simple linear function of total plant surface area (assuming that all surface areas have equivalent rates of evapotranspiration). Finally, reproductive success can be evaluated based on the number of spores produced per morphology and the distance that spores can be dispersed by wind. A simple ballistic model can be used if it is assumed that all sporangia are at the tips of all distal axes and that all sporangia produce an equivalent number of spores of equivalent size and density. The ballistic model is given by the formula $x = HU/T$, where x is maximum distance of spore dispersal, H is a sporangium's height above ground level, U is ambient wind speed, and T is the terminal settling velocity of spores (taken to be 0.15 m/sec). These assumptions and specifications specify that the spore dispersal range is, on average, proportional to the square of plant height.

The relative fitness of each hypothetical variant must be evaluated using basic physics or engineering principles that describe the performance of each task designated to influence growth, survival, and reproductive ability. The ability of each variant to perform one or more of these tasks can then be divided by the maximum performance level in a particular landscape. To compute the fitness of a morphology performing one or more tasks simultaneously f, each task is

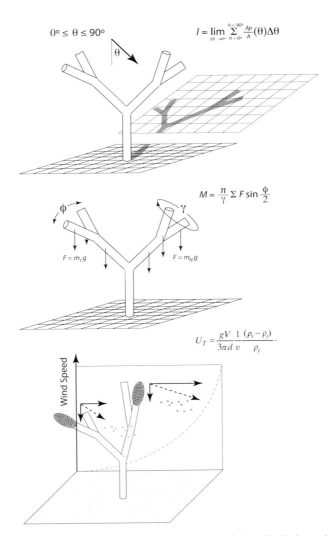

$0° \leq \theta \leq 90°$

$$I = \lim_{\Delta\theta \to 0°} \sum_{\theta = 0°}^{\theta = 90°} \frac{Ap}{A}(\theta)\Delta\theta$$

θ

$$M = \frac{\pi}{\gamma} \sum F \sin \frac{\phi}{2}$$

ϕ

γ

$F = m_1 g$

$F = m_N g$

$$U_T = \frac{gV}{3\pi d}\frac{1}{\upsilon}\frac{(\rho_s - \rho_f)}{\rho_f}.$$

Wind Speed

Figure 8.13. Schematics illustrating how the performance of a hypothetical morphology composed of cylindrical axes is evaluated for intercepting light, maintaining mechanical stability, and dispersing spores (from top to bottom). Light interception (I) is calculated as the sum of the change in the quotient of projected surface area and total surface area (S_p/S) computed as a function of the change in the solar angle ($\Delta\Theta$) for $0° \leq \Theta \leq 90°$ in $1°$ intervals. The maximum bending moment (M) experienced by the basal-most (proximal) axis is computed as the sum of the forces (F) experienced by all of the axes and their orientations as defined by bifurcation and rotation angles (ϕ and γ, respectively; see fig. 8.11). The capacity to reproduce is evaluated as a function of the number of spores produced per sporangium times the number of sporangia per morphology times the maximum dispersal range of spores (which is determined by the terminal settling velocity of spores, U_T). The terminal settling velocity is a function of the volume (V), diameter (d), and bulk density (ρ_s) of the particle (spore) and on three invariant parameters: the acceleration due to gravity (g), and the kinematic viscosity and the density of air (υ and ρ_f, respectively).

assumed to contribute equally and independently to overall fitness such that the fitness of a particular variant f is given by the formula $f = (P_1 + P_2 \ldots + P_N)^{1/N}$, where P is the performance of level of a task and N is the number of tasks being performed. The relative fitness of a morphology in the morphospace will depend on the tasks being performed. However, in each case the relative fitness of each morphology equals f/f, where f is maximum fitness defined by the most fit morphology (the maximum fitness therefore is 1.0 and all other relative fitness values are less than one).

It is important to say that the adaptive walks that will be discussed later in this chapter are not random walks. They are determinate and invariate. Put differently, if a walk starts at the same location in the morphospace and if the walk is directed to locate optimal phenotypes based on the performance of the same set of tasks, it will located the same sequence of phenotypes each time the walk is initiated. This phenomenology emerges because the locations of phenotypes in the morphospace never change. Each phenotype is surrounded by the same set of phenotypes regardless of the set of tasks that define an adaptive walk. In this respect, the morphospace can be envisioned as a crystal lattice in which phenotypic variations are embedded and firmly locked in place. The walks through this space move through the lattice on the basis of the tasks they seek and where they start.

At this juncture it is also good to recall that our purpose is to identify the sequence of morphological variants that results in progressively higher relative fitness as defined by performing one or more designated tasks. All of these "adaptive walks" must begin at the same location in the morphospace, which for the sake of convenience corresponds to the morphology of the most ancient vascular plants, such as *Cooksonia* or *Rhynia* (see fig. 8.12). These plants were predominantly isobifurcate with one or two levels of branching in which most or all terminal axes bore sporangia. Starting from this location, a search algorithm evaluates the relative fitness of all neighboring morphological variants using one, two, three, or four specified tasks as the criteria for evaluating fitness (for example, light interception, light interception and mechanical stability, or light interception, mechanical stability, and water conservation). If one or more neighboring variants have an equivalent or higher relative fitness, the walk proceeds to their locations in the morphospace. This "search," which is free to branch when more than one morphology is located, is reiterated until the relative fitness of the morphologies in the last iteration of a search is higher than that of all surrounding variants. By definition, these morphologies occupy adaptive peaks on a landscape. Finally, fitness landscapes can be "stable" or "unstable." That is, an adaptive walk can proceed through the morphospace wherein the criterion defining relative fitness remains constant until the search for the best morphologies comes to a close. This defines a stable fitness landscape and directional selection. Alternatively, the criterion used to define relative fitness can be changed arbitrarily at any time during a walk to mimic a change in selection. This defines an unstable fitness landscape, which mimics disruptive selection.

Adaptive Walks and Directional Selection

Computer simulations of the morphological transformations maximizing or optimizing functional tasks indicate that comparatively few hypothetical morphologies are capable of maximizing the performance of any one of the four single tasks. In contrast, the number of morphologies capable of optimizing the simultaneous performance of two or more tasks increases as the number of tasks increases. However, the overall (global) relative fitness of these multitasking morphological optima decreases as the number of tasks increases from one to four (table 8.2). Taken at face value, the results of these simulations indicate that adaptive evolution under directional selection is more rapid and easier when selection acts on the ability to perform multiple tasks rather than a single task because the global fitness of competing morphologies differs less among morphologies as the number of tasks increases.

Turning to morphology, the hypothetical morphologies capable

Table 8.2. Summary of the relationships among the number of tasks performed, the number of optimal morphologies with equivalent performance levels, and the relative fitness of these optima

Number of tasks (number of landscapes)	Number of Morphologies (mean ± SE)	Relative Fitness (mean ± SE)
One-task (*n* = 4)	2.50 ± 1.0	35.3 ± 1.80
Two-tasks (*n* = 6)	3.33 ± 1.6	11.6 ± 0.68
Three-tasks (*n* = 4)	6.25 ± 0.5	7.5 ± 0.14
Four-tasks (*n* = 1)	20	2.4

Note. The morphologies with optimal and equivalent performance levels are shown in figs. 8.14–8.17.

of maximizing the performance of any one of the four biological tasks used to quantify relative fitness are, for the most part, comparatively simple in appearance (fig. 8.14). The simplest of these are those that maximize the capacity to conserve water as gauged by minimizing total surface area. These variants are Y-shaped and similar in general appearance to the most ancient vascular plants, such as *Cooksonia*. Thus, water conservation may have been achieved early in the evolutionary history of vascular land plants by virtue of small (and short) morphologies. However, from a mathematical perspective, these Y-shaped phenotypes are trivial, because they are those used to initiate each adaptive walk, whether on a single- or a multiple-task landscape.

In contrast, more elaborate single-task morphologies are those that possess lateral branching systems confined partly or entirely to the horizontal plane and elevated on a single vertical main axis. Some of these morphologies are located in the isobifurcate subdomain, whereas others are found in the anisobifurcate subdomain. Because the latter are reached only by an extensive series of hypothetical morphological transformations, the adaptive walks reaching these morphologies are branched.

The morphologies capable of optimizing the performance of two or more tasks simultaneously are greater in number and arguably more morphologically diverse than those maximizing the performance of a single task (Figs. 8.15–8.17). However, as noted, as the number and

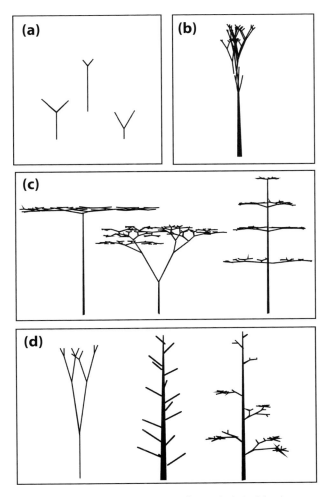

Figure 8.14. Morphologies reached by adaptive walks on single-task landscapes capable of maximizing water conservation (a), spore dispersal (b), light interception (c), and mechanical stability (d).

complexity of these phenotypes increase, relative fitness decreases (table 8.2). In this context it is important to recall that the "currency" by which fitness is measured differs among the different fitness landscapes (for example, the amount of light harvested is in units of energy [watt-hours], whereas reproductive success is gauged by spore number and dispersal range). However, comparisons of relative fitness

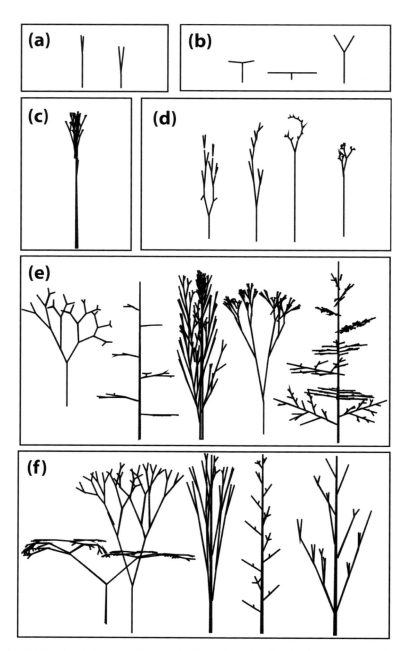

Figure 8.15. Morphologies reached by adaptive walks on two-task landscapes capable of optimizing mechanical stability and water conservation (a), light interception and water conservation (b), mechanical stability and spore dispersal (c), spore dispersal and water conservation (d), light interception and mechanical stability (e), and light interception and spore dispersal (f).

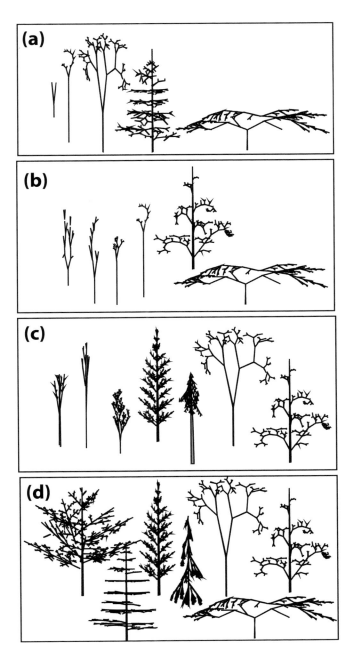

Figure 8.16. Morphologies reached by adaptive walks on three-task landscapes capable of optimizing light interception, mechanical stability, and water conservation (a), light interception, spore dispersal, and water conservation (b), mechanical stability, spore dispersal, and water conservation (c), and light interception, mechanical stability, and spore dispersal (d).

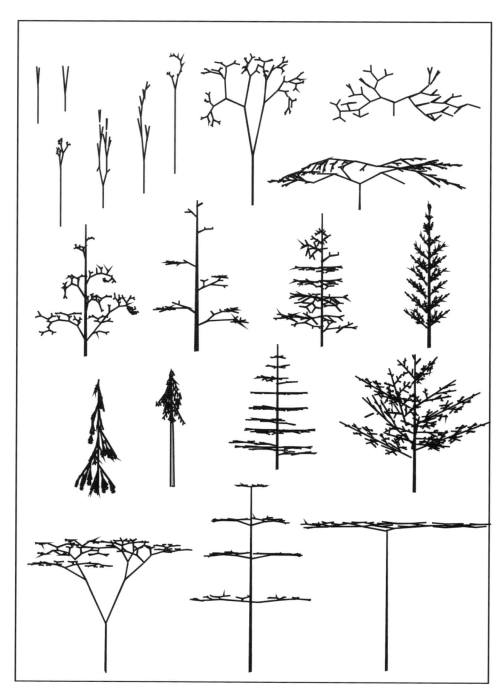

Figure 8.17. Morphologies reached by adaptive walks on the four-task landscape capable of optimizing mechanical stability, water conservation, light interception, and spore dispersal.

across the different landscapes are possible because the fitness of the ancestral *Cooksonia*-like morphology that initiates each adaptive walk on any landscape can be used to normalize the fitness of all other morphological variants within each landscape. Using this protocol, the average relative fitness of morphologies identified as optimizing two or four tasks is 11.6 and 2.4, respectively (table 8.2). In contrast, the average relative fitness of those morphologies capable of maximizing the performance of one task is 35.3. Therefore, as the number of tasks used to define relative fitness increases, the capacity to perform these tasks simultaneous decreases. The decrease in fitness that emerges as the number of simultaneously performed tasks increases explains in part why some of the phenotypes located in multitask-driven walks are the same as those identified by walks driven by fewer tasks. A phenotype that is optimal at performing a single task can have a higher overall, global fitness even when multitasking.

Adaptive Walks and Disruptional Selection

The preceding simulations mimic the hypothetical consequences of directional selection (wherein selection on the performance of one or more functional obligations is sustained throughout each adaptive walk). However, physical environments change over ecological and evolutionary time-scales, often dramatically. It is therefore naïve to assume that selection persistently acts on the performance of any one task or on the performance of any particular combination of tasks, especially over time scales relevant to evolutionary history (10^5 to 10^6 yr.). The consequences of shifting the focus of selection can be simulated by initiating an adaptive walk on one landscape and subsequently changing the landscape one or more times as an adaptive walk proceeds through the morphospace. In this way, the criteria used to quantify relative fitness change over time as gauged by the progress of a walk through the morphospace.

There are no a priori rules for how or when a particular landscape is changed. Therefore, the number of permutations of shifting landscapes is extremely large. Nevertheless, the fossil record and the preceding simulations provide some guidance. As noted, the oldest known vascular land plant fossils are *Cooksonia*-like in their general appearance. Computer simulations also indicate that these morphologies were capable of maximizing water conservation (see fig. 8.14a). Therefore, an adaptive walk on shifting landscape simulations can be initiated using a *Cooksonia*-like morphology as the locus from which other optima on increasingly more functionally demanding landscapes are located.

Two such simulations are shown in fig. 8.18. Each simulation begins on the same single-task landscape, enters two different two-task landscapes, after which both enter the same three-task landscape and come to an end in the four-task landscape. These two simulations are selected because they illustrate an important feature that resurfaces in all similar walks through shifting landscapes. Even though adaptive walks enter the same fitness landscape, they locate different morphological optima, depending on how relative fitness was defined in previous steps of a walk. For example, among the ten morphological optima reached by the two adaptive walks entering the three-task landscapes, only three are the same (fig. 8.18 c and c′), whereas none of the morphological optima reached by the two walks on the four-task landscape is the same (fig. 8.18 d and d′). This may serve as an example of the contingent nature of evolution as dependent upon the physical environment, and thus as an example of "replaying the tape of life" sensu Stephen J. Gould.

Taken at face value, these simulations indicate that the morphologies with the highest relative fitness at any stage in the adaptive evolution of a lineage depend in part on prior selection regimes and genomic competency to adapt, since these two factors define the range of phenotypes that survive and are available for the next round of selection (morphological optimization is historically contingent). Consequently, we should not expect the mechanisms of adaptive evolution to achieve the best conceivable morphologies, but only those that are

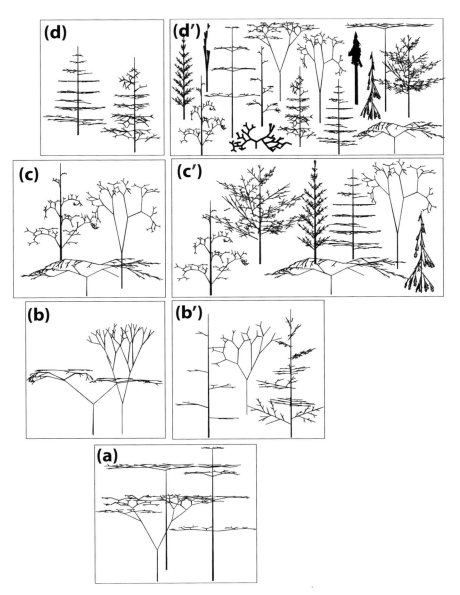

Figure 8.18. Two walks on different shifting landscapes (a–d and a–d') to simulate the consequences of disruptive selection. Both walks begin on a landscape defined by light interception (a). One walk proceeds into a two-task landscape defined by mechanical stability and reproductive success (b). The other walk proceeds into a two-task landscape defined by mechanical stability and light interception (b'). Both walks then enter the same three-task landscape defined by mechanical stability, light interception, reproductive success (c and c'), and conclude in the four-task landscape (d and d'). These simulations indicate that the morphological optima reached by adaptive walks are contingent upon previously achieved morphologies and are thus historically contingent upon previous adaptations.

the best relative to what is available on the basis of past physical and biological events.

Opaque Models and the Telome Theory

Do the preceding simulations shed light on adaptive evolution? The answer to this question is No, at least not directly, because, in the parlance of computer science, these simulations emerge from what is called an opaque mathematical model—that is, these simulations reveal the logical consequences of the assumptions upon which the mathematical model is based. This does not preclude the model from making biologically reasonable predictions, but it does caution against accepting the model's predictions without challenging them with empirical evidence. For example, do the predictions of these simulations agree with trends observed in the fossil record?

In one respect, they do, because some of the morphological transformations identified in some of the walks agree with the predictions of one of the most pervasive and far-reaching theories attempting to explain early land plant evolution, namely, the Telome Theory. First proposed by Walter Zimmerman (1892–1980) and later expounded with vigor by others such as Wilson N. Stewart (1917–2004), the theory postulates that the sporophytes of the most ancient vascular plants consisted of simple paired terminal axes (called telomes) subtended by unbranched axes (called mesomes). These telomes and mesomes are postulated to have been subsequently modified by one or more developmental processes called planation, overtopping, reduction, recurvature, and webbing or lamination (fig. 8.19) to obtain all known extant or extinct plant morphologies. Each of these processes is easily envisioned, although the developmental mechanisms by which they are achieved are not explained. Planation occurs when neighboring axes become oriented in a single plane with respect to the horizontal. Overtopping results from the differential growth in the length of interconnected axes such that some portions of a morphol-

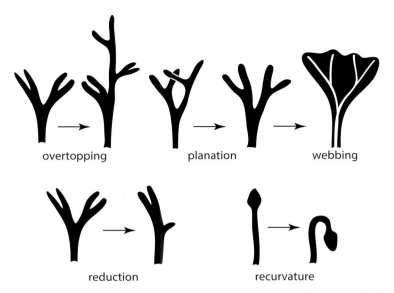

overtopping   planation   webbing

reduction   recurvature

Figure 8.19. The five processes proposed by W. Zimmerman's telome theory to explain how early vascular plants evolved more complex morphologies from simple equally branched structures (upper left). For example, with the exception of the leaves of lycopods, the leaves of all vascular land plants are postulated to have evolved from overtopping, planation, and webbing. From upper left to lower right: Overtopping is the process that gives rise to the unequal growth of some axes that establishes what appears to be a central axis bearing lateral axes. Planation is the process that confines bifurcating axes to one plane of reference. Webbing is the process that intercalates tissues between planated axes. Reduction is the process that gives rise to structures smaller than their developmental precursors. Recurvature is the process that reorients a structure with respect to some frame of reference (in this illustration, an axis bearing a terminal sporangium recurved with respect to the vertical).

ogy are longer than others. Reduction in the size of some telomes can aggregate or positionally subordinate some axes with respect to others. Recurvature involves the differential expansion of one side of an axis with respect to the opposing side. Finally, webbing or lamination is the developmental introduction of tissues between adjoining axes.

Zimmerman proposed that all of the morphological transformations attending the evolution of land plants can be explained by the operation of one or more of these five processes. For example, fern and seed plant leaves are postulated to have evolved by overtopping (to yield sporophytes with a main vertical axis bearing lateral branching systems) and by the planation and webbing of lateral branches (to eventually give rise to leaves). Likewise, the intricate components of the reproductive organs of modern-day horsetails, called sporangiophores, are postulated to have evolved by means of reduction, recurvature, and some degree of fusion or webbing.

Despite its pedagogical (heuristic) utility, the Telome Theory has been criticized for a variety of reasons. One obvious problem is its vagueness regarding the developmental mechanisms responsible for overtopping, planation, and the other processes. Indeed, these terms are purely descriptive and not mechanistic in nature. Another criticism is that the theory never explains why certain morphological transformations occur as opposed to others, nor does it stipulate in many cases the sequence of processes foreshadowing the appearance of a particular morphology. Why should planated and webbed lateral branch systems evolve? Are the leaves of ferns or seed plants functionally adaptive in terms of light interception or some other biological requirement? Did these leaves evolve as the result of the simultaneous operation of reduction, overtopping, planation, and webbing, or did planation and webbing occur after reduction and overtopping? Questions such as these can be answered retrospectively by examining the fossil record, but the Telome Theory sheds little direct light on what morphological trends are expected or what they indicate.

Zimmerman's ideas are nevertheless useful because they provide a lexicon of terms for the morphological transformations observed in the fossil record and for those identified by the computer simulations presented here. For example, adaptive walks on single-task landscapes identify overtopped morphologies with planated lateral branching systems that can maximize spore dispersal, light interception, or mechanical stability (see fig. 8.14). Similar but slightly more elaborate morphologies are reached by walks on multiple task landscapes. This convergence indirectly supports the supposition that overtopping and planation, which have occurred independently in a number of plant lineages, may be functionally adaptive.

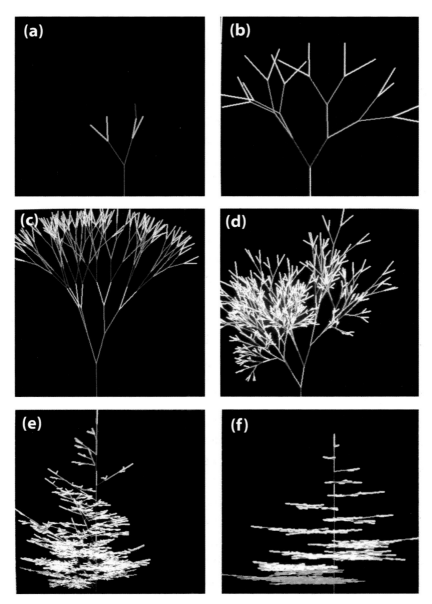

Figure 8.20. Photographs of successive morphologies (a–f) reached in a simulated adaptive walk on a two-task landscape defined by simultaneously maximizing light interception and mechanical stability. The walk begins with a *Cooksonia*-like morphology (a), explores the isobifurcate subdomain of the morphospace where it locates a *Rhynia*-like morphology (b–c), passes into the anisobifurcate subdomain (d), and locates trimerophyte-like morphologies (e) and terminates with a morphology with planated lateral branching systems (f).

This supposition is reinforced when we examine the sequences of morphologies identified by different adaptive walks (fig. 8.20). For example, starting with a simple *Cooksonia*-like phenotype, the adaptive walk on the fitness landscape defined by light interception and mechanical stability identifies a larger, more branched variant. As this particular walk progresses, it identifies still larger and more overtopped phenotypes bearing lateral branching systems that become progressively more planated. Although the webbing or lamination of adjoining axes was not simulated, more sophisticated computer algorithms show that the capacity to intercept sunlight is dramatically improved if some or all of the terminal axes (telome analogues) are interconnected by photosynthetic tissues. Thus, the sequence of morphologies shown in fig. 8.21 e reflects an adaptive sequence of phenotypes.

It is important to bear in mind, particularly after our consideration of development in earlier chapters, that the Telome Theory's conception of developmental processes is naïve and in many cases unacceptable. For example, webbing postulates that each separate vein in a laminated leaf-like structure once had its own apical meristem that somehow became aggregated within a truss of other axes with their own separate apical meristems and that all of these separate identities somehow got reorganized into something even more complicated such as a multiveined leaf. This postulate endorses the notion that separate elements have come together into a single structure while somehow maintaining distinct identities as they continue to get moved around in ways that can be traced through the fossil record like the migrating bones of the vertebrate skull. This conceptualization flouts in almost every respect what has been learned from detailed examinations of how the leaves of lycopods, horsetails, ferns, gymnosperms, and angiosperms develop. The Telome Theory has no basis in fact, and the preceding discussion regarding how its predictions coincide with computer simulations is not intended to endorse the theory but rather to show how a series of morphological transformations may have conferred an adaptive edge.

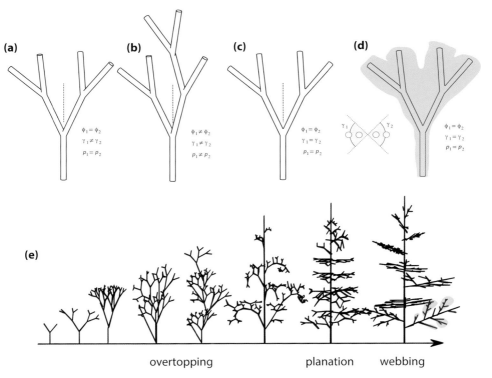

(a)

$\phi_1 = \phi_2$
$\gamma_1 \neq \gamma_2$
$p_1 = p_2$

(b)

$\phi_1 \neq \phi_2$
$\gamma_1 \neq \gamma_2$
$p_1 \neq p_2$

(c)

$\phi_1 = \phi_2$
$\gamma_1 = \gamma_2$
$p_1 = p_2$

γ_1 γ_2

(d)

$\phi_1 = \phi_2$
$\gamma_1 = \gamma_2$
$p_1 = p_2$

(e)

overtopping   planation   webbing

Figure 8.21. Computer simulations that mimic leaf-like structures during an adaptive walk optimizing light interception and mechanical stability. (a) The simulated walk proceeds from the isobifurcate subdomain of the morphospace (defined by equal branching angles, unequal rotation angles, and equal branching probabilities; $\phi_1 = \phi_2$, $\gamma_1 \neq \gamma_2$, and $p_1 = p_2$) into (b) the anisobifurcate subdomain (defined by unequal branching and rotation angles, and unequal branching probabilities; $\phi_1 \neq \phi_2$, $\gamma_1 \neq \gamma_2$, and $p_1 \neq p_2$) and concludes (c) with morphologies bearing lateral branching systems in which all rotation angles are equal ($\gamma_1 = \gamma_2$) that (d) permit the intercalation of photosynthetic tissues to construct a leaf-like structure (shown in green). (e) The sequence of morphologies located by the adaptive walk for maximizing light interception and mechanical stability. The sequence shows the morphological transformation series that increases global relative fitness (from left to right) and mimics overtopping, planation, and webbing (see fig. 8.20). Note that these simulations did not produce webbed structures (shown in green). They only reached the stage at which webbing becomes physically possible.

Enations, Lycopod Leaves, and Some Possible Homologies

The Telome Theory does not explain the evolutionary origins of the leaves produced by lycopods. The fossil record indicates that these leaves likely evolved from what were enations (small scale-like protuberances) on stem-like axes. The development and anatomy of modern-day lycopod leaves also differ from those of the leaves of all other land plant lineages. These and other aspects revealed by the fossil record and the study of extant plants prompted Frederick O. Bower (1855–1948) to propose the Enation Theory for the origin of lycopod leaves. In this context, it is worth further noting that the development of the most ancient kinds of sporangia (called eusporangia) and the development of all leaves bear certain ontogenetic resemblances in that they all involve meristematic initials derived from one or more epidermal cells and one or more hypodermal cells (see, for example, fig. 4.18). These observations prompt the speculation that similar or identical gene regulatory networks were co-opted during the evolution of leaves and sporangia.

The evolutionary origins of roots are far more ambiguous because it is often difficult to determine whether roots are absent as in some extant pteridophytes, are not preserved in the fossil record, or perhaps have not been recognized as such among some fossils. Another factor contributing to the problematic nature of root evolution is the paucity of definitive developmental features that can be used to define accurately and precisely what is meant by "roots." This ambiguity is attributable in part to the fact that roots have evolved many times and therefore in different ways among tracheophytes. For example, root development is typically described as endogenous—that is, the apical meristems of lateral roots develop from cells within a layer of cells called the pericycle, which is near the centrally located vascular tissues. However, the roots of some ferns (such as *Ceratopteris*) emerge from cells in the outermost layer of the cortex just below the epidermis (whose cells divide anticlinally to keep pace with divisions in the hypodermis) in a manner that is essentially indistinguishable

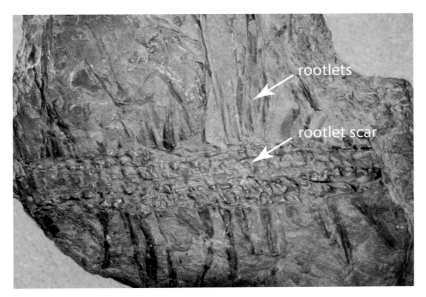

Figure 8.22. Compression fossil of the rhizomatous basal portion of a lepidodendrid tree (an organ genus called *Stigmaria*) showing laterally attached, helically arranged rootlets and scars where rootlets were once attached. On the basis of their arrangement and anatomy (not shown), *Stigmarian* rootlets are interpreted to be highly modified leaves that functioned as roots.

from the ontogeny of eudicot leaves, wherein the initial bulge of the leaf primordium forms as a result of subepidermal and epidermal cell divisions. It is also worth noting that the subepidermal origin of the roots of *Ceratopteris* and some other leptosporangiate ferns is virtually indistinguishable from that of stigmarian roots, which are considered by many workers to be leaf homologs (fig. 8.22).

The Early Paleozoic Explosion

The computer simulations reviewed in this chapter may shed light on yet another feature of land plant evolution—the explosive morphological, anatomical, and reproductive radiation that occurred during the early Paleozoic between 444 Mya and 359 Mya (fig. 8.23). This period of plant evolution, which amounts to roughly 19% of the entire history of vascular land plants, rivals the great Cambrian Explosion

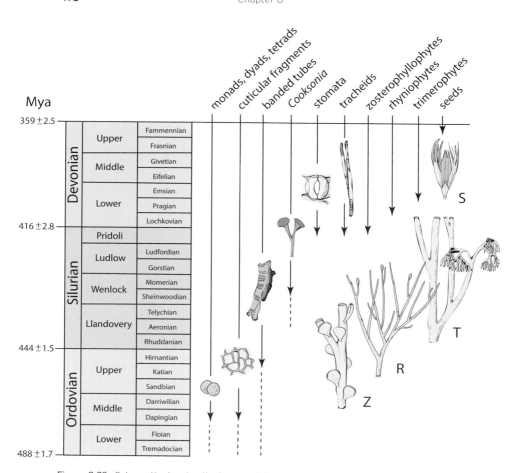

Figure 8.23. Schematic showing the temporal distributions of a few of the key anatomical, morphological, and reproductive innovations that evolved during the Early Paleozoic (from the Lower Ordovician to the Upper Devonian). Solid lines denote unambiguous ranges of occurrence; dashed lines indicate earlier but equivocal fossil remains. The most ancient of these features are spore monads, dyads, and tetrads that extend into the Cambrian and that are believed to be evidence of terrestrial (perhaps nonvascular) plants. Cuticular fragments and cells referred to as banded tubes are also attributed to land plant remains by some authors. Unequivocal land plant remains are reported from the Silurian in the form of fossils referred to as *Cooksonia*. Shortly thereafter, the three great ancestral vascular plant lineages make their appearance, the zosterophyllophytes (Z), rhyniophytes (R), and the trimerophytes (T). The zosterophyllophytes are postulated to have evolved into lycopods. The trimerophytes are postulated to have evolved from the rhyniophytes and to have given rise to all vascular plant lineages with the exception of the lycopods. Seed plants (S) are reported from the Upper Devonian. See chapter 2 for additional information about early land plant evolution.

(between 542 Mya and 517 Mya) during which most of the modern metazoan phyla made their appearance. With the exception of the flowering plants, every major vascular plant lineage made its appearance including the seed plants. Consider, once again, the humble beginnings of the vascular plants, specifically fossils referred to as *Cooksonia*, which consisted of leafless and rootless bifurcating axes bearing terminal sporangia. The remains of these plants date back to Wenlockian strata (between 433 Mya and 427 Mya). Yet, shortly thereafter, by the Lower Devonian, fossils assigned to the zosterophyllophytes, rhyniophytes, and trimerophytes appear and subsequently diversify into lycopods, horsetails, ferns, and gymnosperms by the Upper Devonian (fig. 8.23).

What might account for this remarkable episode during the evolution of the land plants? Recall that the simulations presented in this chapter show that the number of equivalent optimal morphologies is small in single-task landscapes, but that this number increases significantly as the number of functional tasks to be performed simultaneously increases. At the same time, the simulations show that the global relative fitness of optimal morphologies decreases as the number of tasks performed simultaneously increases (see table 8.2). Thus, as the number of tasks increases, the fitness landscape changes from one with a few but very tall peaks to a landscape with many low-lying hills (fig. 8.24). The single most daunting task the first land plants had to perform was to resist desiccation (to survive and reproduce). Modern-day nonvascular plants, such as the mosses, manage to survive by tolerating desiccation. However, vascular plants do not really cope with desiccation. They resist dehydration by virtue of possessing a cuticle and stomata. With this information in mind, it is not unreasonable to speculate that the first land plants evolved on a single-task landscape that had a few *Cooksonia*-like optimal morphologies for dealing with water conservation (see fig. 8.14 a). However, once plants evolved the capacity to produce a cuticle and stomata (features that have been documented for some fossils assigned to the genus *Cooksonia*), the adaptive landscape would have required multifunc-

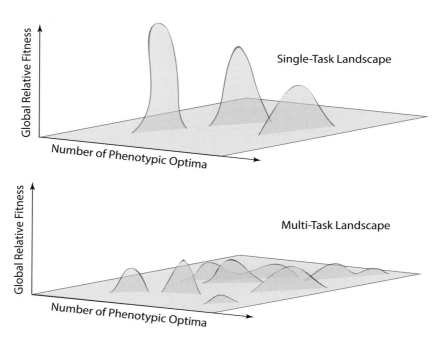

Figure 8.24. Schematic representation of the relationships among the number of tasks a hypothetical morphology must perform simultaneously, the number of functionally equivalent morphological optima, and the global fitness of these optima based on the summary statistics of simulated adaptive walks (see table 8.2). As the number of tasks to be performed simultaneously increases, the number of equivalent optimal phenotypes increases, but the global fitness of these optima decreases. As a consequence, the fitness landscape changes (from having a rough terrain to being a gently rolling hillside) in a manner that permits morphological transformations to explore a larger domain of phenotypes.

tionality and the number of theoretically functionally optimal phenotypes would have increased significantly (see fig. 8.17). This landscape would have had few obstructions to phenotypic transformations because the global fitness of competing phenotypes would have differed little. Extinction would have been a common occurrence, but so too would have been innovations resulting in a rapid diversification of anatomy, morphology, and reproduction.

This is all speculation. It is consistent with the fossil record, but that is an insufficient reason to accept the speculation as being valid. A model can give the right answer for the wrong reasons, and opaque mathematical models are particularly prone to this kind of illusion.

Suggested Readings

McGhee, G. R., Jr. 1999. *Theoretical morphology: The concept and its applications.* New York: Columbia University Press.

Niklas, K. J. 1992. *Plant biomechanics: An engineering approach to plant form and function.* Chicago: University of Chicago Press.

Nobel, P. S. 2005. *Physicochemical and environmental plant physiology.* 3d ed. Amsterdam: Elsevier.

Vogel, S. 1981. *Life in moving fluids: The physical biology of flow.* Boston: Willard Grant.

Wainwright, S. A., W. D. Biggs, J. D. Currey, and J. M. Gosline. 1976. *Mechanical design in organisms.* New York: Wiley.

Wayne, R. 2009. *Plant cell biology: From astronomy to zoology.* Amsterdam: Elsevier.

9

Ecology and Evolution

> The complexity of communities has fascinated naturalists from before
> Darwin, who described it classically.
> —G. EVELYN HUTCHINSON, *The Influence of the Environment* (1965)

Evolutionary ecology is the study of how the interactions among species have changed over the course of Earth's long history. It is clear that no organism or species exists or evolves in isolation. Each organism lives and dies in a population of similar organisms that in turn resides in a community composed of different species that collectively constitute a flora and fauna inhabiting a range of habitats. Further, when organisms adapt to their environment and evolve, they may give rise to new species, or go extinct. In either case, they change their environment, and when the environment changes, organisms must adapt or die. The synergism between organisms and their biotic and abiotic environment was alluded to when discussing the evolution of colonial organisms. In this context, we saw that new network dynamics emerge as cells interact with one other and with the environment they themselves create by virtue of these interactions (see box 7.1). The emergence of new behaviors from the interactions among organisms and their environment is called niche construction, a process whereby

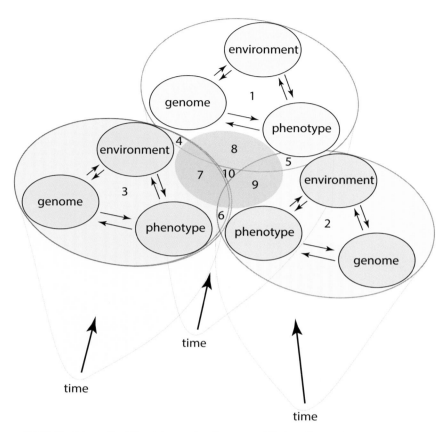

Figure 9.1. Schematic of the niche construction concept that posits organisms interact with their environments and their environments interact with them (shown by small arrows) in ways that create new and progressively more complex interactions (indicated numerically by 1-10). Two additional and important elements are genome-phenotype and genome-environment interactions (epigenetic effects) and the element of time (evolutionary changes in the genomes of interacting organisms).

community dynamics become more complex as new niches emerge as organisms evolve or change their behavior (fig. 9.1). As we have seen, plants change levels of atmospheric gases and modify hydrological and nutrient global fluxes (see fig. 0.2). Likewise, animals build nests and burrow through the soil, while fungi and bacteria decompose organic matter. Indeed, consider the examples of niche construction in terms of the reciprocal interactions among bacteria, fungi, and land plants.

The bacterium *Bacillus thuringiensis* produces toxins that can kill herbivorous insects, the soil fungus *Trichoderma* can kill or outcompete pathogenic fungi and activate a plant's immune system, mycorrhiza such as *Glomus deserticola* extend the capacity of root systems to absorb water and essential elements, the symbiotic interactions between *Curvularia* fungi and temperature-tolerant viruses living in roots can increase a plant's heat tolerance by as much as 20°C, rhizobia living in the roots of legumes manufacture metabolically useful nitrogen, while the cold-tolerant bacterium *Pseudomonas* improves plant growth by fixing atmospheric nitrogen at temperatures as low as 4°C, and *Methylobacterium oryzae* can take up heavy metals thus allowing plants to survive while growing in contaminated soils. Perhaps more telling, research shows that the microbial communities (called microbiomes) associated with plant roots are regulated by plant hormones (such as salicylic acid, jasmonic acid, and ethylene) in ways that foster plant innate immunity. In each case, the interactions among organisms (either unrelated or related genetically and phylogenetically) create new dynamics and establish complex and beneficial community interactions.

Studying the dynamics of niche construction at the level of populations, communities, or an entire habitat is the study of ecology. Studying how these dynamics have changed over the course of Earth's 4.5 billion year history is the study of evolutionary ecology. If there is any real difference between these two disciplines, it is their temporal scale of reference—ecologists study evolution over the course of decades or hundreds of years, while evolutionary biologists study evolution over the course of hundreds of thousands or millions of years. The difference in the temporal scales of reference is important. What appears to be a trend over short spans of time (decades or hundreds of years) may be an insignificant blip when viewed over longer spans of time (thousands or hundreds of thousands of years). Spatial scales are equally important. What appears to be random when viewed over a few hundred square meters may emerge as a pattern or trend when observed over hundreds of square kilometers. A good example is found in the relationship between the number of species found in a patch of habitat N and the area of the patch that is canvassed A. This species-area relationship often conforms to the power-law function $S = \beta A^{\alpha}$, where α and β are numerical constants but only over specified spatial scales and only for specific taxonomic groups of plants and animals. The use of species-area equations has played an important role in examining the consequences of habitat loss, particularly when these equations are used to examine how the number of endemic species changes across different spatial scales.

My goal in this chapter is to review some of the major concepts that apply as much to ecology as to evolutionary biology and to discuss some of the evolutionary implications of the major ecological transformations recorded in the fossil record. Another goal is to wrap up some loose ends by dealing with topics that did not fit comfortably within our previous discussions as well as to explore some topics that required previous discussions to delve into them critically. The tactic is to begin with some useful models for population dynamics, models that consider *intra*specific competition, to move on from there to consider *inter*specific competition, and to increase the scope of the spatial and temporal scales of reference in ways that eventually consider some of the major evolutionary events that changed how species interact with one another.

As in earlier chapters, we will rely from time to time on some mathematical models, and it is good once again to reconsider their uses and limitations before proceeding further. Recall that any mathematical model is an abstraction. Each model attempts to achieve a generalization that helps us understand the world at large. In the process of striving to reach this goal, some elements of reality are ignored or simplified, and one or more assumptions are made. This feature of modeling can make some people uncomfortable, even distrustful of the process of modeling. It is certainly true that a model may give the correct answer for the wrong reasons. However, a scientific model can be empirically tested, and those that are wrong will eventually be identified and abandoned. Also, without models, we would be merely collecting facts and observations, but, without understanding their structure and interconnections, these facts would have no greater meaning. Mathematical models can reveal the structure of data. They are also necessary for the interpretation of data.

Mathematical models are useful in at least two other ways—their logic is transparent and they are heuristically helpful. Verbal arguments and constructs can be opaque and sometimes difficult to follow precisely and accurately. A seemingly good argument may be logically flawed. In contrast, mistakes in mathematics are easy to identify upon reflection, which makes them comparatively easy to correct or abandon entirely. Models are also pedagogically helpful. They provide a "what if" approach to understanding reality. What would a population do if we assume that resources are unlimited? What would happen if we set an upper limit to resource availability? These and many other "what if" questions can be explored using models, and the answers that emerge from even the simplest model can instruct us about our perception of reality. (Notice that I did not say *instruct us about reality*.) Finally, it must be noted that every model contains the seeds of its own destruction. Its ability to make precise predictions makes a model falsifiable. In this way, a model evolves and, if there is sufficient selection pressure, it becomes extinct. This is how science works.

Analogy and Scale

Much of what will be discussed in this chapter lends itself to making comparisons between concepts drawn from different disciplines. Therefore, it is important to bear in mind the merits and the dangers of drawing analogies. Consider the processes that define how different species interact ecologically. Many of these processes are conceptually analogous to the microevolutionary processes that shape allelic diversity reviewed in chapter 3. For example, the arrival of new species into a new habitat by means of speciation or by means of mass migration is conceptually analogous to mutation; the dispersal of individuals among patches within a metapopulation (a regional aggregate of the individual populations of a particular species) is analogous to gene

Table 9.1. Examples of the heritable intrapopulational variation of ecologically important functional traits

Functional trait	Heritability (h^2)	Organism
Fruits per inflorescence	0.56	*Plantago lanceolata*
Inflorescences per plant	0.49	*P. lanceolata*
Yield	0.44	*P. lanceolata*
Seed weight	0.00	*P. lanceolata*
Days to flowering	1.89	*Lolium multiflorum*
	0.55	*Avena fatua*
	0.38	*Festuca microstachys*
Height at maturity	0.67	*L. multiflorum*
	0.48	*A. fatua*
	0.37	*F. microstachys*
Emergence	0.54	*Zea mays*[a]
Seedling weight	0.45	*Z. mays*
Relative growth rate	0.05	*Z. mays*
Dormancy	0.50	*A. fatua*
Seed dormancy	0.13	*Sinapis arvensis*
Germination time	0.00	*Papaver dubium*

[a] Interpopulational distribution of variation.

flow; and the ecological divergence of populations belonging to the same species is conceptually similar to genetic drift or adaptation under selection. The fact that some ecologically important functional traits manifest high levels of heritability also fuels the impulse to draw analogies between microevolutionary processes and ecological processes (table 9.1). As in most cases of analogy, these conceptual similarities cannot be stretched too far. Nevertheless, the analogies between population genetics and ecology are useful at times to understand how different species coexist, or how the population of one species can influence the evolutionary fate of the population of another species. Just as microevolutionary theory treats evolution as a change in allelic diversity as a consequence of mutation, drift, and selection, so too ecological theory can view species diversity as a consequence of demographics governed by immigration, dispersal, and ecological divergence, each of which can be treated statistically.

There are analogies also between ecology and macroevolution as for example the ecological process called succession. If we define succession as *a repeatable sequence of changes in community structure or composition under particular environmental circumstances*, we might be tempted to see similarities with the replacements of species with similar growth habits in the paleofloras once dominated by pteridophytes, gymnosperms, and angiosperms during the Phanerozoic (see fig. 6.8). In this *succession*, forests dominated by progymnosperms at the end of the Devonian were replaced by forests dominated by lycopods and horsetails in the Carboniferous that were successively replaced by forests dominated by gymnosperm trees and more recently by those dominated by angiosperms. Once again, analogies can be treacherous—notice that the use of the word *replaced* does not imply that one type of forest *directly outcompeted* the other. It simply refers to how the taxonomic composition of a forested community changed over evolutionary time.

In the context of ecological succession, it is useful to consider the debate between Frederic Clements (1874–1945) and Henry Gleason (1882–1975). Clements made an explicit analogy between ecological succession and the ontogeny of an individual organism. His perspective, which has been called the pseudo-organismic theory, viewed communities as superorganisms that are born, mature, and eventually die over the course of the deterministic process called succession. Gleason offered a different perspective. He conceived of succession as being far less deterministic and argued that the different distributions of species across neighboring habitats or communities reflected how different species responded individualistically to environmental conditions across gradients. The debate stemming from these two perspectives waged for decades, but it was ultimately resolved in favor of Gleason's point of view, largely owing to the work of Robert H. Whittaker (1920–1980) and John T. Curtis (1913–1961), who provided evidence that the distributions of different species are individualistic and dependent on local environmental conditions.

Another seductive analogy has been proposed by the physicist Per

Bak (1996). Bak drew an analogy between the long-term dynamics of ecosystems and a pile of sand to which more sand is continually added from above. Initially, the pile of sand grows in height as more and more sand is added. However, as the pile gets taller and taller, grains of sand start moving downward and avalanches precipitously occur. The pile never forms another pile (it is always the same ecosystem). It just reaches a maximum critical height (a temporarily stable condition), and then collapses reiteratively (the ecosystem achieves metastable dynamic equilibria). Bak called this phenomenon *self-organized criticality*, and proposed that ecosystems achieve dynamic equilibria when extinctions periodically trigger other extinctions as new species either evolve within them or migrate into them. The result is a constant reorganization of the ecosystem. (The analogy between a pile of sand and the phenomenon called punctuated equilibrium [wherein long periods of evolutionary stasis are punctuated by rapid changes occurring in the fossil record] is also obvious.)

Another facet of Bak's analogy is that the effects of small perturbations (a few gains of sand sliding down a pile of sand) cascade throughout the pile and have large effects (an avalanche). This feature highlights that different spatial scales can be correlated in ways that result in a fractal-like behavior as a result of high interconnectedness. A few grains of sand slide down the hill (a few populations go to local extinction), an aggregate of sand slides downhill (a metacommunity goes to extinction), and the whole pile of sand collapses (the ecosystem gets restructured). Bear in mind that different ecosystems have different degrees of interconnectedness. For example, systems characterized by intense interspecific competition may have less interconnectedness because species are better off being separated spatially. Other systems may have a high degree of mutualistic interactions and thus manifest a high degree of interconnectedness.

Indeed, much of our understanding of ecology depends on our spatial as well as temporal scale of reference. At the largest spatial scale (such as a vast tundra or entire rainforest ecosystem), our predictions about ecological processes may be robust because the noise resulting from localized biotic and abiotic variation and the role of chance are reduced as our scope of reference increases. At smaller scales of reference (such as within a tundra or a rainforest), meso-patterns can be identified that nevertheless manifest variations that make predictions less certain. At the smallest scale, as for example the unit-space occupied by a collection of individuals, predictions based on extrapolation from meso-scale observations may be impossible because chance and uncertainty play very significant roles. This trend reflects statistical certainty regarding large versus small numbers, a feature that mathematicians often refer to the *Law of Large Numbers*. The predictability of a system increases as the number and scope of observations increases. As the scale of the system decreases, predictability declines because random events play a bigger role. An alternative interpretation is that our ability to predict ecological phenomena increases as our spatial scale increases because regularities emerge owing to homeostatic processes that increase in proportion to higher levels of ecological organization. Either way, just as an organism's functional traits are size-dependent, much of our interpretation of ecological phenomenology depends on the spatial as well as the temporal scales used to observe and analyze it (as witnessed by species-area power functions).

Population Dynamics

We reviewed the genetic structure of populations in chapter 3 and saw that population geneticists frequently view evolution mainly in terms of changes in allele frequencies over time. Another useful way of analyzing populations is to consider the dynamics of the growth in the number of individuals within a population because this gives insight into intra-populational competition for limited resources and the extent to which selection acts on variation. This perspective is particularly important when considering sedentary organisms such as the land plants and certain species of algae because these organisms cannot flee from predators and because they are forced to compete for resources with their neighbors that can belong to the same species.

We begin by considering the simplest and therefore the most mathematically elegant model for population growth, the exponential growth model, which is particularly useful when dealing with rapidly growing populations of small organisms such as bacteria or unicellular algae sustained under controlled laboratory conditions. This model stipulates that each individual in a population produces progeny at a maximum rate R and then calculates the number of individuals at time t (denoted as N_t). Starting with an initial population N_o, which for asexually and sexually reproducing species respectively equals 1 and 2, the exponential growth model predicts that the number of individuals in a population will increase exponentially as

$$N_{t=1} = RN_o$$

$$N_{t=2} = RN_1 = R^2N_o$$

$$N_t = R^tN_o$$

$$N_{t+1} = R^{(t+1)}N_o = RN_t,$$

such that the change in the number of individuals over time ($dN/dt = \Delta N_t$) is

$$\Delta N_t = N_{(t+1)} - N_t = (R - 1)N_t = rN_t,$$

where r is the maximum intrinsic growth rate of the population.

Plotting the $\log_{10}$ of the number of individuals predicted by this model against time reveals that even very small growth rates produce astonishingly large numbers of individuals in comparatively short spans of time (fig. 9.2). Intuitively, however, we see that the predictions of this elegant model are limited realistically because the resources that the individuals need within the same population would quickly be exhausted and because the model does not take mortality into consideration. Even if all of the required resources were self-

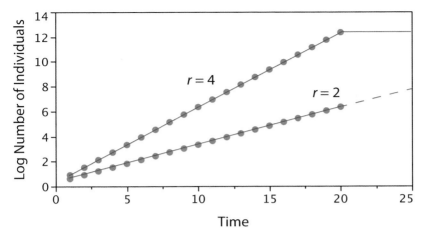

Figure 9.2. Predicted population growth based on the exponential growth model for two populations differing in their intrinsic growth rates (r). Each population began with two individuals. The data points (closed circles) are the log-transformed data for the numbers of individuals in each of the two populations. The extended solid horizontal line represents the trajectory that would occur if we considered the carrying capacity of the environment. The dashed vertical line represents the expected trajectory that would extend to the level of the carrying capacity.

renewing (as in the cultivation of bacteria or unicellular algae within a bioreactor), some limit to population size must exist. This upper boundary condition is called the carrying capacity. It defines the maximum number of individuals that can coexist at any one time in any one location. The carrying capacity is conventionally designated as K.

The classic model that integrates the concept of an environment's carrying capacity with the previous model for population growth is called the Verhulst logistic model, named in honor of its descriptor Pierre F. Verhulst (1804–1849). Using the same notation as before, the Verhulst model can be described mathematically as

$$N_t = KN_0 e^{rt}/[K + N_0(e^{rt} - 1)],$$

where e is the numerical constant approximately equal to 2.71828, which is often called the Euler constant in honor of the Swiss mathematician Leonhard Euler (1707–1783). The number e deals with probability density distributions and is used to perform integral calculus

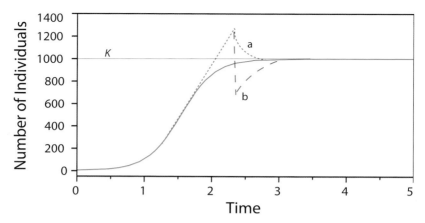

Figure 9.3. Predicted population growth based on the sigmoidal growth model, which incorporates the concept of the carrying capacity (*K*). The solid red line represents the number of individuals in the population, which plateaus at *K* = 1,000 individuals. The dashed red line depicts what would happen if the population growth exceeds the carrying capacity and returns slowly to the carrying capacity (the "from above" return to equilibrium). The dashed green line depicts what would happen if the population exceeds the carrying capacity, undergoes a sudden catastrophic decrease in the number of individuals, and slowly returns to the carrying capacity (the "from below" return to equilibrium). Both of these trajectories would involve significant mortality and high levels of selection.

with exponential functions and logarithms. What is relevant here is that the Verhulst logistic equation can be manipulated to explicitly give us the change in the number of individuals per unit of time:

$$\Delta N_t = rN(K - N)/K = rN(1 - N/K).$$

Graphing the predicted numbers of individuals against time shows that N_t increases exponentially when t is very small, but that N_t increases logistically when t gets progressively larger. The result is that the number of individuals increases slowly at first, accelerates, and subsequently plateaus asymptotically to the numerical value of K. This trend is the classic outcome of a logistic equation that yields a sigmoidal S-shaped plot (fig. 9.3).

Four assumptions are embedded in the Verhulst logistic equation: (1) all individuals are assumed to be identical such that the addition of individuals to the population reduces the rate of increase by the same

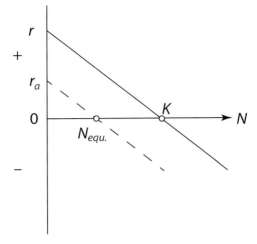

Figure 9.4. Schematic showing the relationships among a population's maximum growth rate (r), actual growth rate, equilibrium population size (N_{equ}), carrying capacity (K), and growth in number (N). The solid lines depicts the situation for the maximum difference between birth and death rates; the dashed line depicts the situation when the death rate is higher than the birth rate.

proportion, (2) r and K are constants, (3) there is no time lag in growth and changes in N, and (4) r is the *maximum* growth rate. The last of these four assumptions is somewhat trivial, because we can solve the logistic equation for the actual rate of growth (r_a), which is variable and a function of r, N, and K:

$$r_a = r(1 - N/K).$$

This equation shows that the actual growth rate decreases linearly with increasing population size and that it is always less than or equal to the maximum growth rate (fig. 9.4). This becomes obvious when we consider that the growth rate of any population depends on the birth and the death rates of individuals and that the *instantaneous rate of growth* is the difference between the instantaneous birth rate per individual b and the instantaneous death rate per individual d. In other words, $r_a = b - d$. The maximum rate of growth r_{max} occurs when $d = 0$, which occurs when the population density is minimal (in other words, when there exists a competitive vacuum). The population reaches an equilibrium ($\Delta N = 0$) when $b = d$, and the rate of growth at any density r_N equals $r_N = b_N - d_N$.

Most models for the growth of the total human population currently predict that we are approaching the condition where $b_N = d_N$. This does not mean that local populations will not change, either by increasing or decreasing in size. It simply means that we as a species appear to have saturated our environment. However, great care has to be exercised in drawing conclusions from this prediction because various factors affect the growth of populations in density-dependent and density-independent ways. Pathogens, for example, often affect populations in a density-dependent manner, whereas climatic factors influence populations typically in a density-independent manner.

Intuition tells us that all four of the assumptions in the logistic model are violated by real populations. For example, it is not uncommon for real populations to overshoot the carrying capacity of their environments, followed by either a slow or rapid (sometimes catastrophic) reduction in the number of individuals until K is established. After overshooting, a population can re-approach the carrying capacity either "from above" or "from below" (see fig. 9.3). Either trajectory can occur if the carrying capacity changes as a population grows in number, or if growth rates change as a function of population density. One historically important example is the Irish potato famine in 1846, which caused the death of roughly one million people and the emigration of 1.5 million people. Another fallacy is that populations that have reached their carrying capacity are described mathematically as static in terms of changing overall numbers. Yet, real populations experience significant losses and gains of individuals even when the total number of individuals varies little (fig. 9.5). Indeed, rates of turnover can be extremely large. For example, in some perennial plant populations, the rates of turnover can be as high as 100% per year, which indicates that selection can be intense.

It is clear that detailed analyses of demographic variables are required when studying population dynamics. Nevertheless, the logistic model is useful as a prelude to understanding the equally elegant

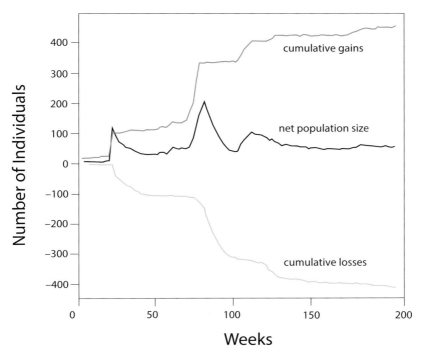

Figure 9.5. Changes in the number of individuals and the cumulative gains and losses in a population of the grass *Agrostis canina* on a copper mine in Wales (data of S. Farrow, Y. McNeilly, and P. D. Putwain reported by A. D. Bradshaw). Note that the net population appears to be stable in number (and thus assumed to be at the environment's carry capacity), and yet the losses and gains continue.

Lotka-Volterra model, named in honor of its codiscoverers Alfred J. Lotka (1880–1949) and Vito Volterra (1860–1940). This model predicts the interactions between two or more species competing for the same resources in the same environment. The simplest case is two species competing for the same resources. Using subscripts 1 and 2 to denote the species, the Lotka-Volterra model is

$$\Delta N_1 = r_1 N_1 (K_1 - N_1 - \alpha_1 N_2)/K_1$$

$$\Delta N_2 = r_2 N_2 (K_2 - N_2 - \alpha_2 N_1)/K_2$$

where α_1 represents the extent to which species 2 competes for resources that would otherwise be garnered by species 1, and where α_2

represents the extent to which species 1 competes for resources that would otherwise be garnered by species 2. The *element of competition* in these equations emerges by a cross term—a variable that breaks the symmetry of a set of equations. Note that the subscripts in the two equations are identical except for one term—N_2 appears in the equation for ΔN_1 and N_1 appears in the equation for ΔN_2. Inspection of these equations further shows that there are only four possible predicted outcomes: (1) the eventual local extinction of species 1, (2) the eventual local extinction of species 2, (3) an unstable periodic equilibrium, or (4) a stable equilibrium. These four possibilities emerge because ΔN_1 and ΔN_2 will not change when $(K_1 - N_1 - \alpha_1 N_2)$ or $(K_2 - N_2 - \alpha_2 N_1)$ equal zero. Thus, if we set $N_1 = 0$, we see that $N_2 = K_2$ or $N_2 = K_1/\alpha_1$, whereas, if we set $N_2 = 0$, we see that $N_1 = K_1$ or $N_1 = K_2/\alpha_2$. It follows therefore that N_1 does not change at any point along the line that joins the two end points $N_1 = K_1$ and $N_2 = K_1/\alpha_1$, and N_2 does not change at any point along the line that joins $N_2 = K_2$ and $N_1 = K_2/\alpha_2$ (fig. 9.6). As was the case for the logistic model, r_1 and r_2 represent the maximum growth rates of the two species. Unlike the logistic model, the instantaneous growths are represented by a family of straight lines that relate r_1 to K_1 and r_2 to K_2 (fig. 9.7).

We can extend the Lotka-Volterra model to accommodate any number of species because, regardless of the number of species, the model has the same general form:

$$\Delta N_i = r_i N_i (K_i - N_i - \Sigma \alpha_{ij} N_j)/K_i,$$

where the subscripts i and j denote species 1 to n. The equilibrium population numbers ($N_{equ.i}$) are achieved when $\Delta N_i = 0$, which is given by the formula $N_{equ.i} = K_i - \Sigma \alpha_{ij} N_j$.

It is always the case that the predictions of a model depend on the assumptions of the model, and it is sometimes the case that some assumptions may not be obvious. For example, in the simplest Lotka-Volterra simulations the interactive terms α_1 and α_2 in $(K_1 - N_1 - \alpha_1 N_2)$ and $(K_2 - N_2 - \alpha_2 N_1)$ are assumed to be constant and first-order terms. This need not be the case. We could have varied the values of α_1 and α_2

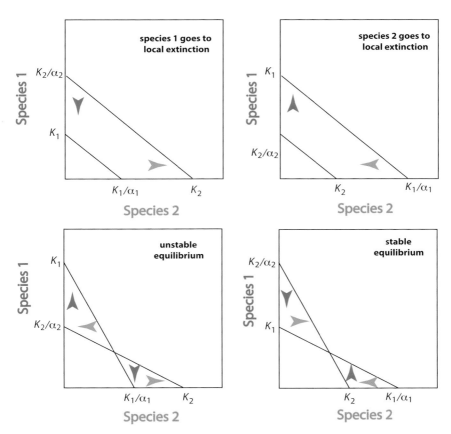

Figure 9.6. Schematics of the four possible outcomes of competition between two species for the same resources predicted by the Lotka-Volterra equations. Clockwise from the upper left: extinction of species 1, extinction of species 2, a stable equilibrium, and an unstable (episodic) equilibrium. The solid diagonal lines in each schematic connecting the points K_1 and K_1/α_1 depict when no change in the numbers of species 1 will occur. The solid diagonal lines in each schematic connecting the points K_2 and K_2/α_2 depict when no change in the numbers of species 2 will occur. The red and green arrows show the direction of change in the numbers of species 1 and species 2, respectively. In the case of the unstable and the stable equilibrium, the point at which the numbers of both species do not change is indicated at the point where the two diagonal lines intersect.

and gotten very different results as would be the case of we had used second-order terms, such as in $(K_1 - N_1 - \alpha_1 N_2 - \alpha_2 N_2)$. Once again, we see that the Lotka-Volterra model is useful heuristically because it sets boundary conditions on the interactions among real populations and species. It also helps us answer "what if" questions that could not be

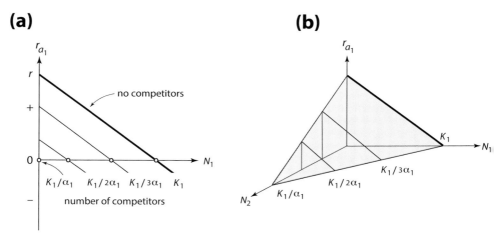

Figure 9.7. Schematic showing the relationships among the maximum growth rate and the actual growth rate of the population of species 1 (r and r_a, respectively), the growth of the populations of species 1 and species 2 (N_1 and N_2, respectively), the carrying capacity of species 1 (K_1), and the number of species 1 competitors (indicated by the numbers in the K_1/α_1 sequence) shown in two and in three dimensions (a and b, respectively). The thick line depicts the relationships among these parameters in the absence of competition. Compare with fig. 9.6.

answered otherwise. In this sense, it is very like the Hardy-Weinberg equation.

Survival of the Individual

The models for the growth of a population and for interspecific competition for resources tell us little or nothing at all about the survival of an individual. That is not their purpose. In this sense, the models just reviewed are as statistical as the equations presented when we discussed population genetics. Yet, the survival of an individual is statistical in the sense that survival depends in part on the number of times an individual is "insulted" by its environment. In other words, survival is an age-dependent parameter. The longer an organism lives, the more times its existence will be challenged.

A very simple mathematical model can be derived from this line of reasoning. If we use the subscripts 1, 2, 3, . . . , i to denote successive challenges, and if we denote the probability of death by each challenge as $P_1, P_2, P_3, \ldots, P_i$, we see that the probability of survival P_s is the sum of $(1 - P_1), (1 - P_1), (1 - P_1), \ldots, (1 - P_i)$, or $P_s = \Sigma(1 - P_i)$. This reasoning assumes that each challenge is an independent event (see Rule 3 in chapter 3). If the probability of death is small with each challenge, the model predicts

$$P_s = 1 - \Sigma P_i = 1 - n\bar{P} = e^{-n\bar{P}},$$

where $\bar{P}$ denotes the mean probability of death. Consider the probability of an owl dying as a result of a particular tree being uprooted by a violent wind storm. Each storm is an independent event. If one such storm occurs every year and if the owl occupies the tree for 5 years, we see that $P_s = e^{-5\bar{P}}$. Notice that this model would not be valid if the probabilities of being uprooted are interdependent. For example, successive wind storms might cause increasing mechanical fatigue in the tree's roots such that the probability of root failure increases with the age of the tree. Under these circumstances, it's a wise owl that moves to live in younger trees.

We can also construct survival models based on the concepts of unit-space and spatial distribution as for example the juxtaposition of neighbors. Consider a hypothetical habitat in which the unit-space is tessellated by triangles and that each triangle can support only one adult among four different species denoted by 1, 2, 3, and 4 (fig. 9.8). In this model, only one of two or more individuals from any of the four species occupying the same unit-space will survive if the individuals differ in their competitiveness, otherwise all individuals die. Further, if the unit-space is occupied by an individual belonging to the most competitive species, that individual will survive regardless of the presence of one or more less competitive individuals from the other species. In this model, the proportion of unit-spaces occupied by surviving individuals depends very much on the initial density of individuals drawn from all four species and on the number of individuals possessing different levels of competitiveness. For example, if we assume that the initial density conforms to a Poisson distribution

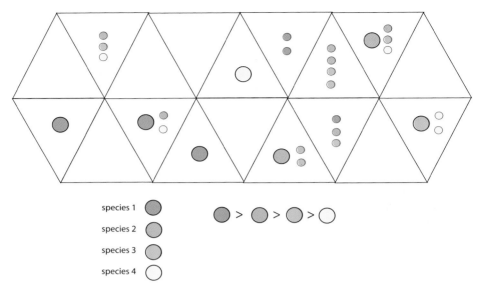

Figure 9.8. Schematic of a unit-space model for plant competition. The "habitat" is tessellated with triangles in which individuals drawn from four species differing in competitiveness (see lower inserts) are distributed initially according to a Poisson distribution. Only one of two or more individuals occupying the same unit-space will survive if the individuals differ in the competitiveness; otherwise all individuals die, and, if a unit-space in occupied by an individual belonging to the most competitive species (species 1), that individual will survive regardless of whether one or more less-competitive individuals co-occupy the unit-space. In this model, the proportion of successfully occupied unit-spaces increases as the number of competing species increases.

(box 9.1), the number of successfully occupied unit-spaces increases in direct proportion with the increase in competing species. That is, the greater the number of competing species, the greater the chance that a unit-space will contain two or more species, but only one of these will belong to the most competitive species.

Why should the two models for survival yield different results? The answer lies in the fact that these two models are based on very different assumptions about competition. The multiplicative model assumes that successive challenges are independent events, whereas the unit-space model assumes that challenges are interdependent. The multiplicative model is well suited to modeling animal survival,

Box 9.1. The Poisson Distribution

Events or objects that occur sporadically in either space or time and that are indepen-dent of one another are distributed in a pattern described by Siméon D. Poisson (1781-1840). Here we will derive the Poisson distribution as a limit to the more familiar binomial distribution.

Consider two mutually exclusive probabilistic conditions (as for example, an organ-ism is either in a unit-space or not) denoted as p and q where $p + q = 1$. The probability of obtaining successive events or "draws" k in n trials is given by the formula

$p^k q^{(n-1)}[n(n - 1)(n - 2), \ldots , (n - k + 1)]/k!$

We now let n grow large and let p grow small, but stipulate that $np = m$, where m equals the mean number of successes (as for example, the number of plants or insects found in a unit-space). Under these conditions, we see that $p = m/n$ and $q = 1 - m/n$. Also, since n is very large, it follows that

$(m/n)^k[1 - (m/n)]^n[1 - (m/n)]^{-k}[n(n - 1)(n - 2), \ldots , (n - k + 1)]/k!$

Notice that

$[n(n - 1)(n - 2), \ldots , (n - k + 1)] \approx n^k,$

$[1 - (m/n)]^{-k}[n(n - 1)(n - 2), \ldots , (n - k + 1)]/k! = m^k/k!,$

$[1 - (m/n)]^n = e^{-m},$

and that

$[1 - (m/n)]^{-k} = 1.$

Therefore, the probability of obtaining successive events k in n trials is given by the elegantly simple formula

$(m^k/k!)e^{-m}.$

Space precludes giving the derivation for the variance of the Poisson distribution σ^2, which equals its mean m.

because it accommodates the behavior of motile organisms. The unit-space model is instructive when dealing with sedentary organisms like the land plants that continue to grow in size. Indeed, unit space models are particularly well suited to modeling a phenomenon called "self-thinning," which we will consider next.

Size-Density Relationships and Self-Thinning

The structure and dynamics of populations and communities are profoundly influenced by the consequences of the interactions of three
fundamental ecological processes: (1) competition for resources and
space, (2) the effect of an organism's body size on resource and space
use, and (3) the effect of population density on growth and mortality.
Traditionally, two approaches have been used to study these interactions. One focuses on theoretical models and empirical measurements of abundance, spacing, survival, mortality, and recruitment as
functions of body size in relatively undisturbed natural populations
and communities, where competition and the elimination of individuals is complicated by a number of effects such as asymmetries in resource supply and mortality rates. The second approach focuses on the
structure and dynamics of organisms in highly controlled laboratory
or field settings, as for example where plants of nearly identical age
are grown as crops. Studies of such simplified agricultural systems
have led to theoretical and empirical insights into a process called
self-thinning wherein plants belonging to the same species compete
with one another and mortality results in a temporal trajectory of
decreasing population density as a function of increasing plant size.
Such studies frequently (although not invariably) report that plants
grow with minimal mortality until they reach a critical density N_{crit} at
which point the growth of a limited number of individuals is possible
only if resources are made available by the death of other individuals
belonging to the same species.

In such resource-limited environments, the scaling of resource use
with body size typically results in a tradeoff between population density and body size. This tradeoff is frequently (albeit not invariably)
described by the formula

$$N_{crit} = \beta M^{\alpha},$$

where N_{crit} is the maximally packed number of individuals with an average mass M that initiates self-thinning, β is a normalization constant,

and α is a scaling exponent. It is generally but not invariably observed that α = −¾, which can be derived easily (box 9.2). When plotted on logarithmic axes, this equation yields a linear relationship, where β is the log-log Y-intercept and α is the log-log slope (fig. 9.9). The initial "seeding" density is often less than N_{crit}, as for example when propagules are dispersed over great distances in a tundra. In these cases, plants will typically grow with minimal mortality until they increase in size and start to compete for resources and space, after which they follow a "self-thinning" log-log linear trajectory, with some individuals dying and thus freeing up resources and space that surviving individuals can use. If the initial density is sufficiently low, plants will never reach a critical thinning density and they will mature to some more or less uniform maximum size (see fig. 9.9).

Some of the evolutionary implications of the self-thinning trajectory come to light once we notice that maximum body size increased over the early course of land plant evolution. As noted in chapter 8, the most ancient vascular land plant sporophytes, such as *Cooksonia*, were diminutive and used wind to disperse their propagules. These sporophytes were probably morphologically and physiologically similar to the sporophytes of modern-day mosses. Consequently, they likely experienced little or no density-dependent self-thinning. By Lower Devonian times, the sporophytes of most vascular plants had evolved a rhizomatous growth habit in which older parts decomposed, while the growing tips spread out to explore and garner resources in an expanding unit-space (fig. 9.10). (It is interesting to speculate about the consequences of somatic mutations or spontaneous chromosome doubling that could, in theory, give rise to a new species with this kind of growth habit.) A modern-day analog may be seen in the lycopod *Huperzia lucidula*. On the basis of permineralizations found in the Rhynie Chert, some Lower Devonian plant communities were thickly occupied and self-thinning may have been intense, especially in older parts of a community. The advent of central root systems and secondary growth, which evolved by Upper Devonian times, changed the global ecology in many ways, not the least of which was the appear-

Box 9.2. A Derivation of the Self-Thinning Rule

The self-thinning phenomenon conforming to $N_{crit} = \beta M^{\alpha}$ is described by the following model for the relationship between critical plant density N_{crit} and plant body mass M_T and establishes the relationship between plant density and canopy radius r for any population of plants of more or less uniform size.

At low densities (or when the size of mature plants is small in comparison to the unit-space), the canopies of neighboring plants do not make contact and plant-plant interferences are minimal or nonexistent (fig. B.9.2 a). At higher densities (or when the size of mature plants is large in comparison to the unit-space), the canopies of neighboring plants touch and begin to intersect (fig. B.9.2 b). At this point, the critical plant density equals the total number of plants divided by the unit-space such that

$$N_{crit} = n_A n_B / (4 n_A n_B r^2) \propto 1/r^2,$$

where n_A and n_B are the number of plants in the orthogonal dimensions of the unit-space. Bio-mechanical theory and observations indicate that, on average, plant body biomass is propor-tional to the 8/3 power of any reference dimension of plant size such as plant height or canopy radius. Thus, $M_T \propto r^{8/3}$, from which it follows that $r^2 \propto M_T^{6/8} \propto M_T^{3/4}$ and that

$$N_{crit} \propto 1/M_T^{3/4}.$$

In general, the area of any canopy is $A_{can} = l_{can}^2$, where l is the length of one side of the tessellated arrangement, so that $l = 3r + fr = r(3 + f)$, where f is some fraction of canopy radius. It follows that the critical density of any uniformly occupied unit-space is proportional to l^2 such that

$$N_{crit} \propto 1/l^2 \propto 1/\bar{r}^2 \propto 1/\bar{M}^{3/4},$$

where $\bar{r}$ and $\bar{M}$ denote the average canopy radius and plant mass when plant sizes are not uniform.

This derivation can be extended into the third dimension of canopy height by considering the volume of mutually shaded leaves created when two neighboring canopies (which is approx-imated here as having a cylindrical geometry) intersect along their height. This volume equals canopy height, h, multiplied by two times the area of the segment created by the intersecting circular canopies (fig. B.9.2 c-d). The area of each segment, A_{seg}, is given by the formula

$$A_{seg} = \left(\frac{\pi \theta r^2}{360°} - \frac{r^2 \sin \theta}{2} \right) = r^2 \left(\frac{\pi \theta}{360°} - \frac{\sin \theta}{2} \right),$$

where θ is the angle defining the area of the segment, which will increase as r increases (fig. B.9.2 d). The volume created by two intersecting canopies therefore is given by the formula

$$V = 2hr^2 \left(\frac{\pi \theta}{360°} - \frac{\sin \theta}{2} \right).$$

This volume is a direct function of the arrangement of plants in the unit-space, which is a function of plant density. Note that the distance between neighboring plants d is a constant for any arrangement of plants such that the chord created by two intersecting canopies will be located at a distance $d/2$ provided $r < d$ (fig. 9.2.1 c-d). Given this geometric arrangement, we see that $r = d/[2\cos(\theta/2)]$. Inserting this relationship into the formula for A_{seg} gives

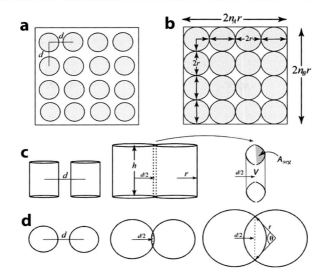

Figure B.9.2. Geometric relationships among planting distance, d, and canopy radius, r, and height, h. a. Polar view of equally distanced plants (canopies indicated by shaded circles). Neighboring canopies do not intersect provided the planting density is low (or that mature plants are sufficiently small in comparison to density). b. At higher plant densities, neighboring canopies begin to make contact and self-thinning begins. At this juncture, the total number of plants equals $n_A n_B$ and the planted area equals $4 n_A n_B r^2$, where n_A and n_B are the numbers of plants in the orthogonal dimensions of the planted field such that the critical plant density $N_{crit} = n_A n_B / 4 n_A n_B r^2 \propto 1/r^2$. c. Side view of the canopies (with radii r) of two neighboring plants at a fixed distance d. As canopies increase in size, their canopies begin to intersect (middle diagram). The intersecting volume of neighboring canopies, V, equals twice the area of the segment of each circular intersecting canopy, A_{seg}, multiplied by canopy height, h. d. Polar views of neighboring plants showing that the location of the chord between the two intersecting canopies in c is always located at $d/2$, whereas the area of the segments defined by the chord is a function of the angle θ.

$$A_{seg} = \left(\frac{d}{2 \cos(\theta/2)} \right) \left(\frac{\pi\theta}{360°} - \frac{\sin\theta}{2} \right),$$

where θ is an angle in degrees and d is a numerical constant. Integration of this formula with respect to θ gives

$$\frac{\partial A_{seg}}{\partial \theta} = \frac{d^2}{16} \sec^3\theta \left(\frac{8\pi\theta}{360°} \sin\theta + \frac{4\pi}{360°} \cos\theta + \cos 2\theta - 3 \right),$$

which shows the rate at which the area of intersecting neighboring canopies increases (as does the volume of intersecting canopies) as the function of the initial distance between neighboring plants. Further, because d^2 is plant density, the extent to which the canopies of neighboring plants intersect is a function of the critical density N_{crit} and maximum plant size (as gauged by canopy radius or height).

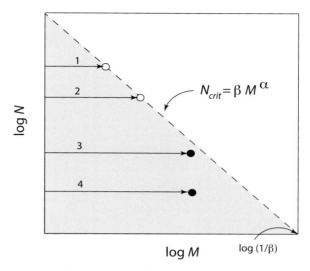

Figure 9.9. Schematic of the traditional self-thinning model for crops. In this hypothetical case, plants are grown with four different initial densities (numbered 1–4 in order of decreasing density) and subsequently harvested after the same fixed time when plants have matured. Initially, all plants grow at the same rate (horizontal arrows). At the two highest densities (1–2), plants reach their critical population densities (N_{crit}) at different body masses (O). At the two lowest densities (3–4), plants never reach their N_{crit} and mature at the same maximum size (●). Note that this scenario is easily changed to cope with naturally growing plants if we assume that the lowest densities reflect low seedling densities (perhaps due to early predation) and that adults never reach critical densities because of adult mortality or determinate growth in size (as in annuals).

ance of the tree growth habit and an ensuing increase in community densities and stature. As discussed in chapter 8, computer simulations indicate that the evolutionary transformation from equal branching to unequal branching architectures in tandem with increasing body size and height undoubtedly resulted in more intense selection and self-thinning as more and more species were packed into the same community space and competed for space, light, water, and other resources.

Competition and Patchiness

Thus far, we have been discussing competition, either between members of the same population or between members of different species.

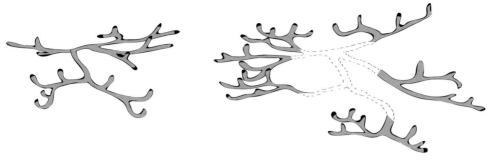

Figure 9.10. Schematic of a rhizomatous growth habit with a "rolling ball of yarn" strategy for garnering resources. This strategy entails growing at the tips of rhizomes and rotting at the ends of the oldest portions of the plant. This results in fragmentation of a single individual into many genetically identical individuals, and it provides a method of locomotion much like a rolling ball of yarn leaves some thread behind it. Fossil plants found in Lower Devonian sediments, such as *Rhynia gwynne-vaughnii* and *Aglaophyton major* (which should be named *A. majus* to be grammatically correct), very likely possessed this growth habit.

The logistic curve reviewed earlier indicates implicitly that members of the same population will experience selection and that variation among them will result in differential mortality. The Lotka-Volterra equations predict local extinction in two out of four hypothetical cases. In each case, competition increases nonlinearly as the system approaches saturation. High mortality is predicted even if we make the absurd assumption that all individuals belonging to the same population (or belonging to two different species) are identical. This prediction emerges because resources are always finite and the law of supply and demand cannot be contravened.

Indeed, the mathematical parameters in the Lotka-Volterra equations were quickly used by ecologists as heuristic devices to define the end points of a spectrum of competition. Species with high rates of growth and that produce many progeny and that live in unpredictable or quickly changing environments were defined as r-selected species (r for intrinsic growth rate), whereas species existing at densities close to the carrying capacity of their environment and that produce few progeny and that live in predictable and stable environment were defined as K-selected species (K for carrying capacity selection). Curi-

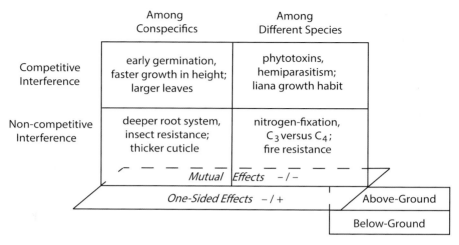

Figure 9.11. Matrix giving examples of above- and below-ground competitive interference and noncompetitive interference among individuals of the same species (conspecifics) and individuals of different species.

ously, however, *K*-selection technically does not exist. Environments do not *select for* large, slow-growing organisms that produce few progeny per year. They *permit* organisms with these traits to exist. In other words, the characteristics of *K*-selected organisms reflect circumstances in which *r*-selection is negligible. Although a life-history paradigm has replaced the *r*- and *K*-selection paradigm, the notion of competition among opportunistic species and among equilibrium species is instructive when comparing an annual weed to an oak tree, or a mouse to an elephant.

Yet, species requiring much the same of their environments coexist. How can this be? One explanation lies in the distinction between two forms of competition—competitive interference and noncompetitive interference. Competitive interference occurs when organisms consume resources that would be otherwise available to their neighbors. Noncompetitive interference occurs when organisms consume resources that are not available to their neighbors (fig. 9.11). This distinction shows that the ability to garner resources and the ability to compete with other organisms are not necessarily correlated.

For example, a plant may acquire more water by growing a deeper root system and thus tap into a resource that is unavailable to its more shallow-rooted neighbors. This partitioning of the soil profile could even decrease competition for water among some neighbors, although competition for mineral nutrients or light may increase as a consequence of more vigorous growth or increases in population densities. Likewise, a particular phenotype among conspecifics may acquire insect resistance by virtue of a spontaneous mutation and thus grow larger than normal, but not necessarily at the expense of its neighbors. Indeed, conspecifics can differ in their ability to convert resources into biomass by virtue of differences in water- or nutrient-use efficiency, and these differences can affect growth rates in the absence of above- or below-ground direct interference.

Another explanation for coexistence is the role of spatial heterogeneity and time. Throughout this book, we have been discussing populations as if they have sharp exclusionary boundaries that remain fixed in place. Yet, empirical observations show that real populations typically lack sharply defined boundaries and that these boundaries move, even in the case of land plants, because spores, seeds, and fruits can be transported considerable distances. We also know that animals and plants (by means of spores, seeds, or fruits) can migrate among populations and that this can increase genetic and phenotypic diversity within the previously established genotypic spectrum of a population. This fluidity does not make the concept of a population illusionary. It simply means that populations must be understood in terms of their spatial patchiness, temporal fluctuations, and their life-history characteristics (fig. 9.12). For example, a species can maintain itself in an unfavorable location by means of random dispersal from a favorable location, or by offsetting the disadvantage of competition with the advantage of reproductive superiority, which is dependent on the density of its competitors.

By making three assumptions, we can formalize mathematically how the fluidity of populations helps to explain the coexistence of species requiring much the same of their environments: (1) the envi-

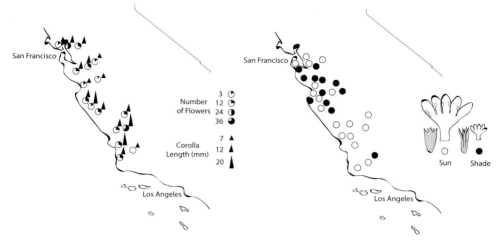

Figure 9.12. Regional interpopulational differences in the number of flowers per head, corolla length, and sun- versus shade-adapted colonies of the California gily (*Gilia achilleaefolia*) in the central California area between San Francisco and Los Angeles. The diagram illustrating the differences in the size and shape of flowers produced by individuals in sunny and shaded habitats is drawn to scale. Data taken from Grant (1954).

ronment consists of unit-spaces ("patches"), (2) the carrying capacity of each patch is reached rapidly once colonized, and (3) dispersal has no effect within a patch. These assumptions give rise to a dynamics between local extinction and colonization such that the change in the proportion of patches *p* occupied by a species on the regional scale is given by the formula

$$\Delta p = cp(1 - p) - ep,$$

where *c* is the colonization rate of empty patches, and *e* is the extinction of occupied patches. Assuming that the probability of colonizing patches is proportional to the proportion of patches already occupied, the equilibrium condition ($\Delta p = 0$) is given by the formula $p_e = 1 - e/c$. Note that this model deals with metapopulations (a collection of populations composed of the same species) and that the persistence of a species in the same region requires $c/e \geq 1$. It also states explicitly that *space* in the form of patches is a *resource*, which makes intuitive sense especially when dealing with sedentary organisms such as plants.

Note further that some patches may be sinks in which reproductive rates are low and mortality rates are high and that others may be sources in which reproductive rates are high and mortality rates are low. In this manner, a species may persist in locations that are not intrinsically conducive to its survival. This scenario must also consider spatial scale as well as the demographics within individual populations. Empirical studies have shown that environmental dissimilarity (variance) tends to increase as a function of distance, sometimes in a fractal-like manner. Under these circumstances, environmental heterogeneity will also tend to increase with distance. Thus, a sink-source model must consider the ability of animals to migrate successfully, or the ability of the propagules of different kinds of plants to reach different distances. (Maximum seed dispersal distances on the order of 1–20 km have been reported for a large number of angiosperm species [see Cain, Milligan, and Strand 2000].)

The evolutionary implications of dispersal models pertain as much to the past as to the present. We can easily imagine Earth's terrestrial landscape occupied by the first land plants as a quilt of patches reached by wind-dispersed spores just as today's bryophytes and pteridophytes colonize new sites by means of spores (fig. 9.13). In this context, it is interesting to consider the extent to which populations diverge under the effects of mutation, genetic drift, and migration of individuals among populations. Theoretical treatments have considered this with the assumption that migrants are a random sample of the entire metapopulation (an island model), or that migration is between adjacent populations (one- or two-dimensional network models). Despite these restrictive assumptions, these models have heuristic value because they show the combined effects of population density and migration rates. Sewall Wright was one of the first to explore the consequences of migration. He showed that the extent to which populations have genetically diverged from one another can be calculated as the normalized variance of gene frequencies F_n, which is the correlation between two gametes drawn at random from each population:

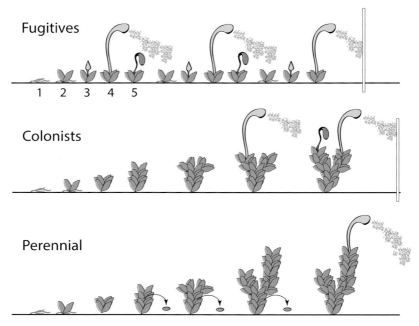

Figure 9.13. Schematic comparisons among three moss life history strategies (fugitives, colonists, and perennials). Fugitive species finish their life cycles within one year and disperse to other ephemeral habitats. Their gametophytes grow quickly and tend to be small. They also tend to produce large numbers of small spores. Fugitives are archetypal r-selected species. 1–5 denote different stages in the lifecycle of an individual: 1. Protonema. 2. Production of antheridia and archegonia. 3. Early growth of sporophyte. 4. Mature sporophyte releases spores. 5. Sporophyte senesces. Colonists have a life span of a few years. They tend to be larger and to grow more slowly than fugitive species. Their habitats are also more stable and last longer than those of fugitives. The vertical rectangles indicate when the habitat for a fugitive and a colonist are no longer suitable. Perennials are long lived and reproduce both sexually and asexually (arrows denote the release of asexual propagules). Perennials also tend to produce fewer but larger spores. Adapted from During (1979).

$$F_n = (1 - m)^2/[2N - (2N - 1)(1 - m)^2],$$

where m is the rate of migration and N is population density. This simple formula shows that (1) migration rates have little effect on interpopulational divergence when population densities are low, (2) genetic divergence will occur little or not at all when densities and rates are high, and (3) divergence will occur at high densities if migration rates are low (fig. 9.14).

Another explanation for coexistence of species comes from the

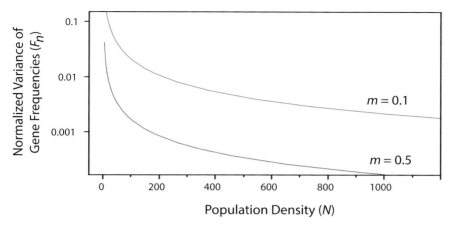

Figure 9.14. Graph showing the relationships among migration rates (*m*), population density (*N*), and the extent to which populations will diverge genetically (F_n) if the migrants are selected at random from the metapopulation as a whole. At low population densities, populations will diverge at comparatively high rates, even at comparatively high migration rates. At high population densities, populations will not diverge even at high migration rates.

recognition that we can think of competition in two different ways: (1) competition among individuals in the same population (intraspecific competition), and (2) competition among individuals belonging to different species (intraspecific competition). Intra- and interspecific competition can have opposite effects on the persistence and resource use of a population. In the absence of another species, individuals occupying a particular patch can initially maximize resource use and individual fitness (fig. 9.15). As the population grows in number and expands in its range, the resource gradient expands to include less optimal conditions and individual fitness in these regions declines. Additional growth in population size expands the extent to which the resource gradient is further explored and overlaps with the peripheral growth of neighboring populations of other species. Individuals of the two or more species coexist in these overlapping regions because they exploit less optimal (and less contested) resources. This scenario, which is consistent with viewing populations as pancakes draped over adaptive peaks in one of Sewall Wright's fitness landscapes, ultimately results in an equilibrium condition in which the intensity of intra-

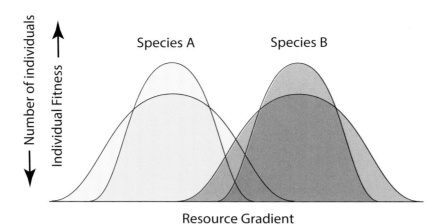

Figure 9.15. Schematic showing the hypothetical relationships among the number of individuals, individual fitness, and resource gradient occupancy of two species. When a patch
is initially colonized, individual fitness is maximized and resource use is optimized. As the
populations grow in number, individual fitness declines and resource use expands across
the resource gradient, which permits individuals of both species to coexist in marginal
resource gradients (indicated by overlapping curves).

specific competition equals the intensity of interspecific competition
within overlapping regions. Clearly, genetic and phenotypic differences among individuals will restrict the extent to which a resource
gradient can be explored by different species such that a precise balance between these two forms of competition may never be achieved.
However, we can imagine conditions in which the taxonomy of your
neighbors matters less than their behavior.

Finally, we cannot easily dismiss the concept of *neutral morphology* when considering species coexistence. This concept, which is an
analog to the neutral theory of molecular evolution, proposes that
natural selection is indifferent to some phenotypic differences such
that some diverse morphologies can coexist. For example, it can be
used to explain why vastly different unicellular and colonial forms of
planktonic organisms coexist because these organisms are continuously multitasking and require much the same resources. Recall from
the previous chapter that computer simulations show that the number
of equally fit (optimal) phenotypes increases as the number of tasks

an organism performs simultaneously increases. The concept of neutral morphology can be applied equally well and for the same reasons to other organisms, such as trees or understory plants. This concept does not argue that species coexistence implies a lack of competition. It simply posits that different morphologies are equally competitive.

Novelty and Adaptive Radiations

Earlier, an analogy was drawn between the concept of ecological succession and the three great paleo-megafloristic transitions that occurred during the Phanerozoic, wherein a pteridophyte-dominated paleo-megaflora was replaced by a gymnosperm-dominated paleomegaflora that was replaced in turn by communities dominated by flowering plants (see fig. 6.8). Indeed, a fourth floristic transformation can be added, since the pteridophyte communities that dominated the Late Silurian and Devonian were undoubtedly preceded by communities composed of bryophyte-like plants and lichens. Regardless of the number of transformations, in each case, the fossil record records a significant increase in species origination and extinction rates with an overall increase in the number of species belonging to the new megaflora. Taken at face value, it is tempting to think of each of these successive floristic turnovers as an *adaptive radiation*. Yet, upon some reflection, this view is flawed, or at least problematic.

Consider the two most widely espoused conceptions of the word *adaptation*, one that emphasizes history and another that emphasizes adaptive currency. Darwin's formulation of evolution is a historical theory about adaptation and diversification. Organisms "fit" into their environments by virtue of their "contrivances" (Darwin's words), which allow them to function as coherent and better adapted phenotypes as determined by natural selection. This Darwinian construct obtains a definition for adaptation that specifies that *an adaptation is a consequence of natural selection*. In other words, an adaptation is any functional and heritable trait that enhances fitness. This conception of adaptation does not explain adaptation. To do this, we must uncover the reason(s) why a functional trait came into being as well as why it has been maintained. Note further that this definition fits neatly into the *Weltanschauung* of the Modern Synthesis because the process that gives rise to an adaptation can be understood in terms of changing allele frequencies. This is the *historical* perspective of what an adaptation is.

An alternative conception of adaptation ignores the historical perspective and emphasizes current usage and maintenance. Here,

---

Box 9.3. Evolutionary Stable Strategies

The concept of evolutionary stable strategies (ESS) is generally credited to John Maynard Smith. However, it emerges directly from classical game theory (CGT). Both ESS and CGT attempt to determine mathematically which among alternative strategies is optimal when players compete for the same prize. The difference between ESS and CGT is that the CGT assumes that players are cognizant of their actions and that they behave rationally. ESS does not make these assumptions. It assumes only that each player (organism) has a strategy (determined genetically and developmentally), and lets evolution "test" alternative strategies for their ability to confer high probabilities for survival and reproductive success. The "goal" of a player therefore is to produce as many of its kind as possible. The "payoff" to a player is in units of relative fitness. Here, we will explore two examples of a game, one example drawn from CGT and another from ESS.

The example from CGT is one of individual decision making under uncertainty. The gist of the problem is that a choice must be made from among a set of actions, $A_1, A_2, \ldots, A_n$, but the payoff for each action depends on a condition or "state of nature," $c_1, c_2, \ldots, c_n$. The player is aware that one of the conditions is a true state of nature, but has no idea about the probabilities of the various conditions, or even if it is meaningful to think about probabilities. This problem can be framed in concrete terms by considering a person preparing an omelet with six eggs among which one *may be* rotten and potentially poisonous. The person has broken five eggs into a bowl without encountering a bad egg and must now decide whether to break the sixth egg into the bowl. There exists a set of only three actions (strategies): break the egg into the bowl, break the egg into a saucer, or leave the sixth egg alone. The following lists the choices and the consequences:

| | Conditions | |
Actions	Good egg	Poisonous egg
Break into the bowl	Six-egg omelet	Five ruined eggs
Break into a saucer	Six-egg omelet	Five-egg omelet
Leave the egg alone	Five-egg omelet	Five-egg omelet

In this problem, the person may know the probability of finding a bad egg out of a random sample of six eggs and may elect to take the risk of breaking the egg into the bowl if the probability of finding a bad egg is deemed to be very low. In general, if a probability distribution an adaptation is said to exist on the basis of how a trait functions in the present day when its functionality is assessed using some sort of optimality or game theory modeling that posits a set of potential strategies among which one is better than the rest. What counts most in this context is the extent to which the currency of a particular strategy is robust against mutation. In other words, an adaptation is defined as an evolutionary stable strategy (box 9.3). This perspec-

---

is already known (if a person knows the probability of finding a rotten egg out of six random draws), the problem outlined here can be transformed into the mathematical domain of decision making under risk. Specifically, if the conditions $c_1, c_2, \ldots, c_n$ have probabilities $p_1, p_2, \ldots, p_n$, where the sum of probabilities is

$$\sum_{j=1}^{n} p_j = 1, \ p_j \geq 0,$$

the expected utility or gain from performing act A_i equals $u_{i1}p_1 + u_{i2}p_2 + \ldots + u_{in}p_n$ and the act that is expected to yield the maximum utility (gain) will be chosen. We can think of this as a game between the person and nature. The person has a set of strategies $(A_1, A_2, \ldots, A_n)$, whereas nature has a set of conditions $(c_1, c_2, \ldots, c_n)$. The payoff to the person for the strategy pair (A_i, c_j) is u_{ij}, and if the person knows that nature is employing the mixed strategy $(p_1c_1, p_2c_2, \ldots, p_nc_n)$, the person should adopt the strategy that is best against the mixed strategy (that is, the best strategy to deal with the a priori probability distribution). In the example of the six-egg game, if a person thinks there is even a slim chance that the sixth egg might be spoiled, washing a small saucer would be the best strategy.

Turning now to ESS, consider two species of annual flowering plants, one that grows aggressively (species 1) and one that grows only half as aggressively (species 2). Both species shed their seeds at the same time into the same area, both compete for the same resources R (space, water, sunlight, and soil nutrients), and both incur a cost C for competing with another plant. The payoffs for the two competing species are summarized as

	Meets species 1	Meets species 2
Species 1	$R/2 - C/2$	f_1R
Species 2	f_2R	$R/2$

where f denotes some fraction of R such that $f_1 > f_2$. As in the previous example, this payoff matrix depends on the probabilities that seeds from the two species land in the same unit-space. It also depends on the mean number of seeds produced per plant and on previous iterations of the game of competition because the results of the previous growing season must be fed back into the next competitive interactions the following year. In the simplest case, if $f_1 \approx f_2$ and $C > R$, the mathematics ends in an evolutionary stable strategy with a mix of the two strategies where $R \approx C$ and the fitness of the two species is equal. Note that fitness is defined (in this game) both in terms of the survival of an individual plant and its reproductive output per year.

tive was entertained in chapter 8 when we modeled the evolution of branching patterns and identified phenotypic optima for the performance of one or more functional tasks. The appearance of evolution in this modeling (the adaptive walks on landscapes) was an illusion in the sense that it was unnecessary to identify optima, although these walks helped to reveal elements of historicity that could be compared to the fossil record.

A pluralistic view about adaptation could be advocated such that questions concerning the origins of adaptations would adopt the historical definition, whereas questions about the maintenance of adaptations would adopt the currency definition. Upon reflection, however, an attempt to determine whether something in the fossil record amounts to an adaptive radiation requires both definitions and thus both modes of analysis, particularly since evidence for species diversification is necessary but not sufficient for claiming that a sequence of historical events constitutes adaptive evolution. Consider, for example, the emergence of the first vascular floras during the early Paleozoic, which involved the convergence of not one but many functional traits, each of which had an individualized historicity and its own currency (a cuticle, stomata, sporopollenin-rich spore walls, and vascular tissues). In turn, consider the evolution of the seed plants in light of the fact that a "seed" is more than an "integumented megasporangium" (among other things, the evolution of the seed also involves a reduction in the number of functional megaspores and some sort of sperm delivery system such as a pollen tube). The same admonition can be directed toward the evolution of the flower, which involved a constellation of phenotypic innovations among serially homologous body parts.

When viewed critically, each of the three major paleo-megafloristic radiations involved a convergent juxtaposition of individual functional traits that produced an innovation such as the seed and the flower. Each of these innovations was subsequently evaluated by passing through the gauntlet of selection. Thus, each innovation *became* an adaptation *after* its appearance. It is not certain whether all or most of the functional traits in each of these events preexisted in the ancestral lineage from which the innovation arose. That some of these traits did preexist is beyond doubt. The ancestors to the first vascular plants had spores with sporopollenin-rich walls; the ancestors to the gymnosperms had megasporangia; the ancestors to the flowering plants had seeds. When viewed in this way, each of the functional traits drawn upon to synthesize an innovation was both an adaptation in the sense that it conferred a selective advantage and an exaptation in the sense that the beneficiary of the advantage was the ancestor and not the descendant. If one elects to use the word *novelty*, we see that in each case it was the innovation that was novel and not its individual functional traits. The key points are that the historical definition and the currency definition are not interchangeable, one cannot be reduced to the other, and both are required when evaluating the fossil record. Understanding history and currency is also required to understand the ecology of extant organisms in terms of how they "fit" into their environments and how their "contrivances" function.

Biogeography and Continental Drift

One of the major goals of ecology is to understand the factors that influence the geographic distributions of species. This study of the spatial distributions of all organisms and the search for explanations for these distributions is called biogeography. Alfred Russel Wallace, who along with Darwin proposed natural selection as the driving force of evolution, was one of the most famous biogeographers. He and other naturalists, such as Alexander von Humboldt (1769–1859), quickly observed that different parts of the world were characterized by distinct and different species assemblages that could be segregated into six major biogeographic regions. Three of these coincided more or less with the continents of South America (the Neotropical region), Australia (and the Australian region), and Africa south of the Sahara (the Ethiopian region). The other three regions are the Nearctic (which includes North America north of the Mexican escarpment and the

Antilles), the Palearctic (which includes the land mass north of the
Himalayas and Africa north of the Sahara), and the Oriental (which includes the land mass south of the Himalayas, the Philippines, Borneo,
Sumatra, Java, and other islands east to the Celebes). Although the
boundaries of these six regions are debated, few who have traveled the
world would disagree that the organisms characterizing these regions
are assembled in distinctive and recognizable fauna and flora. It must
be noted also that the boundaries of these six regions are established
by barriers to dispersal—a tall mountain range, a narrow isthmus, an
ocean, or a desert—that have been steadily eroded by human activity.

The study of biogeography has recognized a number of trends or
"rules" based on reoccurring patterns in all or most of the six biogeographic regions. For example, body volume with respect to body
surface area tends to increase among warm-blooded animals living in
progressively colder climates, presumably because body surface area
tends to decline as volume increases (Bermann's rule). Conversely,
the appendages of warm-blooded animals tend to be longer or larger
in area in warmer climates (Allen's rule). Among vascular plants, the
diameters of water-conducting cells tends to decrease with increasing
latitude or elevation, whereas the number of cross bars in the end
plates of vessel members tends to decrease in progressively drier habitats. Both of these trends make sense when viewed from a hydraulic
perspective because narrower vessel members are less prone to cavitation and the elimination of cross bars reduces the resistance to water
flow. However, like most "rules," exceptions to each of these trends
are well known because selection acts on the entire phenotype whose
functional traits are integrated in ways that cannot be individually
dissected.

One of the more dramatic biogeographic trends is the noticeable
increase in species diversity as one travels from the higher latitudes
to the equator, or when one travels from high to low elevations
(fig. 9.16). A number of theories have been proposed to explain these
tendencies. Broadly speaking, these fall into one of three categories:
(1) biotic explanations claiming species interactions (competition,

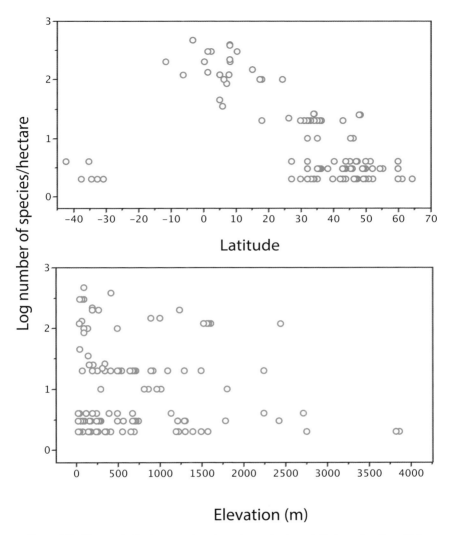

Figure 9.16. Changes in the log$_{10}$ number of species per hectare plotted as a function of latitude (upper graph) and elevation (lower graph). Overall, species numbers increase toward the equator and toward lower elevation. Data from Cannell (1982).

predation, specialization, etc.) account for the latitudinal diversity gradient, (2) historical-evolutionary elucidations revolving around evolutionary time or rates account for the higher diversity in the tropics, and (3) climate-area hypotheses that draw attention to how climate equitability affects resource utilization. No theory has received wide acceptance, and it is likely that multiple ecological and historical factors have influenced the biodiversity of different kinds of organisms.

It is strange, however, that people do not reverse the question and ask, "why do species numbers decline with increasing elevation or latitude?" because, from the perspective of a botanist, the answer would be, "because it gets progressively colder and at some point water freezes in tracheids and vessels." Clearly, this answer is simplistic because many factors contribute to the biogeographic distribution of any species. But consider three facts: (1) the forests found at the highest latitudes or elevations consist predominantly of conifers, (2) tracheids are on average less susceptible to cavitation than vessels, and (3) conifer woods are primarily composed of tracheids. Since angiosperms are the most species-rich land plant clade, the constraints imposed by freezing water on the latitudinal and elevational gradients in plant biodiversity must be important.

No discussion about biogeography would be complete without mentioning continental drift (fig. 9.17). Much of eighteenth- and nineteenth-century biogeography was governed by the unwavering assumption that the continents were always located where they are today. This conceit focused discussions about biogeographic patterns on dispersal mechanisms such as animals or plants rafting across a geographic barrier, such as an isthmus or an ocean. Yet, the notion that the continents might have moved was put forward as early as 1596 by the Flemish cartographer and geographer Abraham Ortelius (1527–1598), who is widely recognized as the author of the first modern atlas (entitled *Theatrum Orbis Terrarum*), originally printed in 1570. Ortelius noted certain similarities between the western coast line of north America and the eastern coast line of Africa. Many others did as well. These included Alexander von Humboldt, who drew attention to this in the first two volumes of his epic and widely read *Kosmos* (published between 1845 and 1847), and Alfred Wallace, who wrote in 1889, "It was formerly a very general belief, even amongst geologists, that the great features of the earth's surface, no less than

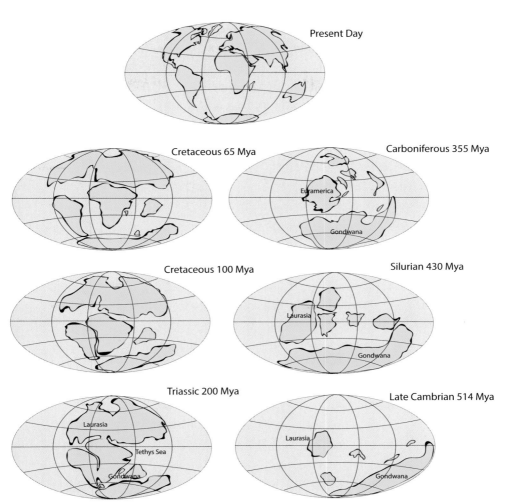

Figure 9.17. Schematic showing the positions of the major land masses at various times in Earth's history. These schematics indicate that the supercontinent called Pangaea consisted of two supercontinents called Laurasia and Gondwana, which existed from roughly 510 Mya to 180 Mya, and that Gondwana formed before Pangaea.

the smaller ones, were subject to continual mutations, and that during the course of known geological time the continents and great oceans had again and again changed places with each other." These and other speculations were known to the German geologist Alfred Wegener (1880–1930), who in 1912 was the first to put forth the idea that the continents are fragments of what was once a single landmass called Pangaea (see fig. 9.17). Wegener was the first to use the phrase continental drift. His hypothesis was rejected largely for two reasons. First, there was no known mechanism that could account for the migration of the continents, and, second, his estimates of the rate of continental drift were off by two orders of magnitude (the accepted rate of the separation of the Americas from Europe is ≈2.5 cm/year, whereas Wegener's estimate was 250 cm/year). Unfortunately, Wegener died before his hypothesis became accepted—that is, when the theory of plate tectonics was introduced in 1958 by the Australian geologist Samuel W. Carey (1911–2002) and when seismological evidence for plate tectonics was published ten years later by the American geologist Jack E. Oliver (1923–2011).

The evidence supporting Wegener's theory of continental drift is overwhelming. For example, fossils of freshwater reptiles such as *Mesosaurs* are found in South Africa and Brazil, whereas the biogeography of the southern beeches (*Nothofagus*) is explained best by the breakup of a supercontinent. However, the evidence for continental drift did not always come easily. Consider the unlikely and ultimately tragic collaboration between the British paleobotanist Marie C. Stopes (1880–1958) and the Antarctic explorer Robert F. Scott (1868–1912). While at the university of Manchester between 1904–1907, Stopes was studying coal balls and the university's collection of *Glossopteris* fossils in part to prove the existence of Gondwana and Pangaea (fig. 9.18). She speculated that fossils of this seed fern would be found in Antarctic sediments, since these plants appear in the fossil record near the beginning of the Permian around 299 Mya. Their distribution across several detached land masses led the Austrian geologist Eduard Suess (1831–1914) to speculate that there once was a single supercontinent. Stopes met Scott during one of his fund-raising lectures in 1904 and convinced him to look for and collect *Glossopteris* fossils during his 1912 Terra Nova expedition. Scott and his party died during the expedition, but *Glossopteris* fossils from the Queen Maud Mountains were recovered near his body. The disjunct distribution

Figure 9.18. Compression fossils of the organ genus *Glossopteris* in Permian sediments from Australia. In the personal collection of the author.

of these fossils added yet another line of evidence supporting the existence of Pangaea that in turn supported the theory of continental drift. Today, both theories are universally accepted, at least by the scientific community.

Global Climate Changes

Glossopteris went extinct by the end of the Permian roughly 252 million years ago along with an estimated 90% of all contemporary species during one of five major extinction events that changed the global climate and Earth's great ecosystems (fig. 9.19). There is still much debate about the causes of these five extinction events. Some may have been sudden such as the great Cretaceous-Paleogene extinction roughly 65 million years ago caused by an asteroid collision that eliminated an estimated 75% of all contemporary species. Other extinction events such as the Permian-Triassic event roughly 250 million years ago that killed *Glossopteris* might have been caused by sustained volcanic and tectonic activity (called flood basalt eruptions) such as those that produced the Siberian Traps, which cover over 2,000,000 square kilometers (770,000 square miles). Still other extinction events appear to conform to a press-pulse model in which extinction is a sudden catastrophic event (the "pulse") that is preceded by a long-term pressure on the global ecosystem (the "press"). In this model, the catastrophe is the final blow that kills off an already weakened ecosystem. What can be said with certainty is that these major extinction events dramatically changed Earth's abiotic conditions that in turn significantly changed the composition of Earth's flora and fauna.

The existence of mass extinctions was not anticipated by Darwin or his geological contemporaries owing to the influence of Lyell's widely read *Principles of Geology*, published in 3 volumes between 1830 and 1833. Charles Lyell (1797–1875) argued for geological uniformitarianism—the notion that geological processes observed to operate today are steady and slow and the same as those processes operating throughout Earth's history. This conception undoubtedly

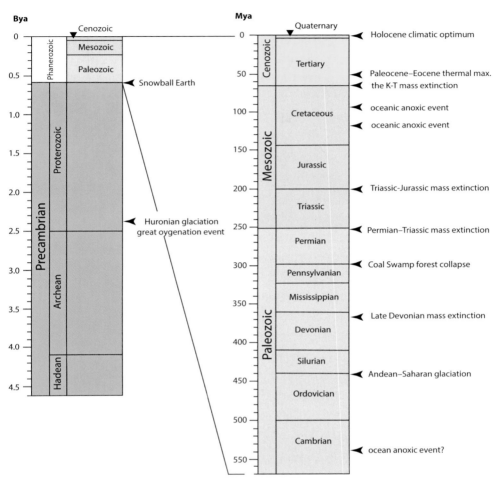

Figure 9.19. Events that had significant effects on Earth's climate and ecosystems. Among these are the five mass extinction events that occurred during the Ordovician-Silurian (≈450-440 Mya), the late Devonian (≈375-360 Mya), the Permian-Triassic (≈251 Mya), the Triassic-Jurassic (≈200 Mya), and the Cretaceous-Paleogene (66 Mya).

influenced the 22-year-old Darwin, who was given a copy of the first edition of Volume 1 by the captain of the *HMS Beagle*, Robert Fitz-Roy (1805–1865). Indeed, it is fair to say that Darwin's gradualistic conception of evolution is attributable in no small measure to the influence of Lyell's ideas as opposed to the concept of catastrophism as advanced by the French paleontologist and anatomist Georges

Cuvier (1769–1832), whose anatomical comparisons among living and extinct animals provided conclusive proof that extinction was commonplace in the fossil record. Regardless of the worldviews of the nineteenth century, it is abundantly clear that Earth's long history has been punctuated many times by global catastrophes and that the endgame of evolution is death.

Plants clearly played a role in changing our planet's climate. Consider the Huronian glaciation period that extended from 2.4 to 2.1 billion years ago during the early Proterozoic. This period of time is the oldest known glaciation event during which the entire surface of the globe was covered in ice. It occurred shortly after the great oxygenation event caused by the evolution of photosynthetic cyanobacteria-like organisms. The rise in oxygen and the corresponding decrease in atmospheric methane and CO_2 (both of which are greenhouse gases) undoubtedly contributed to a decrease in average temperature. The location of the supercontinent Pangaea and increased silicate weathering of newly formed basalts were additional contributing factors.

The importance of the greenhouse gases can be fully realized once we consider that a change in the average temperature of the Earth of 10°C is estimated to make a difference between a fully glaciated Earth and an ice-free Earth. By the end of the Proterozoic, roughly 542 million years ago, Earth's climate started to warm. Isotopic data indicate that the average global temperature was 22°C, or roughly 14°C higher than today's average (fig. 9.20), in part because atmospheric CO_2 concentrations were significantly higher than today's levels (fig. 9.21). Nevertheless, Earth experienced significant periods of glaciation, one of which may have contributed to the collapse of the great Coal Swamp forests. Indeed, global climate has undergone numerous periods of cooling and warming in part because of what is called *climate forcing*. This refers to the difference between the radiant energy received at Earth's surface and the longwave infrared radiation from the troposphere going back into space. Depending on the balance between the two, Earth either warms or cools. Along with aerosols, the so-called greenhouse gases (methane, nitrous oxide, water vapor, and carbon

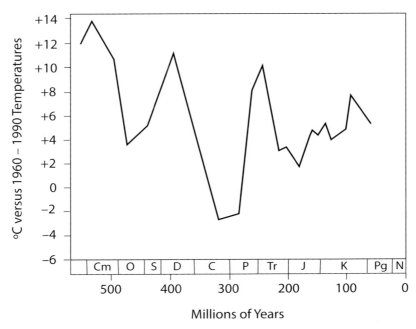

Figure 9.20. Estimates of Earth's average temperature referenced to average 1960 to 1990 temperatures. Data are averages taken from a number of primary sources. Cm = Cambrian, O = Ordovician, S = Silurian, D = Devonian, C = Carboniferous, P = Permian, Tr = Triassic, J = Jurassic, K = Cretaceous, Pg = Paleogene, N = Neogene.

dioxide) absorb certain wavelengths of longwave radiation, which is directed back to Earth's surface and thus contributes to climate warming.

A number of other factors are important to climate change as for example plate tectonics and the Milankovitch cycles. The creation of mountain ranges by tectonic activity can increase the weathering of soils and rocks, which can act as a CO_2 sink that in turn can reduce global temperatures. Volcanic activity can have the opposite effect by emitting CO_2. Earth's axial precession is equally important. The axial precession is the movement of the rotational axis of a planet or moon in a circle around the ecliptic pole. Earth's axis slowly moves through its precession once every 26,000 years during which the angle between the rotational axis and the normal to the orbital plane slowly

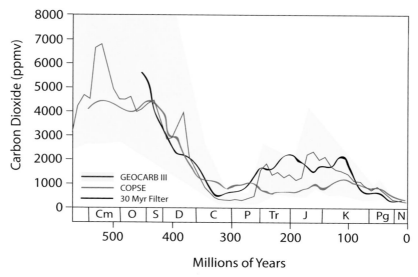

Figure 9.21. Changes in CO_2 during the Phanerozoic have been significant and played an important role in global climate changes. The highest concentrations are estimated to have been between 4,500 and 6,900 ppmv. These predate terrestrial plant life, during which solar output was more than 4% lower than today. With the ascendency and diversification of the land plants and the evolution of leaves and roots during the early Paleozoic, CO_2 levels dropped, reaching a local minimum during the late Carboniferous and early Permian times when the great coal deposits were formed. During the Mesozoic, CO_2 levels increased again as the sun gradually approached modern levels of radiance output. The last 200 million years include periods of considerable warmth and episodes of glaciation. During some intervals, sea levels rose dramatically as for example during the Cretaceous. Toward the close of the Paleogene, atmospheric CO_2 concentrations reached present-day levels. Adapted from Royer (2006). Cm = Cambrian, O = Ordovician, S = Silurian, D = Devonian, C = Carboniferous, P = Permian, Tr = Triassic, J = Jurassic, K = Cretaceous, Pg = Paleogene, N = Neogene. COPSE = co-incident probabilistic climate change weather data for a sustainable environment; GEOCARB III = geologic carbon cycle version III.

oscillates between 22.1° and 24.5° on a 41,000 year cycle. The Serbian geophysicist and astronomer Milutin Milanković (1879–1958) proposed that changes in Earth's precession, eccentricity, and axial tilt are responsible for climate changes in geological timeframes. His theory of orbital forcing is supported by the fact that many observed climatic periodicities statistically fit well with the orbital periods. Nevertheless, there are also significant problems with this theory. For example,

the ice age cycles in the Quaternary coincide with a 100,000 year periodicity over the past million years, which cannot be explained easily
by Milankovitch cycles. Certainly, one contributing factor is changes
in the sun's luminosity resulting from the changes in the movement
of solar fluid and the frequency and size of sunspots. These changes
occur in roughly 11 year cycles. One of the more recent manifestations
of this phenomenon is called the Maunder minimum, a 70-year-long
period between 1645 and 1715, which is referred to as a mini ice age.

There is no question that Earth's climate has changed, that it is
changing now, and that it will continue to change. Climatologists can
model with great accuracy the effects of virtually every contributing
variable that influences climate, with the exception of CO_2. The effects of this greenhouse gas on climate are difficult to predict because
of uncertainties concerning Earth's carbon cycle and its effects on
rainfall and temperature. However, reliable models have excluded all
of the other contributing variables as drivers of currently observed
changes in global climate, albeit not in time-frames of a geological
scale, whereas detailed measurements of atmospheric levels of CO_2
indicate that this greenhouse gas is increasing in concentration and
that the average global temperature is increasing. It is ironic that this
rise in atmospheric CO_2 levels is the result in part of burning coal that
was produced hundreds of millions of years ago by complex plant communities of species that are now all extinct. Whether human activity
is entirely or mostly responsible for the climatic changes occurring
now remains controversial among nonscientists. However, there is
little doubt that human activity has contributed to changes in climate
and to Earth's biosphere. This is hardly surprising when we recall the
concept of niche construction, which was introduced earlier on. Our
very existence has changed the biotic and abiotic environment in uncountable ways whose long-term consequences can only be guessed at.
By constructing our niche globally, we have poisoned the environment
and extirpated other species whose ecological importance can only
be judged in the future when it is too late to remediate the conse-

quences. While it is true that extinction is an inevitable consequence of evolution, we are the only species known to us that is cognizant of what extinction means, and the only species that has some significant control over extinction during our undoubtedly short geological stay on this planet.

> The law of the past cannot be eluded!
> The law of the present and future cannot be eluded!
> —WALT WHITMAN, *To Think of Time* (1855)

> The world is a fine place and worth fighting for
> and I hate very much to leave it.
> —ERNEST HEMINGWAY, *For Whom the Bell Tolls* (1940)

Suggested Readings

Bak, P. 1996. *How nature works*. New York: Springer.

Cain, M. L., B. G. Milligan, and E. Strand. 2000. Long-distance seed dispersal in plant populations. *Am. J. Bot.* 87: 1217–27.

Cannell, M. G. R. 1982. *World forest biomass and primary production data*. London: Academic.

Carlquist, S. 1979. *Ecological strategies of xylem evolution*. Berkeley: University of California Press.

During, H. J. 1979. Life strategies of bryophytes: A preliminary review. *Lindbergia* 5: 2–18.

Gould, S. J., and N. Eldridge. 1977. Punctuated equilibria: The tempo and mode of evolution reconsidered. *Paleobiology* 3: 115–51.

Grant, V. 1954. Genetic and taxonomic studies in *Gilia*. IV. *Gilia achilleaefolia. Aliso* 3: 1–18.

Grime, J. P. 1979. *Plant strategies and vegetation processes*. Chichester: Wiley.

Harper, J. L. 1982. *Population biology of plants*. London: Academic.

Holyoak, M., M. A. Leibold, and R. D. Holt, eds. 2005. *Metacommunities: Spatial dynamics and ecological communities*. Chicago: University of Chicago Press.

Levin, S. 1999. *Fragile dominion: Complexity and the commons*. Reading, MA: Perseus.

Luce, R. D., and H. Raiffa. 1957. *Games and decisions: Introduction and critical survey*. New York: Wiley.

MacArthur, R. H., and E. O. Wilson. 1967. *The theory of island biogeography*. Monographs in Population Biology, Vol. 1. Princeton, NJ: Princeton University Press.

Pianka, E. R. 1988. *Evolutionary ecology*. New York: Harper.

Pielou, E. C. 1991. *After the Ice Age: The return of life to glaciated North America*. Chicago: University of Chicago Press.

Royer, D. L. 2006. CO_2-forced climate thresholds during the Phanerozoic. *Geochim. Cosmochim. Acta* 70: 5665–75.

GLOSSARY

The terms included in this glossary were selected using one or more of three criteria: (1) the comparative difficulty of understanding the concept a term conveys, (2) a truncated definition offered in the text so as not to interrupt the flow of the text, and (3) whether a term has multiple definitions, among which one was used in this book. Taxonomic names are not included in this glossary.

Allopolyploidy—A form of polyploidy (the possession of two or more sets of chromosomes) resulting from the mating of two species. The result of hybridization. (See autopolyploidy.)

Allosteric—The binding of a molecule to a site on a protein that differs from the active site and that alters the configuration or the activity of the protein.

Alternative Splicing—A process whereby a single gene can encode for multiple kinds of proteins by virtue of a spliceosome excluding or including different exons from pre-mRNA before mRNA translation.

Anticlinal—Having an orientation at a right angle to some plane or surface of reference. (See periclinal.)

Antheridium—A multicellular structure consisting of a stalk and a sterile layer of cells surrounding spermatogenous cells produced by an embryophyte. (See archegonium.)

Apetalous—Lacking petals.

Apomixis—The formation of a viable embryo without the fusion of sperm and egg to produce a zygote.

Aptation—An **exaptation** (a trait or set of traits that provides access to a previously inaccessible niche that originated as an adaptation to a previous accessible niche, or as a neutral mutation) or an **adaptation** (a trait that has resulted as a consequence of selection acting on previous manifestations of the same traits in the same or similar niche).

Archegonium—A multicellular structure consisting of a neck, venter, and egg cell produced by an embryophyte. (See antheridium.)

Autopolyploidy—The spontaneous doubling of the nuclear genome that results in an organism with twice the original set of paired chromosomes. (See allopolyploidy.)

Body Plan (*Bauplan*)—The phenotypic architecture shared among members of the same taxon. It is also the developmental motif used to construct the phenotype characterizing a particular group of organisms.

Boolean Network—A network composed of Boolean variables (those that have only two possible states or conditions such as "true" or "false, "yes" or "no", "+" or "-", and "0" or "1") whose states are determined by other variables in the network.

Carrying Capacity—The largest population of a species that the environment can sustain.

Chiasmata—When matching segments of the non-sister chromatids of homologous chromosomes exchange genetic materials during meiotic crossing-over. Sister-chromatids may also cross-over. However, this process does not result in a genetic change in the cells resulting from meiotic division.

Chromatid—One copy of a newly replicated chromosome that is joined to its companion ("sister chromatid") by a centromere. Companion chromatids become separated during anaphase in mitosis and during anaphase II in meiosis.

Cis-Regulatory Element—A noncoding DNA segment that regulates the transcription of nearby genes. Cis-regulatory elements typically function as binding sites for transcription factors.

Clade—A large group of organisms descended from a common ancestor including the ancestor. A monophyletic group or organisms consisting of a nested set of interrelated (extinct and extant) lineages.

Cladistics—A methodological approach to infer phylogenetic relationships among organisms based on the distribution of shared character states and the number of character transformations required to resolve relationships among taxa. The result of a cladistic analysis is a cladogram that represents a parsimonious evolutionary hypothesis about the phylogenetic relationships among the organisms included in the analysis.

Coenocyte—A body plan consisting of a single multinucleated cell capable of indeterminate growth in size. (See colonial and multicellular.)

Colonial—A body plan consisting of aggregated cells that may cooperate yet maintain their individuality. (See coenocyte and multicellular.)

Conspecific—Individuals belonging to the same species.

Constitutive Exons—Exons that are always (or almost always) included in mRNA after the alternative splicing of pre-RNA.

Crossing-Over—The exchange of genetic material between homologous chromosomes occurring in prophase I during meiosis. Crossing-over typically occurs when the matching ("homologous") DNA segments of homologous chromosomes break and reconnect (homologous crossing-over). Nonhomologous crossing-over can occur when similarities in DNA segments are mismatched along the lengths of homologous chromosomes, which affects many loci simultaneously and can result in a significant chromosomal mutation.

Darwinian Fitness—The genomic contribution of an individual to the genomic composition of the next generation compared to the average contribution of members in the same population.

Differentiation—The development of specialized cells, tissues, organs, or body parts from simpler antecedent cells, tissues, organs, or body parts.

Dikaryon—Possessing two individual nuclei per cell.

Dioicous—A unisexual gametophyte incapable of self-fertilization. Not the same thing as dioecious, which pertains to a diploid sporophyte that produces either microspores or megaspores. (See monoicous.)

Dissociation (Developmental)—The loss of correlation(s) in the timing or the course of development in one or more cell-types, tissues, organs, or organ systems of an organism.

DNA-Binding Domain—A folded protein domain or a folded nucleic acid sequence that contains one or more motifs capable of recognizing double- or single-stranded DNA. These domains are structurally involved with DNA repair, storage, replication, or modification (as for example in methylation), or they function in the regulation of gene expression patterns (as for example as components of transcription factors).

Dynamical System—A model that describes the temporal behavior and evolution of a system as the smooth action of real components (a continuous dynamical system) or integers representing real components (a discrete dynamical system).

Endemic (Species)—A species that is unique to a particular geographic location. A species that is found nowhere else.

Epigenetics—The study of phenotypic variations resulting from genomic responses to external or internal environmental factors as opposed to phenotypic variations resulting from genomic mutations. A heritable and stable change in phenotypic expression that is independent of differences in DNA sequence.

Epistasis—The effect of one or more (modifier) genes on the expression of one or more other genes. The mutations of epistatic genes have different effects in combination than individual mutations.

Eudicots—The largest of three clades of angiosperms characterized by species with dicotyledonous characteristics that produce tricolpate pollen grains (pollen grains with three grooves, called colpi, running parallel to the polar axis of the grain). The other two angiosperm clades are the monocotyledons and the basal angiosperms.

Eukaryote—A unicellular or multicellular organism consisting of cells possessing a nucleus and other membrane-bound organelles, such as the mitochondrion and the chloroplast. (See prokaryote.)

Euploidy—A genome consisting of three or more complete sets of the haploid chromosome number.

Eusporangium—A spore-producing structure produced by most extant pteridophytes developing from two or more initial cells, possessing a sporangial wall two or more cells thick, and producing a comparatively large number of spores. With the exception of some late-divergent ferns, all extant pteridophytes produce eusporangia. (See leptosporangium.)

Evolution—Organic heritable change; descent with heritable modifications from ancestors.

Evolutionary Stable Strategy—A stratagem by which a population in a specified environmental context cannot be usurped by any other stratagem that is initially rare. An axiom emerging from game theory.

Evolvability—The capacity of a natural system to change adaptively, or more generally the ability of an organic system to evolve.

Exaptation—A shift in the functionality of a trait resulting from selection operating on the function of a previously adaptive trait under the conditions of a different environmental context. (See aptation.)

Game Theory—The study of mathematical models designed to predict the consequences of cooperation and conflict among players that must follow proscribed rules of behavior. Also, the study of interactive decision-making. The use of game theory in biology has been extended to consider decision-making in organisms other than humans (such that the stipulation that decision be made rationally is relaxed).

Exon—A coding DNA segment or its RNA-translated segment that is retained in the RNA molecule for protein translation. (See intron.)

G (–Value) Paradox—The inconsistent relationship between organismal complexity and the number of protein-coding genes (hence the G). Across different organisms, organismal complexity appears to not increase in proportion to the increase in the number of protein-coding genes. This paradox is closely related to the C-paradox (the failure of organismal complexity to increase in proportion to cellular DNA content).

Gene Regulatory Network—A set of genes or transcription factors interacting with one another and with internal and external cellular signals to control gene expression patterns and the course of development.

Genotype—The entire set of genes within an organism. The term is often, but inappropriately, used to specify all nuclear genes. (See phenotype.)

Heterochrony—A temporal displacement of the appearance or development of one or more organs with respect to other organs. A phyletic change in the timing of development in a descendant with respect to the timing of development of its ancestor.

Heterospory—The production of two different kinds of spores that develop into unisexual gametophytes, one type that produces eggs (megagametophyte) and another type that produces sperm (microgametophyte). (See homospory.)

Homeo-Domain (Homeobox)—A DNA segment (typically 180 base pairs long) that encodes homeo-domain proteins that are transcription factors regulating development in eukaryotes.

Homeotic Gene (Homeosis)—Genes that regulate the development of body parts by means of regulating transcription factors, or other genes that regulate gene regulatory networks. Homeosis is the transformation of one body part into another kind of body part resulting from the mutation of a homeotic gene.

Homodimer—A molecule consisting of two identical molecular units.

Homology—A similarity between organic structures or traits resulting from the inheritance of the same structure or trait from a common ancestor.

Homospory—The production of spores that develop into bisexual gametophytes that produce eggs and sperm. (See heterospory.)

HOX Genes—A group of related genes that encode homeo-domain proteins that function as transcription factors affecting the developmental identities of segments along the head-tail (anterior-posterior) axis of animal embryos. HOX genes are characterized by containing a homeobox sequence (≈ 180 base pairs long) that binds to DNA. Like MADS-box gene mutations, HOX-box gene mutations (homeotic mutations) result in the abnormal developmental expression of cell-, tissue-, or organ-types. (See MADS-box genes.)

Hypercapnia—Excessive concentrations of carbon dioxide in the bloodstream resulting from an inadequate supply of oxygen for respiration.

Intron—A noncoding DNA segment or its RNA-translated segment that is removed (spliced) from the RNA molecule before protein translation. (See exon.)

K–**Selection**—Selection of individuals in populations at or near their carrying capacity typically selecting for few, large, slowly developing individuals that produce few progeny. (See *r*-selection.)

Karyotypic—Referring to the appearance or condition of the chromosomes in a nucleus. (See genotype.)

Intrinsically Disordered Protein—A protein that lacks a single equilibrium three-dimensional conformation under normal physiological conditions and that can take on multiple conformations depending on its physiological context. Many transcription factors contain intrinsically disordered protein domains.

Leptosporangium—A spore-producing structure developing from a single initial cell, possessing a one-cell-thick wall, and producing a comparatively small number of spores. A distinguishing feature of the leptosporangiate (sometimes classified as the Polypodiidae) ferns. (See eusporangium.)

Life-History Strategy—A set of traits adapted to a specific kind of environment that involve quantitative features such as reproductive age, fecundity, and life span.

Ligand (**Binding**)—A molecule that binds to another molecule and forms a complex serving a biological function. In protein ligand binding, the ligand is typically a small signaling molecule that binds to a target protein.

Linkage Disequilibrium—A deviation in the frequencies of alleles at different loci from that which would be expected from the random, independent assortment of alleles at different loci. Linkage disequilibrium can result from mutations, genetic drift, a change in the rate of recombination, or selection.

Logic Gate—A device that operates using Boolean functions such as "and" and "or."

Macroevolution—Evolution at or above the species level. Typically defined in terms of evolutionary events or processes affecting higher taxonomic groups. (See microevolution.)

Macromutation—A dramatic genomic alteration that results in the sudden appearance of a phenotype that significantly differs from its antecedent.

MADS-Box Genes—Genes that contain the conserved MADS-box sequence ($\approx$ 168 to 180 base pairs) encoding the DNA-binding MADS domain. MADS-box genes control important aspects of plant gametophyte, embryo, seed, and fruit development. In animals, these genes are involved in cell proliferation and differentiation. Like HOX gene mutations, MADS-box gene mutations (homeotic mutations) result in the abnormal developmental expression of cell-, tissue-, or organ-types.

Meiosis—The type of cell division that reduces the number of chromosomes by one half. (See mitosis.)

Meristem—A region of nondifferentiated and self-propagating cells (initials) that gives rise to new cells that subsequently differentiate into one or more tissues or tissue systems.

Metapopulation—A geographically disjunct set of populations consisting of members of the same species.

Microevolution—Pertaining to evolution below or at the species level. Temporal changes in the allele frequencies in a population. (See macroevolution.)

MicroRNA—A small noncoding RNA molecule occurring in eukaryotes that silences RNA molecules and is thus involved in the post-transcriptional regulation of gene expression.

Mitosis—The type of cell division in which the chromosomes within the nucleus replicate and become separated into two identical sets of chromosomes (barring mutation). (See meiosis.)

Monogenic mutation—A phenotypic alteration resulting from the mutation of a single gene.

Monoicous—A bisexual gametophyte capable of self-fertilization. Not the same thing as monoecious, which pertains to a diploid sporophyte that produces microspores and megaspores. (See dioicous.)

Monophyletic—One or more taxonomic groupings of organisms descending from a last common ancestor that includes all the descendants of the ancestor. (See polyphyletic.)

Multicellular—A body plan in which somatic cells adhere to one another after cell division, and establish physiological integration and cooperation. (See coenocyte and colonial.)

Multilevel Selection Theory (Group Selection)—A theory positing selection acting at the level of a group rather than acting at the level of the individual components of the group. The theory maintains that groups (composed of cells, organisms, species, etc.) have a functional organization that experiences selection in a manner analogous to selection acting on the individuals constituting groups.

Native (Species)—A species that is either **indigenous** (found in a particular geographic location and in other geographic locations) or **endemic** (found only in a particular geographic location).

Neontology—The study of living organisms. (See paleontology.)

Niche—The sum of an organism's interactions with its biotic and abiotic environment. The role an organism plays in its environment.

Nucellus—The megasporangium of a seed plant.

One-Step Meiosis—A rare process wherein meiosis is achieved in a single cell division event before chromosome replication. One-step meiosis achieves a greater degree of genetic variability than two-step meiosis.

Ontogeny—The developmental sequence of events over the life history of an individual.

Optimization—The selection based on one or more criteria of the most efficient or effective functional component (from a set of alternatives) for operating in a larger system. The procedure for finding the maximum of a mathematical function.

Paleontology—The study of ancient, typically extinct organisms. (See neontology.)

Oxidant—An element or molecule that removes ("accepts") an electron from another element or molecule. (See reductant.)

Pangenesis—Darwin's theory of heredity wherein heritable particles residing in somatic cells can be influenced (altered) by the environment, be conveyed to sex cells, and thus affect the phenotype of the next generation.

Panmictic (Panmixia)—The random mating of individuals within a population. The absence of sexual selection. The outmoded theory of panmixia was consistent with Lamarckian inheritance.

Paraphyletic—One or more lineages descending from a last common ancestor that do not include all the descendants of the ancestor.

Parenchymatous—Consisting of a ground tissue composed of thin-walled, more or less isodiametric polyhedral living cells. Consisting of a morphologically undifferentiated tissue composed of cells with plasmodesmata on all or most adjoining cell walls.

Parsimony (Parsimonious)—An optimality criterion that posits the best phylogenetic (cladistic) hypothesis is the one that involves the fewest character state transformations (the fewest evolutionary changes).

Parthenocarpy—The formation of a fruit without benefit of the formation of embryos or seeds. The result is a seedless fruit.

Periclinal—Having an orientation parallel to some surface or plane of reference. (See anticlinal.)

Pericycle—The outermost layer of cells (derived from the procambium) of the vascular strand (stele) of a root.

Phenotype—The set of all observable traits that collectively distinguish one individual organism from another. (See genotype.)

Phylogeny—The evolutionary history of a lineage or clade.

Plate Tectonics—A theory emerging from the concept of continental drift proposing that Earth's mantel is composed of (seven or eight major) plates, which move over the lithic layer surrounding the core as a consequence of convection in the mantel near the Earth's core. Subduction at convergent plate boundaries carries plate materials into the mantle. In turn, the loss of crust is compensated for by the formation of new oceanic crust, which drives sea-floor spreading.

Pleiotrophy—The effect of one gene on multiple traits. Antagonistic pleiotrophy occurs when the effects of one gene are beneficial with regard to one or more traits and detrimental to one or more other traits.

Pleuromitosis—A type of mitotic division in which the spindle microtubules form as separate half spindles at an angle to one another, or lie side by side of one another. Different forms of pleuromitosis occur in the foraminiferans, sporozoans, and parabasalian flagellates.

Polyphyletic—A group of organisms, lineages, or clades that does not descend from the same common ancestor. (See monophyletic.)

Post-translational Modification—The modification of a protein after ribosomal translation of a mRNA molecule.

Pre-mRNA—Precursor mRNA. The mRNA molecule that is transcribed directly from DNA before introns are removed by a spliceosome to form the mRNA molecule for protein translation.

Primordium—A body part in its earliest recognizable or observable form.

Prokaryote—A unicellular or multicellular organism consisting of cells lacking a nucleus and other membrane-bound organelles such as the mitochondrion and the chloroplast. (See eukaryote.)

Proteasome—A protein complex that degrades damaged or unneeded proteins. Proteasomes occur in eukaryotes, Archaea, and some eubacteria.

Proteome—The entire set of proteins that a genome produces.

Protists—A large polyphyletic group of ecologically diverse eukaryotes consisting mainly of unicellular and colonial organisms.

Quantitative Trait—A trait that various continuously (in a non-discretized manner). Typically, a trait that results from the interactions of two or more genes.

_r_–Selection—Selection of individuals in populations well below their carrying capacity typically selecting for many, small, and rapidly developing individuals that produce many progeny. (See K-selection.)

Reductant—An element or molecule that loses ("donates") an electron to another element or molecule. (See oxidant.)

Regulatory Gene—A gene that controls a developmental process by regulating the turning on or off of structural genes that manufacture proteins. (See structural gene.)

Riboswitch—A regulatory messenger RNA (mRNA) molecule that binds to a small molecule resulting in a change in the production of the protein encoded by the mRNA molecule.

Second-Order Term—A mathematical variable or function of a variable that results in a higher order of complexity, as x^2 versus x.

Self-Thinning—The progressive decrease in plant density (number of individuals per unit area) and progressive increase in average plant size over time.

Sexual Selection—A mode of selection in which members of one gender select mates from the other gender.

Species Selection (Species Sorting)—A macroevolutionary trend or pattern resulting from differential speciation or extinction rates.

Spliceosome—A molecular structure (typically found in the nucleus) composed of five small nuclear RNA molecules and a range of protein factors that removes introns from pre-mRNA molecules.

Sporic Meiosis—The formation of haploid spores by the meiotic division of diploid cells (called sporocytes). (See zygotic meiosis.)

Stele—All of the tissues in a root or stem that develop from the procambium.

Structural Gene—A gene that encodes a protein. (See regulatory gene.)

Synapomorphy—A trait shared by two or more taxa that was present in the most recent common ancestor of the taxa but that was not a trait of the ancestor of the most recent common ancestor.

Trait—A characteristic or distinguishing heritable phenotypic feature of an organism.

Transcription Factor—A protein that binds to a specific DNA sequence and thereby influences the transcription of nearby genes.

Transposon—A DNA segment that can change its location within a chromosome or among chromosomes in the same cell. There are two classes of transposons: retrotransposons (that can be described as "copy and paste," which involves mRNA intermediates) and DNA transposons (that can be described as "cut and paste," which does not involve mRNA intermediates).

Ubiquitination—The post-translational modification of a protein resulting from the binding of the protein to ubiquitin (a small regulatory protein that is found in the cells of eukaryotes). Ubiquitination can result in the degradation of the bound protein, the alteration of the protein's location within a cell, or an alteration in the effect or activity of the protein.

Vicariance—The physical separation of conspecifics belonging to a population as a result of an abiotic or biotic barrier.

Weismann's Doctrine (Barrier)—The proposition that heritable materials are transmitted by sex cells to the next generation and that their transmission is excluded from somatic cells to the next generation. In more contemporary terminology, Weismann's doctrine posits that heritable information is conveyed by DNA and RNA to protein structure and function and not vice versa.

Zygotic Meiosis—The zygote undergoes meiotic division after it is formed by the fusion the egg and a sperm cell. The formation of haploid cells following the formation of the zygote. (See sporic meiosis.)

INDEX

A page number followed by the letter *f* refers to a figure or its caption, and a page number followed by the letter *t* indicates a table.

Archaeopteryx, 157–58

Archaeosperma arnoldii, 137–38

Archean, 30f; fossil cells from, 49, 50f, 51; hydrothermal vents during, 57; ocean chemistry of, 39, 41; oxygenic prokaryotes of, 62

archegonium, 99–102, 100f, 101f; development of, 267f; of *Equisetum*, 266f; evolution of, 105–8, 107f; seed plants and, 135; of unisexual vs. bisexual gametophyte, 109

archetype, 377. *See also* body plans

ARF (auxin response factor), 232–33, 233f

Aristotle, 260

AS. *See* alternative splicing (AS)

asexual reproduction: in angiosperms, 144, 145; by apomixis, 285–86; biological species concept and, 285–86, 330; coexisting with sexually reproductive population, 199, 337–39, 341, 345, 371; exploration of new habitats and, 199; of first land plants, 97–98; by hybrids unable to sexually reproduce, 281; of many successful species, 336–37; in matrix of life cycles, 337, 338f; modes of, 338t; plants exclusively or predominantly using, 285; rarity of, 339, 341; self-compatibility and, 109; with sexual alternative, 75, 78; by sporophytes of modern pteridophytes, 127–28; unicellular organisms with phase of, 75, 416–17, 417f; of well-adapted population, 78

Asplenium, 289, 290, 290f

Asteraceae: dominated by annuals, 366; inflorescences of, 309, 364

asteroid collisions with Earth: abiotic syntheses and, 38; Cretaceous-Paleogene mass extinction and, 13, 528

atmosphere: of early Earth, 34–35; greenhouse gases in, 530–31; reducing, 35, 36. *See also* carbon dioxide; oxidizing atmosphere; water conservation

ATP (adenosine triphosphate): abiotic synthesis of, 38; from cellular respiration, 63; from glycolysis, 41, 42f; in photosynthesis, 59f

ATPases, in IAA-mediated signal transduction, 232, 233f, 234, 235f

ATP-dependent transport, 229

autocatalytic molecules, 45–46, 47f

autogenous hypothesis, 64f, 67–68

autopolyploidy: evolution of meiosis and, 81, 342, 343f; genetic variation due to, 170; speciation through, 287–89, 288f

autozygous genotype, 174

AUX1 co-transporter, 229, 230f

AUX1 gene, 229

AUX-IAA gene, 232–33, 233f

auxin, 227, 229; developmental patterning and, 398; GRAS proteins and, 249. *See also* indole-3-acetic acid (IAA)

auxin response factor (*ARF*), 232–33, 233f

Bacillus subtilis, two forms of, 411

bacteria: biological species concept and, 285; indole-3-acetic acid in, 228; plants and fungi interacting with, 150, 484–85. *See also* cyanobacteria; Eubacteria

bacteriochlorophylls, 54–55; biosynthesis of, 55–56, 56f; hydrothermal systems and, 57; light absorption by, 57–58

bacteriorhodopsin, 53

Bak, Per, 489–90

banded iron ore formations, 30f, 32, 32f

banded tubes, 110, 478f

membrane-bound protobionts, 36, 46–48

membranes: of organelles, 67; of pro-karyotes, 66, 67

Mendel, Gregor, 17

Mendelian genetics, 17–20, 155, 159–60, 161, 162

mesomes, 470

messenger RNA (mRNA), 45; microRNA and, 242–43; riboswitches and, 242–43. *See also* alternative splicing (AS)

metabolism, 31. *See also* glycolysis; pentose phosphate pathway; photosynthesis

metazoans, number of cell types in, 422, 424–25, 424f, 425f, 426–27, 428f, 429t

methane: of early atmosphere, 35, 530; as greenhouse gas, 530–31

microevolution: defined, 325; in Modern Synthesis, 24

microgametophytes, 128, 130–31; of Carboniferous gymnosperms, 139–40; of seed plants, 133, 135

microphylls, 124

microplasmodesmata, 83

micropyle, 133, 134f; Devonian structures related to, 137, 139; of Jurassic gymnosperms, 146; pollination droplet emerging from, 136f

microRNA (miRNA), 25, 241, 242–43

microspheres, 47–48

microsporangia, 129–30; of stamen, 140

microspores, 128–29, 130; heterozygosity enhanced by, 131; pollen grains developing from, 140; of seed plants, 135

microtubules: cell plate orientation and, 90; future cell wall's position and, 398–400, 400f; of phragmoplast, 72; proteins associated with, 396; of spindle apparatus, 70

migration: to cope with environmental change, 198–99; genetic variation due to, 170, 511, 513–14, 515f; between populations, 511, 513–14

Milanković, Milutin, 532

Milankovitch cycles, 531–33

Miller, Stanley L., 36, 44

missing links, 157–58, 159f

mitochondria: aerobic prokaryotes similar to, 63; endosymbiotic hypothesis and, 63–69, 64f, 65t

mitosis, 70–74, 71f; asymmetric, 400, 401; cell adhesion and, 397–98; cytokinesis and, 70, 72, 397–98; dynamical patterning modules and, 397–400, 400f; motility of unicellular organisms and, 84; phragmoplastic, 102; preferential sister chromatid segregation in, 405; symmetric, 398–400, 400f; of unicellular eukaryotes, 75, 78. *See also* cell division rates

mitotic nondisjunction, 287, 288f, 289

mitotic spindles: persistent, 102; self-assembly of, 90

Mnium, 109f

Modern Synthesis, 21–25, 22t, 155, 160–62, 161t; adaptation and, 518; biological species concept in, 272; gradualism in, 24, 326–27; speciation in, 21–22, 22t, 24, 25, 161–62, 172, 272, 326–27; Weismann's doctrine and, 336

monogenic speciation, 306, 308–11

Mooers, Arne O., 421–22

Morgan, C. Lloyd, 271

morning glory (*Ipomoea purpurea*), 187

morphospace: adaptive walks in, 447–49, 460–70, 462t; matrix of functional combinations and, 452; for simple sporophytes, 455–61, 456f

origin of life: four properties required
for, 48; self-organization and, 89–
90; stages of, 35–36. *See also* abiotic
organic synthesis
Ortelius, Abraham, 524
Oryza, 253
Oscillatoria, 83
overtopping, 470–71, 471f, 472, 474,
475f
ovules, 133, 134f, 135; of angiosperms,
140–41, 141f; developmental
regulation of, 141f; of Devonian
fossils, 136–37, 137f, 139; of gymnosperms, 138, 146; pollination
droplets and, 136f
Owen, Richard, 257–58, 377
oxidizing atmosphere: abiotic organic
synthesis and, 35; banded iron ores
and, 30f, 32; creation of, 6–8, 7f, 32,
62, 530
oxygen: levels over geological time, 7f;
in organisms, 39
oxygen released by photosynthesis,
6–8, 7f; banded iron ores and, 30f,
32; changing the ancient Earth, 31;
mechanism of, 61

paedomorphosis, 314–15, 315t
PAGE4 protein, 250
paleo-polyploids, 178
Pangaea, 525f, 526, 528, 530
pangenesis, 19, 159–60
parallel logic circuits, 221–23, 222f
parapatric speciation, 286–87, 286f,
300; subspecies and, 293
paraphyletic group: bryophytes as, 117,
123; defined, 117
parasitic fungi, 149–50
parasitism, reduced phenotypes in, 10,
12f
pareidolia, 5
parenchymatous tissue construction,
389f, 390, 391t

Pascal's Triangle, 164–65, 165f
Pascher, Adolf, 387
patchiness of populations, 511–13,
512f
peas (*Pisum sativum*): Mendel's studies
of, 17–18, 19f; single-gene mutations in, 22, 23f
pectin methylesterase (PME), 236, 237f,
238
pectins, 6, 236, 238–39; dynamical
patterning modules and, 397–98
penetrance of a gene, 169
Penstemon, hybrid speciation in, 295
pentose phosphate pathway, 31, 39–41,
40f
Pentoxylales, 146
peramorphosis, 314, 315t
perennial mosses, 514f
perfect flowers, 142
perianth, 140
pericarp, 86f
pericycle, 476
peripatric speciation, 274f, 275, 276,
286, 304–5
peroxisomes, 102
petals, 140; apetalous flowers and, 22,
308; regulation of development of,
141f
phagocytosis, and multicellularity, 418
Phanerozoic, 93, 94f; great floristic
transitions in, 349f, 360, 517,
520–21
phenotype: assumed to emerge simply
from genotype, 25; conflicting
requirements for, 206–7; defined,
10; mapping to genotype, 189–94;
selection acting on, 188–89
phenotypic divergence of species, 331,
332f, 333
phenotypic plasticity: of angiosperms,
355, 372–73; assimilated into
development, 403; environmental
conditions and, 9–10, 9f, 192–94,